Advances in Neuroethics

Advances in neuroscience research are bringing to the forefront major benefits and ethical challenges for medicine and society. The ethical concerns related to patients with mental health and neurological conditions, as well as emerging social and philosophical problems created by advances in neuroscience, neurology and neurotechnology are addressed by a specialized and interdisciplinary field called neuroethics.

As neuroscience rapidly evolves, there is a need to define how society ought to move forward with respect to an ever growing range of issues. The ethical, legal and social ramifications of neuroscience, neurotechnology and neurology for research, patient care, and public health are diverse and far-reaching — and are only beginning to be understood.

In this context, the book series "Advances in Neuroethics" addresses how advances in brain sciences can be attended to for the benefit of patients and society at large.

Veljko Dubljević • Allen Coin
Editors

Policy, Identity, and Neurotechnology

The Neuroethics of Brain-Computer Interfaces

Springer

Editors
Veljko Dubljević
Department of Philosophy and Religious Studies
North Carolina State University
Raleigh, NC, USA

Allen Coin
North Carolina State University
Raleigh, NC, USA

ISSN 2522-5677 ISSN 2522-5685 (electronic)
Advances in Neuroethics

ISBN 978-3-031-26800-7 ISBN 978-3-031-26801-4 (eBook)
https://doi.org/10.1007/978-3-031-26801-4

This Springer imprint is published by the registered company Springer Nature Switzerland AG
The registered company address is: Gewerbestrasse 11, 6330 Cham, Switzerland

Contents

An Introduction to Policy, Identity, and Neurotechnology: The Neuroethics of Brain–Computer Interfaces

Allen Coin and Veljko Dubljević

Brain–Computer Interface (BCI) technology is a promising and rapidly advancing research area. It was initially developed in the context of early government-sponsored futuristic research in biocybernetics and human–machine interaction in the United States (US) [1]. This inspired Jacques Vidal to suggest providing a direct link between the inductive mental processes used in solving problems and the symbol-manipulating, deductive capabilities of computers, and to coin the term "Brain-Computer Interface" in his seminal paper published in 1973 [2]. Recent developments in BCI technology, based on animal and human studies, allow for the restoration and potential augmentation of faculties of perception and physical movement, and even the transfer of information between brains. Brain activity can be interpreted through both invasive and noninvasive monitoring devices, allowing for novel, therapeutic solutions for individuals with disabilities and for other non-medical applications. However, a number of ethical and policy issues have been identified in context of the use of BCI technology, with the potential for near-future advancements in the technology to raise unique new ethical and policy questions that society has never grappled with before [3, 4]. Once again, the US is leading in the field with many commercial enterprises exploring different realistic and futuristic applications of BCI technology. For instance, a US company named Synchron recently received FDA approval to proceed with first-in-human trials of its endovascularly implanted BCI device [5].

In this volume, we explore the landscape of thought on the ethical and policy implications of BCI technology, with a deliberate focus on the North American context. In many ways, the US has shaped—and will continue to shape—the field of BCI research with ample funding, public support, and a permissive regulatory environment. The development of commercial BCI technology in the US has

A. Coin · V. Dubljević (✉)
NC State University, Raleigh, NC, USA
e-mail: allen_coin@ncsu.edu; veljko_dubljevic@ncsu.edu

V. Dubljević, A. Coin (eds.), *Policy, Identity, and Neurotechnology*, Advances in Neuroethics, https://doi.org/10.1007/978-3-031-26801-4_1

important ramifications for the worldwide adoption of this technology. In fact, two factors have led to the internationalization of specifically North American ethical and policy oversight: First, overall globalization of health issues and technological products as well as the internationalization of manufacturing for the North American market; and second, simultaneous multinational testing and approval [6].

As we continue to evolve into a global society, it is clear that health-related technologies such as BCIs, and the science that supports them, know no borders. Whether working to foster innovation in product development, or investigating a potential health risk from a product, the concerns, challenges, and needs are shared by public health officials and the public around the world. The US regulatory environment is clearly one of the leading and influential models for health technology adoption across the globe (along with the EU regulatory environment, which arguably has been better researched in regard to neuroethics scholarship [7]).

Companies providing new health technologies such as BCIs test their products in places and on people around the world and use those data to support marketing applications both in the US and in other countries. With the revolution in information technology, data acquisition and sharing have become easier than ever. As a result, regulatory agencies have had to make their marketing application review processes more efficient and compatible with both industry practices and the new ways data is acquired and shared. As the industries the US Food and Drug Administration (FDA) oversees have become global, many regulatory systems around the world have become increasingly competent and sophisticated. Some have modeled themselves on the US, and the FDA maintains its very strong position as a world leader in the regulatory promotion and protection of public health.

In this volume, we focus on policy oriented ethical analysis of clinical and commercial applications of BCI in the North American context and provide an up-to-date overview of the ethical, social, and policy implications of BCI. Since BCI is a rapidly advancing field of research, it is vital for ethicists and policymakers to stay current on the recent developments and the state-of-the-art in the field. The following 14 chapters specifically consider past developments in BCI, current technological capabilities, and current commercial aspirations that will guide the future of this promising technology.

While prior work in neuroethics may have contributed to some of the major issues covered in this volume (see, e.g., [7]), no single edited volume has exclusively focused on the intersection of ethical and policy issues in BCI technology within the North American regulatory context. This book may serve as an introductory textbook into the neuroethics of BCI or as a resource for neuroscientists, engineers, and medical practitioners to gain additional insight into the ethical and policy implications of their work. We seek to paint a rich and detailed picture of the field of BCI ethics with contributors from various fields and backgrounds with experience in both academia and the commercial sphere.

To accomplish our aforementioned aims, we have structured the book into three parts, outlined below.

1 Overview of Part One: The Past, Present, and Future of BCI Technology

In Part One, we present unique perspectives on the history of BCI, the current state-of-the-art of the technology, and considerations about potential future developments and applications of this technology. In Chap. 2, Mai Ibrahim explores aspects of the history of BCI through the application of a feminist, posthumanist lens, considering specifically the potentialities of the convergence of BCI technology with virtual reality (VR). In Chap. 3, Paul Tubig and Frederic Gilbert present an analysis of one of the most salient concerns about the application of current BCI technology, namely the impact BCIs can have on personality, identity, agency, autonomy, authenticity, and/or self (PIAAAS). The chapter draws on their prior work [8–10] to outline the motivations patients may have for pursuing neural implants and the seriousness of changes in PIAAAS that can occur in patients with BCIs. In Chap. 4, Adam Fry and colleagues from the Icahn School of Medicine at Mount Sinai and Synchron explore the potential for the use of recently developed stent-based endovascular electrode array technology [11–13] to mitigate some ethical concerns regarding BCIs—specifically the safety, health risks, permanency, and informed consent process of the more common intracranially inserted invasive BCI devices. In Chap 5, Surjo Soekadar and colleagues discuss the likely merger of BCI with elements of another promising and rapidly advancing technology, artificial intelligence (AI). The authors build upon prior work [14–16] to provide an overview of current trends and technological challenges associated with the combination of BCI and AI. Additionally, the authors discuss the ethical and societal implications of applications of AI-enhanced BCI both in medical/therapeutic contexts and for non-medical use that could elevate human cognition beyond a "normal" baseline. In Chap. 6, Elisabeth Hildt builds upon prior work [17, 18] to consider the ethical and policy implications of the future of one application enabled by BCI, namely brain-to-brain interfacing (BBI). Hildt considers the implications of a futuristic scenario in which BBI technology expands into everyday contexts, leading to issues around autonomy, shared agency, identity, and privacy.

2 Overview of Part Two: Ethical and Philosophical Issues

In Part Two, we present unique ethical and philosophical perspectives on BCIs, the social concerns that arise from widespread implementation of this technology, and considerations about essential features of the human experience that may be negatively or positively affected by either scientific research or BCI product design. In Chap. 7, Abigail Lang and colleagues build upon prior work [4] to present a scoping review that systematically categorizes and synthesizes all relevant academic publications pertinent to the ethical, legal, and social implications (ELSI) of BCI technology published before 2020, and after 2016, the year in which the last such scoping review was conducted [3]. The authors report that, in 2020, almost as many relevant academic papers discussing BCI ethics had been published in the years since 2016

($n = 34$) as had previously been identified in all years prior to 2016 ($n = 42$). Lang and colleagues' analysis serves as a useful overview and introduction to this rapidly growing body of knowledge. In Chap. 8, Brielle Lillywhite and Gregor Wolbring present the results and analysis of a survey conducted exploring the perceptions of science, technology, engineering and math (STEM) undergraduates on BCI's impact on the ability for users to have a good life. Drawing from prior work [19, 20], the authors argue that their survey results reveal a techno-optimistic perception of the impact of BCIs among future STEM professionals. In Chap. 9, James Hughes adds to a body of work on moral enhancement [21] and social issues around "cyborgization" [22] by discussing the potential for brain stimulation and BCI to modulate moral emotions, cognition, and behavior. Hughes considers the possibility that healthy individuals could soon use BCI to inhibit or boost moral thoughts and emotions and proposes a model of six virtues that could be targets for such neuromodulation. In Chap. 10, Andreas Schönau and Rajesh P. N. Rao consider the ethical, moral, and social justice implications of brain co-processors: BCI interfaces that utilize AI to convert brain activity into brain stimulation patterns for restoring or augmenting brain function. The authors draw on their considerable experience in experimental BCI work [23–25] to review different kinds of current brain co-processors, potential future applications of the technology, and the resulting ethical issues that may arise.

3 Overview of Part Three: Legal and Policy Implications

In Part Three, we present unique, practical perspectives on BCIs, the regulatory landscape that provides the framework for implementation of this technology, and considerations about commercialization and downstream consequences for law and policy. In Chap. 11, Robert H. Blank summarizes the policy implications arising from BCI. The chapter draws on his prior work [26, 27] and extensive expertise in "brain policy" to contextualize the complex regulatory framework, national culture, and legal system of the US, including public funding of BCI research by the National Institutes of Health (NIH), development of military applications through the Defense Advanced Research Projects Agency (DARPA), and efforts to regulate the technology by the FDA. In Chap. 12, Marc Blitz and Woodrow Barfield draw on prior work [28, 29] to provide an analysis of the legal challenges presented by the advent of BCI devices that can aid in the formation and retrieval of memories in humans, including potential implications of the technology in US constitutional law. In Chap. 13, Nadine Liv and Dov Greenbaum introduce the nascent subfield of cybersecurity and neuroscience, "cyberneurosecurity." The authors draw on prior work [30] to advocate for the regulation of the many ethical and policy issues that arise from widespread use of BCI technology, including privacy concerns, manipulation of neural signals, physical and psychological harms resulting from hacking, and self-hacking on the part of BCI users. In Chap. 14, Michael Pflanzer presents an in-depth introduction to the FDA's guidance and permissive stance toward regulating BCI and argues that the resulting de facto deregulation of BCI leads to risks for the safety and effectiveness of the rapidly advancing technology. Finally, in Chap. 15, Adam Molnar and colleagues from Neurable, a company producing noninvasive BCI devices for commercial use, draw on their extensive experience in research and development in BCIs [31] to provide an insider's perspective into ethical decision-making within a private neurotechnology company and how those decisions are informed by different stakeholders including those within the organization, its end-users, the general public, and policy makers.

4 Conclusion

As can be gleaned from the discussion above, there is already a rich body of work on the ethical and regulatory aspects of BCI technologies, and the chapters contained in this volume build upon that body of knowledge to provide an up-to-date overview of the most salient ethical and policy considerations of BCI. This book is the first such venue (to our knowledge) where academic researchers of ethics and policy are presented alongside industry representatives conducting experimental work in both wearable (Neurable) and implantable (Synchron) BCI technologies as equal participants "at the table." We hope to encourage further dialogue around ethics and policy of BCI between experts in academia, public policy, and in the private commercial sphere. That said, it is important to ensure that ethics keeps pace with the rapid engineering advances in BCI and that policy does not lag behind, as each iteration of BCI technology carries with it unique challenges based on the type of interface, the population of users, and specific security concerns which may arise in the future.

The general picture painted about the future of BCI by the unique collection of contributions to this volume is one of cautious optimism. However, considering the unique potential for future applications of the technology, including for human enhancement purposes [32], it is vital for society to continue to enable important ethics and policy work in order to anticipate and mitigate any potential deleterious effects. Consider as an analogy the example of genetic material (e.g., blood) collected and preserved before the advent of genetic sequencing. At that time, science and technology were not advanced enough to glean more than general information (e.g., blood type) about the individual that deposited the said genetic material. However, with the advent of DNA sequencing, precise markers and detailed identification of the individual became not only possible but came into widespread use in both legal (e.g., law enforcement) and commercial (e.g., establishing ancestry) applications. It is possible that, in a similar vein to this analogy, brain data collected by today's BCI technologies, and currently used for training algorithms, may be used to identify or even influence individuals in the future. Even though in this volume we try not to be "alarmist" about the future of BCI, we believe it is uncontroversial to assert that more structured conversations between industry, academia, and government are necessary, and that such dialogue will help to ensure a future for BCI that is beneficial to society.

References

1. Kübler A. The history of BCI: from a vision for the future to real support for personhood in people with locked-in syndrome. Neuroethics. 2020;13:163–80. https://doi.org/10.1007/s12152-019-09409-4.
2. Vidal JJ. Toward direct brain-computer communication. Annu Rev Biophys Bioeng. 1973;2:157–80.
3. Burwell S, Sample M, Racine E. Ethical aspects of brain computer interfaces: a scoping review. BMC Med Ethics. 2017;18:60. https://doi.org/10.1186/s12910-017-0220-y.
4. Coin A, Mulder M, Dubljević V. Ethical aspects of BCI technology: what is the state of the art? Philosophies. 2020;5:31. https://doi.org/10.3390/philosophies5040031.
5. Han JJ. Synchron receives FDA approval to begin early feasibility study of their endovascular, brain-computer interface device. Artif Organs. 2021;45:1134–5. https://doi.org/10.1111/aor.14049.
6. US Food and Drug Administration. FDA Globalization. 2022. https://www.fdagov/international-programs/fda-globalization.
7. Friedrich O, Wolkenstein A, Bublitz C, Jox RJ, Racine E, editors. Clinical neurotechnology meets artificial intelligence. Cham: Springer; 2021. https://doi.org/10.1007/978-3-030-64590-8.
8. Gilbert F, Tubig P. Cognitive enhancement with brain implants: the burden of abnormality. J Cognit Enhancem. 2018;2:364–8. https://doi.org/10.1007/s41465-018-0105-0.
9. Gilbert F, O'Brien T, Cook M. The effects of closed-loop brain implants on autonomy and deliberation: what are the risks of being kept in the loop? Camb Q Healthc Ethics. 2018;27:316–25. https://doi.org/10.1017/S0963180117000640.
10. Gilbert F, Cook M, O'Brien T, Illes J. Embodiment and estrangement: results from a first-in-human "intelligent BCI". Trial Sci Eng Ethics. 2019;25:83–96. https://doi.org/10.1007/s11948-017-0001-5.
11. Raza SA, Opie NL, Morokoff A, Sharma RP, Mitchell PJ, Oxley TJ. Endovascular neuromodulation: safety profile and future directions. Front Neurol. 2020;11:351. https://doi.org/10.3389/fneur.2020.00351.
12. Sefcik RK, Opie NL, John SE, Kellner CP, Mocco J, Oxley TJ. The evolution of endovascular electroencephalography: historical perspective and future applications. Neurosurg Focus. 2016;40:E7. https://doi.org/10.3171/2016.3.FOCUS15635.
13. Oxley TJ, Yoo PE, Rind GS, Ronayne SM, Lee CMS, Bird C, et al. Motor neuroprosthesis implanted with neurointerventional surgery improves capacity for activities of daily living tasks in severe paralysis: first in-human experience. J Neurointerv Surg. 2021;13:102–8. https://doi.org/10.1136/neurintsurg-2020-016862.
14. Soekadar SR, Haagen K, Birbaumer N. Brain-computer interfaces (BCI): restoration of movement and thought from neuroelectric and metabolic brain activity. Underst Complex Syst. 2008;2008 https://doi.org/10.1007/978-3-540-74479-5_11.
15. Soekadar SR, Birbaumer N, Slutzky MW, Cohen LG. Brain-machine interfaces in neurorehabilitation of stroke. Neurobiol Dis. 2015;83:172–9. https://doi.org/10.1016/j.nbd.2014.11.025.
16. Soekadar SR, Witkowski M, Gómez C, Opisso E, Medina J, Cortese M, et al. Hybrid EEG/EOG-based brain/neural hand exoskeleton restores fully independent daily living activities after quadriplegia. Sci Robot. 2016;1:eaag3296. https://doi.org/10.1126/scirobotics.aag3296.
17. Hildt E. What will this do to me and my brain? Ethical issues in brain-to-brain interfacing. Front Syst Neurosci. 2015;9:17. https://doi.org/10.3389/fnsys.2015.00017.
18. Hildt E. Multi-person brain-to-brain interfaces: ethical issues. Front Neurosci. 2019;13:1177. https://doi.org/10.3389/fnins.2019.01177.
19. Sample M, Aunos M, Blain-Moraes S, Bublitz C, Chandler JA, Falk TH, et al. Brain-computer interfaces and personhood: interdisciplinary deliberations on neural technology. J Neural Eng. 2019;16:063001. https://doi.org/10.1088/1741-2552/ab39cd.

20. Wolbring G. Why NBIC? Why human performance enhancement? Innovation Europ J Soc Sci Res. 2008;21:25–40. https://doi.org/10.1080/13511610802002189.
21. Hughes JJ. Moral enhancement requires multiple virtues: toward a posthuman model of character development. Camb Q Healthc Ethics. 2014;24:86–95. https://doi.org/10.1017/S0963180114000334.
22. Hughes J. Citizen cyborg: Why democratic societies must respond to the redesigned human of the future. 2004.
23. Rao RPN. Brain co-processors: using AI to restore and augment brain function. Handbook of Neuroengineering. 2021. https://doi.org/10.1007/978-981-15-2848-4_32-1.
24. Bell CJ, Shenoy P, Chalodhorn R, Rao RPN. Control of a humanoid robot by a non-invasive brain-computer interface in humans. J Neural Eng. 2008:5. https://doi.org/10.1088/1741-2560/5/2/012.
25. Rao RPN. Brain-computer interfacing: An introduction. 2011. https://doi.org/10.1017/CBO9781139032803.
26. Blank RH, Burau V, Kuhlmann E. Comparative health policy. Bloomsbury Publishing; 2017.
27. Blank RH. Cognitive enhancement: Social and public policy issues. 2015. https://doi.org/10.1007/978-1-137-57248-6.
28. Barfield W, Williams A. Law, cyborgs, and technologically enhanced brains. Philosophies. 2017;2:4. https://doi.org/10.3390/philosophies2010006.
29. Blitz MJ. Searching minds by scanning brains. 2017. https://doi.org/10.1007/978-3-319-50004-1.
30. Cyberbiosecurity GD. An emerging field that has ethical implications for clinical neuroscience. Camb Q Healthc Ethics. 2021;30:662–8. https://doi.org/10.1017/S096318012100013X.
31. Jantz J, Molnar A, Alcaide R. A brain-computer interface for extended reality interfaces. ACM SIGGRAPH 2017 VR Village, SIGGRAPH, 2017. https://doi.org/10.1145/3089269.3089290.
32. Coin A, Dubljević V. The authenticity of machine-augmented human intelligence: therapy, enhancement, and the extended mind. Neuroethics. 2021:14. https://doi.org/10.1007/s12152-020-09453-5.

Part I

The Past, Present, and Future of BCI Technology

Posthuman Subjectivity in BCI-VR Entanglement

Mai Ibrahim

1 Introduction

The idea of minds connecting to machines has been a staple of science fiction movies and fantasy books. Today, however, with the unprecedented advancements in neuroscience, this has become a reality. Brain–computer interfaces (BCIs) are devices that connect the brain to a computer and decrypt brain activity [1]. The brain activity is then analyzed and translated into commands which in turn allows us to communicate with our surrounding environment or others who are physically or virtually present around us [2]. For example, through BCI, a user can move a ball on a computer screen by only imagining the movement [3].

Technologies like BCIs cast doubt on the understanding of the human as an independent isolated entity and question several long-held assumptions about the understanding of subjectivity. These include, for instance, the understanding of the human as a singular, self-contained, unchanging entity that is non-relational and non-transversal (Wilde & Evans, as cited in [4] p 793; Braidotti, as cited in [5] p 6). The intimate relationship between humans and technology today demands a renewed understanding of what it means to be human. Posthumanism is a growing field of study with diverse, often conflicting, theorizations and definitions that is concerned with decentering of the human due to its enmeshment in a wide array of technical, medical, informatic networks that cannot be overlooked or ignored [6]. Posthumanism should not be confused with transhumanism, which takes the enhancement of the human condition using advanced technologies as its focus [7]. Within the posthuman landscape, it has become imperative to rethink the notion of subjectivity as "emergent rather than given, distributed rather than located solely in consciousness,

M. Ibrahim (✉)
Department of Communication, Rhetoric, and Digital Media, North Carolina State University, Raleigh, NC, USA
e-mail: mwabdelk@ncsu.edu

V. Dubljević, A. Coin (eds.), *Policy, Identity, and Neurotechnology*, Advances in Neuroethics, https://doi.org/10.1007/978-3-031-26801-4_2

emerging from and integrated into a chaotic world rather than occupying a position of mastery and control removed from it" [8]. This understanding reveals that the human being is "a heterogeneous subject whose self-definition is continuously shifting, and that exists in a complex network of human and nonhuman agents and the technologies that mediate between them" [9].

This chapter builds on the understanding of the human being as an entangled subject that is embedded in an ever-shifting network of relations by exploring the integration of brain–computer interface (BCI) in virtual reality (VR) technology. The coupling of BCI and VR is a growing area of interest that promises great potential since BCI can provide a powerful tool for communication and control of virtual reality environments. Simultaneously, and with the rapid development of electronics, optics, and motion tracking technologies, VR is becoming a popular technology [10] that promises to further expand the scope of BCI.

By drawing on some canonical posthumanist and feminist works by Karen Barad and Bruno Latour, this chapter postulates that the human being is part of an intertwined network of human and nonhuman entities with perpetually shifting boundaries. I theorize how the entwinement of the human with the brain–computer interface and VR could transform the understanding of subjectivity. In addition, and by building on Barad's agential realism framework, I critique the existing model of subjectivity and propose to reframe the human–BCI-VR relationship as a posthuman subjectivity, a form of subjectivity that underscores the mutual shaping and perpetually shifting boundaries of the human. This chapter has three sections and a conclusion. The first discusses the history and development of brain–computer interfaces (BCIs). In the second section, I discuss virtual reality (VR) and some of the applications that result from the convergence of BCI and VR, namely therapy and rehabilitation, as well as gaming. In the last section, I provide a theoretical discussion of the posthuman and introduce an alternative view to the conventional understanding of subjectivity through reframing it in light of the co-constitution of the human with BCI and VR as posthuman subjectivity.

2 History of Brain–Computer Interfaces

Although BCI research has advanced at an unprecedented rate in recent years, the history of BCI is not recent. The origins of BCIs can be traced back to the 1920s when Hans Berger, a German scientist, showed that the human brain was producing electrical signals that could be measured using electroencephalography (EEG) (Berger 1929, as cited in Nam et al., as cited in [3] p 2). The development of EEG brought to the fore the idea that the brain could be used as a communication channel [3]. Berger hoped to be able to record the fluctuations of the electrical current from the human skull. In an experiment on a man with a bone defect at the forehead, Berger succeeded in showing that variations in electrical current could be recorded with electrodes placed on the skin [1].

The works of José Delgado and Eberhard Fetz in the 1960s introduced a rudimentary version of BCIs as we know them today [11]. Delgado developed an implantable chip that was used to simultaneously stimulate the brain and send electrical signals to the brain, permitting the human subject to move freely [11]. In a similar well-known experiment, he used a "stimoceiver" that sent electrical signals to a bull's brain and manipulated its aggressive impulses [11]. In the same spirit, Fetz showed that monkeys can control their brain activity to obtain food rewards [11]. His experiment demonstrated that the activity of a single neuron in the monkey's cortex could be acclimatized to control a needle's movement which was coupled to the neuron's firing rate. The monkeys learned to move the needle until it reached a certain threshold, which resulted in receiving a reward [11].

It wasn't until 1973 when Jacques Vidal, a Belgian researcher, coined the term "Brain-Computer Interface" in his seminal paper (Vidal 1973, as cited in Nam et al., as cited in [3] p 2). He defined BCIs as the use of brain signals to control external tools such as computers or prosthetic devices [3]. Vidal's paper called attention to the "evoked responses" of the brain upon exposure to sensory, visual, auditory, or somesthetic stimuli [2]. In his article, which remains influential to date, Vidal introduced a system that translates EEG signals into computer control signals [12] and delineated all necessary elements to build a BCI [1]. It is worth noting that the first real-time BCI came to light in the 1990s [3]. The basis of clinical trials, however, was founded when Sterman and colleagues undertook some experiments that found that "instrumental training of sensorimotor rhythms in cats increased seizure thresholds" (Sterman et al. 1969, as cited in Chaudhary et al., as cited in [12] p 513). Subsequent case studies using similar techniques in humans with epilepsy resulted in a reduction of grand-mal seizures (Sterman et al. 1972, as cited in Chaudhary et al., as cited in [12] p 514).

The beginning of the twenty-first century witnessed a radical expansion in BCI research that had become a field of its own [3]. Central to BCI research today is the distinction made between active, reactive, and passive BCI. Active BCI revolves around the presumption that users can manipulate their brain activity and consequently are able to give commands to a brain-controlled device [2]. Being in an angry or relaxed state, for instance, is revealed through brain activity. Similarly, imagining the movement of a particular body part can be picked up by electrodes which results in the movement of one's wheelchair in a particular direction or controlling one's avatar in video games [2]. This is to say that conscious manipulation of brain activity results in controlling the environment [2]. While in active BCI users manipulate their brain activity to issue commands, in reactive BCI, the application generates stimuli that give rise to changes in brain activity when the user pays attention to them [2]. Such stimuli are usually presented on a computer but can also be presented auditorily or by touch [2]. In passive BCI, the user is not required to evoke a particular imagery, provide a command through their brains, or focus on an external stimulus; instead, brain activity is deployed to induce changes in the environment [2]. Users are simply observed, and their brain activity is measured. Ideally, users are not aware of being measured to prevent the transformation of passive BCI into active BCI in case users become cognizant of their mental states [2]. Affective BCIs are subcategories of passive BCIs which monitor the users' affective states such as joy or fear in order to create applications which respond to those states (Mühl et al. 2014, as cited in Nam et al., as cited in [3] p 4).

Today, a wide range of BCI applications are being developed and explored, ranging from lie detection and security, to gaming, education, and art, as well as mental and physical augmentation of human capacities [11]. Invasive (implanted) as well as non-invasive (outside the body) BCIs are developed for patients suffering from disorders that restrict their ability to fully interact with the world, such as patients with locked-in syndrome [13]. Similarly, some BCI devices have moved into clinical trials for the purpose of providing artificial vision or bionic eyes, while others are used to help restore human hearing through cochlear implants [14]. In addition, BCIs are being deployed for non-medical purposes such as gaming and human–technology interfaces [13]. Coin and Dubljević [14] contend that it seems probable that "advances in BCI technology in the near future could allow for the enhancement of vision, hearing, and mobility beyond the capabilities of a normal human" (Coin & Dubljević, as cited in [14] p 286). Such expansive applications have resulted in increased interest in BCI technology outside the BCI field such as in human–computer interaction, as well as for the purposes of scientific research [3].

3 Virtual Reality

A natural companion to BCI is virtual reality (VR) [15], which could be broadly defined as "a technological reproduction of the process of perceiving the real" [16]. To understand the BCI-VR alliance, it is necessary to briefly touch on the development of VR technology. The term "VR" was coined by Jaron Lanier when he founded VPL Incorporated, although many track this back to Ivan Sutherland [17]. In his 1965 paper titled *The Ultimate Display*, Sutherland outlined the model for human–computer interface that has inspired the thinking about virtual environments (Sutherland, as cited in [18] p 964). Sutherland's idea was that the computer display could produce a simulation of the physical world [18]. As Ken Hillis argued in *Digital Sensations* [16], the development of VR has been guided by first "an ongoing (Western) motivation to alter conceptions of space" (p 1–2). It wasn't however until 1970 when Sutherland and a team of researchers developed the first head-mounted display system (Rheingold, as cited in [18] p 964) which acted as a starting point for the takeoff of VR technology in later years in areas such as art, robotics, military [18], and gaming.

VR typically functions through bracketing the external physical world and supplanting it by a virtual simulated representation that is distinguishable from yet dependent on the real. It uses "computerized and behavioral interfaces to simulate the behavior of 3D entities such as people, places and objects in a virtual environment (VE)" (Hudson et al., as cited in [19] p 459). Central to the VR experience is the feeling of immersion which is used in different contexts, yet mostly refers to games and virtual reality [20]. Immersion can be defined as "a feeling of isolation from the real world" where performing tasks can increase feelings of immersion [21]. Stated differently, immersion is the manipulation of the user's sensory experience through computer graphics, haptics, sounds, affective computing, and advanced user interfaces that work on increasing the sense of presence [22]. It has to do with what Bollmer and Suddarth [23] explain as "a phenomenal sense of absorbing presence elsewhere" which is produced through a necessary coupling between the body and the technology [23].

VR has become a prominent medium deployed for a range of different purposes. Ranging from shows that aim to attract large audiences, such as *Björk Digital*, to others which force the viewers to witness a violent scene, such as Jordan Wolfson's 2017 *Real Violence* [24], and still others whose purpose is to foster empathy, such as *Project Syria*, a 2013 VR work directed by journalist Nonny de la Peña, that simulates the bombing of Aleppo [25], VR is a medium that is gaining more prominence lately. Despite its abundant representation in popular media and depictions in books demonstrating its great success, there is no clear direction as to where this technology may be headed. Yet, the deployment of VR in a wide array of fields engendered utopian predictions that VR will create new understandings of space and ability to travel to unreachable destinations (Cheong, as cited in [26]), as well as foster new social and cultural relations that produce a sense of empathy (Milk, as cited in [26]). There are, on the other hand, some skeptical considerations including concerns of loss of identity and self (Batchen, as cited in [26]) and a "culture of increasing surveillance and voyeurism" [16].

Notwithstanding those disparate perspectives on VR, its powerful potential can't be underplayed. The companionship between VR and BCI is one of the notable areas demonstrating this. Friedman [15] posits that both technologies complement each other given their powerful capacities. BCI introduces new ways for manipulation and control in VR, while VR provides rich feedback for BCI in a safe environment [15].

4 The Convergence of BCI and VR in Therapy & Gaming

BCIs have been adopted as novel technology that merged with VR producing new ways of interaction in virtual environments. BCI makes it conceivable to explore this natural integration of the brain and VR [27]. For instance, BCIs can be used for navigation in VR as well as controlling avatars and the virtual worlds in full [27]. There are several prototypes that exist today offering innovative ways for users to "navigate in virtual scenes or manipulate virtual objects solely by means of their cerebral activity, recorded on the scalp via electroencephalography (EEG) electrodes" (Lécuyer et al., as cited in [28] p 66).

At the same time, VR offers a safe and controlled environment that provides more insight into brain responses [28]. It is perceived as a new "feedback modality for BCIs and a safe test bed to provide subjects with the opportunity to train and test the BCI applications before using them in reality" (Leeb et al., as cited in [29] p 30). The recent development of VR head-mounted devices (HMD) has "paved the way to the commercialization of combined BCI+VR technology" (Cattan et al., as cited in [30] p 2). The HMD design supports the embedding of the EEG electrodes required for BCIs [30]. Such advancements have attracted the attention of VR and BCI communities alike given the powerful potential that this convergence could result in. For instance, Cattan et al. [30] point out that the Neurable Company has announced a product that combines an HTC Vive, which is referred to as an active device, with an EEG cap [30].

Many BCI-VR applications, mostly based on active and reactive BCIs discussed earlier, have been proposed. In a BCI-VR system, users interact with virtual environments by regulating their brain activity, whereas in a reactive BCI-VR system, VR generates stimuli which prompts changes to the brain activity that results in active controlling of objects in the virtual environment (Kathner et al., Faller et al., & Beveridge et al., as cited in [29]). In addition, the interaction between BCI and VR technology produces new tasks such as navigation to change the view or the control and manipulation of virtual objects in the virtual environment. When it comes to navigation in the virtual world, BCI permits users to control the camera position by using brain signals such as "left- or right-hand motor imagery (MI) or two steady-state visual-evoked potentials (SSVEPs) at different frequencies" (Lécuyer et al., as cited in [28] p 66). For instance, MI BCIs have been used to rotate in a virtual room and move along a virtual street, demonstrating that BCI "could be used to control locomotion events in a CAVE-like [Cave Automatic Virtual Environment] setting" (Friedman et al., as cited in [27] p 101). As for the manipulation of virtual objects, Lécuyer et al. [28] state that most BCIs are based on "P300 or SSVEP signals" where virtual objects deliver "a stimulus that triggers a specific and recognizable brain signal that draws the user's attention to the associated object to select and manipulate it" (Lécuyer et al., as cited in [28] p 66). Such BCIs allow users to control devices such as a virtual TV or lamp by providing the users with a range of commands, each of which is associated with a specific activity (Bayliss, as cited in [28]).

The convergence of BCI and VR technology proves to be of mutual benefit due to the multimodal and multisensory experience gained through VR, which provides a more immersive experience, while BCI could work as a channel to manipulate virtual scenarios as opposed to standard controllers. The addition of BCI to VR technology reduces the need for physical interaction with the environment [2]. This is to say that the use of VR technology comes with promises of "immersion and the creation of interactive, explorable scenes" for the field of BCI [2]. To further explore those possibilities, I emphasize the discussion of converging BCI and VR for therapy and rehabilitation as well as gaming, as two domains that highlight the significance of this convergence.

4.1 Therapy

Therapy and rehabilitation are core areas that have been strongly influenced by the development of both BCI and VR in recent years. In one of their pilot studies, Salisbury et al. [31] assessed the efficacy of VR training used for cognitive rehabilitation for individuals suffering from stroke, brain neoplasm, and anoxic injury. The study focused on treatments associated with certain functions such as cognitive flexibility and memory, among others. While BCI is deeply entrenched in this area, VR is rapidly catching up since it affords a high degree of immersion and feelings of presence that cannot be available otherwise, providing the opportunity to explore new therapeutic approaches that cannot always be implemented in reality [2]. Miloff et al. [32], for instance, underscore the increasing effectiveness of VR in exposure therapy for phobias. In a similar work, Dunn and colleagues [33] review several studies on the role of VR in the treatment of phantom limb pain which note a reduction in pain.

In a different study, Vourvopoulos et al. [34] developed NeuRow, which is a self-paced BCI neurogame in an immersive VR environment with the purpose of motor imagery training (MI). Motor imagery training, which is "the mental rehearsal of movement -without any muscle activation- and is a mental ability strongly related to the body or 'embodied' cognition" (Hanakawa, [35] as cited in Vourvopoulos et al., as cited in [34] p 1), can provide novel communication means to neurologically impaired patients [34]. The game uses "multimodal stimulation through vision, sound and vibrotactile feedback and delivered through a VR Head Mounted Display" (Vourvopoulos et al. 2019, 1). The fusion of BCI and VR provides an illusion of movement to individuals with low levels of motor control. In their designed game, the BCI-VR task involves rowing a boat solely through mental imagery with the purpose of collecting flags (Vourvopoulos et al., as cited in [34] p 6). Similarly, Škola and Liarokapis [36] examine the effect of VR embodiment on motor imagery, a prevalent BCI paradigm that "requires users to consciously replay bodily motor action" (Vourvopoulos et al., as cited in [34] p 59). In their study, the motor imagery BCI training was implemented in a VR environment (Škola and Liarokapis, as cited in [36] p 60). They employed VR to create virtual embodiment "showing the person's hands executing the detected imagined movements" [2]. The results demonstrated that the group trained using embodied VR exhibited higher accuracy [36].

4.2 Gaming

Games can be defined as "a mental contest, played with a computer according to certain rules for amusement, recreation, or winning a stake" (Zyda, as cited in [22] p 25). The integration of BCIs within games appears to be promising and is thus "one of the oldest domains for which the combination of BCI and VR technology has been discussed" [2]. BCI and VR technologies are considered "excellent candidates for enhancing the possibilities of entertainment and satisfaction in video games" since both enhance immersion and consequently increase feelings of amusement (Cattan et al., as cited in [30] 1). VR technologies are suitable companions to BCIs, which are in turn considered efficient interaction devices. Additionally, it has become widely accepted that VR is a competent medium for understanding and enhancing BCI technologies (Lécuyer et al., as cited in [28] p 66).

Immersive games work with three different inputs, namely: visual, auditory, and mental [20]. According to Cattan et al. [30], since BCI works to "transform 'mental' signals into input commands, such an interface may play a unique role in the mentalization process involved in the feeling of immersion" (Cattan et al., as cited in

[30] p 2). In their study, they highlight that BCI integrates well with games that require concentration and thinking such as strategy, puzzles, simulation, among others, while simulation and adventure games work best for VR.

Coyle and colleagues [37] designed a spaceship game with the purpose of testing BCI's users' ability to control objects on the screen. In the game, the user is required to control the location of a spaceship using motor imagery to avoid asteroids. A key advantage of the game is that while users can manipulate the position of the spacecraft, "the asteroids are timed to be consistent with the timing of standard motor imagery trials and appear on either right or left of the screen so that data during game play can be triggered and labeled" (Coyle et al., as cited in [37] p, 1).

5 The Emergence of the Posthuman

The rapid surge in the development of technologies in general and the merging of humans with technologies in particular has resulted in questioning what it means to be human today. A vast body of work has explored the meaning of posthuman which is a highly contested term with multiple definitions and understandings. Some of those accounts have gained attention due to their version of the posthuman as a "disembodied, high-tech, transcendental subjectivity" [4]. The posthuman is oftentimes portrayed as utopian in a sense that would potentially allow us to overcome, for instance, age and class [4]. Nick Bostrom [7] postulates that being posthuman means possessing general capacities that exceed the maximum attainable capacities by current human beings. Other times, the posthuman is depicted as dystopian in the case of an apathetic and inhuman cyborg [4]. Both views present different understandings of what posthuman is.

Broadly speaking, the posthuman is related to the radical deconstruction of the human which began in the 1960s. Posthumanism does not grant primacy to the human, and by deconstructing the human, it seeks to destabilize and reinvestigate dualisms such as human/machine, life/death, and natural/artificial [38]. In Philosophical Posthumanism, Francesca Ferrando [38] explains that the term "posthuman" has become a key concept that is used to show the urgency of redefining the notion of the human, following the "onto-epistemological, as well as scientific and bio-technological developments" (Ferrando, as cited in [38] p 1). Cary Wolfe [6] posits that the posthuman is not the evolution of the human into something new; it opposes transhumanism and the traditional understanding of humanism which lead to the intensification of the human. Today, there is a "new reality" where "the human occupies a new place in the universe, a universe now populated by what I am prepared to call nonhuman subjects" (Wolfe, as cited in [6] p 47). This understanding underscores the fact that the human needs to be understood as one actor among other human and nonhuman actors.

Katherine Hayles [8] argues that while the concept of the posthuman has multiple articulations, there are some foundational assumptions by which the term could be defined. Central to those is the posthuman view that the body is a prosthesis that can be manipulated or extended, and that the human being can be united with intelligent machines (Hayles, as cited in [8] p 3). A posthuman perspective holds that there are "no essential differences or absolute demarcations between bodily existence and computer simulation, cybernetic mechanism and biological organism, robot teleology and human goals" [8]. That said, Hayles [8] points out that the creation of a posthuman does not necessarily entail bodily interventions or that the human be transformed into a cyborg per se. Rather, a defining characteristic of the posthuman is the "construction of subjectivity" (Hayles, as cited in [8] p 4). We are a "posthuman collectivity, an 'I' transformed into the 'we' of autonomous agents operating together to make a self" [8]. The posthuman subject is, simply put, is a set of diverse components [8].

The concept of subjectivity has often been described as "posthuman" through the human's relation to the animal, the machine, or the environment [4]. The understanding that I want to draw on here postulates that new forms of subjectivity emerge in our relationship and co-production with technology [8]. For Hayles [8], technology reveals a more networked and co-produced understanding of the human. In the following section, I discuss how the human-BCI-VR enmeshment introduces a new understanding of subjectivity, namely posthuman subjectivity. This conceptualization of posthuman subjectivity can describe the human-BCI-VR relation as a "fluid, horizontal and relational experience between human and machine" [4].

6 Posthuman Subjectivity in the Human-Technology Entanglement

A longstanding Western assumption is that subjectivity is constructed in relation to the ownership of a biological body [39]. However, this traditional understanding of subjectivity that is bound to corporeality needs to be revisited since "any transformation in our dominant model of the body will or should effectuate a transformation in our dominant model of subjectivity" [9]. In other words, and considering the understanding of the human as a part of a shifting network of relations, it has become necessary to reevaluate the classic notion of subjectivity associated with a corporeal human body. In this section, I call this assumption into question and propose to understand posthuman subjectivity as that which arises from a mutual reciprocation of entwined entities previously defined as distinct. By drawing on Bruno Latour and Karen Barad, I want to apply the human-BCI-VR entanglement as a case study to argue that subjectivity is not exclusively human and that the human is not an independent entity. Instead, by discussing several examples, I will first build on Latour to demonstrate that the distinction between humans and nonhumans cannot be maintained. In other words, the rigid boundaries between human users, BCI, and VR technologies cannot be upheld within this human-nonhuman network. Subsequently, I build on Barad's agential realism to suggest that subjectivity is intra-actively constituted through the entanglement of the human users, the BCI, and the VR and is not limited to the corporeal human body.

Latour [40] criticizes the dualism between nature and society or subject and object that provide the foundation for modernist thought. He rejects the separation arguing that it has never existed in pre-modernism and highlights the need to rethink dualism because while modernity invokes such segregation, this is not actually the case. Humans and technology are always entangled. He redefines the social to include nonhumans, which range from a doorknob to technologies and scientific apparatuses. Therefore, for Latour [41], everything is part of the social construction of the world, including technology.

Latour [42] introduced the concept of "translation" which does not refer to the shift from one thing to another, rather, translation is about mediation and "the creation of a link that did not exist before and that to some degree modifies the original two" (Latour, as cited in [42] p 179). So, in discussing whether guns kill people or people kill people, Latour points out that neither is responsible for the killing, but someone else is, namely "a citizen-gun, a gun-citizen" (Latour, as cited in [42] p 179). An actor with a gun in his hand is not the same person as an actor without one. Therefore, a more productive approach to study technoscientific practices includes what Latour [42] labels as "symmetry" where subjects and objects have a symmetrical effect on each other. Translations are symmetrical in that "you are different with a gun in your hand; the gun is different with you holding it. You are another subject because you hold the gun; the gun is another object because it has entered into a relationship with you" (Latour, as cited in [42] at p 179). This is to say that the concept of symmetry blurs the dichotomy between subjects and objects and instead focuses on their connections and associations within the given network.

When we look at the human-BCI-VR from Latour's perspective, it becomes plausible to abandon the subject-object dichotomy. It seems imperative to overcome the understanding of the human user as a subject that operates and controls the BCI-VR technology as an object. From this Latourian perspective, we are more attentive to "the understanding of collectives" [42]. For instance, a human actor does not autonomously move the ball on a computer screen by imagining its movement nor does the BCI induce the motion. Similarly, individuals with low-level motor abilities do not row the boat independently, nor does the VR head-mounted display. Instead, "responsibility for action must be shared among the various actants" (Latour, as cited in[42] at p 180). The human user and the BCI-VR are together responsible for the action. Outputs, such as moving the ball or rowing a boat, are a result of the simultaneous interaction of the different individual actors in this network. As Latour [42] writes, "action is a property of associated entities" (Latour, as cited in [42] p 182).

Every individual actor in any given network could only be afforded action by the other actors. For instance, Yeh et al. [29] proposed a BCI-VR car racing game which allows multiple users at different locations to mentally play the game. The car's movement is prompted by electromyographic (EMG) activity; therefore, the game is "based on a hybrid BCI control (EEG plus EMG)" [29]. Consequently, the car racing game is "a property of the whole association of entities" that includes the players, EEG, EMG, massive amounts of data, servers, cloud-computing-based signals, etc. [29]. Action, accordingly, is not a human characteristic but the relations and associations of the different actors [42]. This is not to claim that either BCIs or

VR can create outputs independently of the human user; rather, the different outcomes are a co-production of the human-BCI-VR entanglement.

While this understanding of actors as a part of a more extensive network is plausible, the nature of interaction between the individual actors remains somewhat murky in Latour's framework. In other words, while Latour makes it clear that no one acts alone, his framework seems to fall short when it comes to highlighting the way action itself is produced between the actors. Barad, on the other hand, surveys the nature of this interaction, which she calls "intra-action." In Barad's [43] agential realism framework, matter is a dynamic and intra-active actor. Barad posits that matter is about engagement with a new understanding of the world where bodies and the environment are intra-actively co-constituted. Intra-action, unlike the conventional notion of interaction, presumes the prior existence of independent entities and is understood as "the mutual constitution of objects and agencies of observation within phenomena" ([43] p 197). That is to say, while interaction is concerned with separate entities that interact, intra-action is about the inseparability and mutual emergence of the subject and object through their interaction [44]. In this way, Barad (2007) challenges dichotomies between the human and nonhuman, material and discursive, natural and cultural and stresses that subjects and objects are intertwined as they intra-act with one another. Intra-action conceptualizes that "it is the action between (and not in-between) that matters" ([44] p 14). Entities, therefore, Barad posits, do not exist independently, but intra-act with one another [44].

We are not unitary human subjects independent of our interaction with "others," where "others" include all human and nonhuman agents. Rather than understanding that either "me" or the "BCI" or the "VR" is individually responsible for the actions, we need to understand that action stems from the intra-action between myself, the BCI and the VR resulting in a distributed subjectivity. Such a "relationality implies a more rhizomatic relationship between human and machine, as each is dependent on the other in order to emerge as an entity" ([45] p 376). For instance, central to the integration of BCI and VR with the human user is the way in which cognition and brain signals are distributed between the human and the technologies displacing a single unified human subjectivity, rendering it posthuman. Ma et al. [46] discuss gaze input, which solely depends on eye movement, as an input method of interaction in VR applications. They designed and implemented "a hybrid gaze-based text-entry system in VR for high-speed typing" which combined an SSVEP-based BCI with eye-tracking module in virtual reality head-mounted display resulting in the detection of the eye gaze direction and text entry (p 264). Similarly, Lécuyer et al. [28] discuss the way some games involve controlling a 3D character within a virtual environment where the objective "is to gain one-dimensional control of the character's balance on a tightrope using only the player's EEG" (Lécuyer et al., as cited in [28] p 67). These examples demonstrate that neither the user nor the BCI or VR technologies could act independently of the other. The human, as well as the act of text entry in the first example and controlling the virtual character in the second are all interdependent, creating a posthuman subjectivity. In such instances, the boundaries disintegrate, and human-technology is understood as enmeshed. There exists a dynamic and inseparable intra-action between both [4].

The intra-active relationship between the human and technology complicates notions of a single body and single subjectivity given the various actors (Sundén [47] as cited in Wilde & Evans, as cited in [4] p 792). By focusing on the entanglement of the users with BCI-VR and the way action is distributed, it becomes difficult to demarcate the boundaries which consequently results in engendering a posthuman subjectivity. Hayles [8] stresses that our bodies do not end at our skin. Referring to her VR experience, she implies that "in these systems, the user learns, kinesthetically and proprioceptively, that the relevant boundaries for interaction are defined less by the skin than by the feedback loops connecting body and simulation in a technobio-integrated circuit" (Hayles, as cited in [8] p 27). This understanding is not about "leaving the body behind but rather of extending embodied awareness in highly specific, local, and material ways that would be impossible without electronic prosthesis" (Hayles, as cited in [8] p 291). It points to the ways in which subjectivity transcends bodily boundaries through technology. The human-BCI-VR entwinement demonstrates that the rigid distinction between biology and technology collapses. In this co-constitutive relationship, it becomes difficult to identify where the human body ends, and the technology begins. What is being created here demonstrates a "posthuman subjectivity that blends the embodied materiality" of the human user with the technical and informational BCR and VR technologies [4]. This posthuman understanding demonstrates the impossibility of any separation between the "self," "other," and "environment" [4]. The "I" is no longer an isolated subjectivity but is rather "constructed and experienced in a permeable fluidity with a range of different external stimuli" ([45] p 376). Subjectivity in this sense is fragmented rather than unified and no longer pertains to the human "I" but is rather the result of the interaction between the different intra-acting actors. Put differently, subjectivity is the product of mutual intra-actions among various interdependent subjectivities. This is to say that the body is always interacting with the world, and therefore changes in the body can change subjectivity itself. In other words, "insofar as subjectivity is tied to the body…that a shift in the formulation of bodies can have a direct implication on the formulation of subjectivity" [9].

7 Conclusion

In discussing the human-BCI-VR entanglement, it becomes pertinent to underscore that the idea of an independent human subject no longer stands. In this chapter, I have attempted to describe how this human–technology relationship complicates the understanding of the subject as a unified entity and necessitates a re-thinking of the notion of the self as distributed with dynamic and perpetually shifting permeable boundaries. This requires a shift from the limited understanding of subjectivity that is bound to corporeality to a more encompassing conceptualization of subjectivity, namely posthuman subjectivity. Subjectivity should rather be re-defined as "the notion that the human being is not an independent and autonomous entity with clear cut boundaries but a heterogeneous subject whose self-definition is continuously shifting, and that exists in a complex network of human and nonhuman agents and the technologies that mediate between them" [9]. If we maintain the misguided understanding of the human as an autonomous standalone subject, as Hayles [8] contends, our relationship to technology will remain that of real life on the one hand and an illusion on the other, concealing the influence of the development of technologies [8].

References

1. Kübler A. The history of BCI: From a vision for the future to real support for personhood in people with locked-in syndrome. Neuroethics. 2019;13(2):163–80.
2. Nijholt A. Brain art: brain-computer interfaces for artistic expression. Cham: Springer; 2019.
3. Nam CS, Nijholt A, Lotte F. Brain–computer interfaces handbook: technological and theoretical advances [Internet]. 1st ed. Boca Raton, FL: Taylor & Francis, CRC Press; 2018. https://www.taylorfrancis.com/books/9781351231947
4. Wilde P, Evans A. Empathy at play: Embodying posthuman subjectivities in gaming. Convergence. 2019;25(5–6):791–806. https://doi.org/10.1177/1354856517709987.
5. Braidotti R. Affirmative ethics, posthuman subjectivity, and intimate scholarship: a conversation with rosi braidotti. In: Strom K, Mills T, Ovens A, editors. Advances in research on teaching. Bingley: Emerald Publishing Limited; 2018. p. 179–88. https://doi.org/10.1108/S1479-368720180000031014/full/html.
6. Wolfe C. Introduction: What is posthumanism. In: What is posthumanism? Posthumanities series. Minneapolis, MN: University of Minnesota Press; 2010. p. 12–35.
7. Bostrom N. Why I want to be a posthuman when I grow up. In: The transhumanist reader: classical and contemporary essays on the science, technology, and philosophy of the human future. Chichester: Wiley-Blackwell; 2013. p. 28–53.
8. Hayles NK. How we became posthuman: virtual bodies in cybernetics, literature, and informatics. Chicago, IL: University of Chicago Press; 1999.
9. Sharon T. Human nature in an age of biotechnology. Dordrecht: Springer; 2014. (Philosophy of Engineering and Technology; vol. 14). https://doi.org/10.1007/978-94-007-7554-1.
10. Yao Z, Wang Y, Yang C, Pei W, Gao X, Chen H. An online brain-computer interface in mobile virtual reality environments. ICA. 2019;26(4):345–60. https://www.medra.org/servlet/aliasResolver?alias=iospress&doi=10.3233/ICA-180586
11. Rao RPN. Brain-computer interfacing: an introduction. First paperback edition. Cambridge/New York/Melbourne/New Delhi/Singapore: Cambridge University Press; 2019.
12. Chaudhary U, Birbaumer N, Ramos-Murguialday A. Brain–computer interfaces for communication and rehabilitation. Nat Rev Neurol. 2016;12(9):513–25. http://www.nature.com/articles/nrneurol.2016.113
13. Coin A, Mulder M, Dubljević V. Ethical aspects of BCI technology: what is the state of the art? Philosophies. 2020;5(4):31. https://www.mdpi.com/2409-9287/5/4/31
14. Coin A, Dubljević V. The authenticity of machine-augmented human intelligence: therapy, enhancement, and the extended mind. Neuroethics. 2021;14(2):283–90. https://doi.org/10.1007/s12152-020-09453-5.
15. Friedman D. Brain-computer interfacing and virtual reality. In: Nakatsu R, Rauterberg M, Ciancarini P, editors. Handbook of digital games and entertainment technologies. Singapore: Springer Singapore; 2015. p. 1–22. https://doi.org/10.1007/978-981-4560-52-8_2-1.
16. Hillis K. Digital sensations: Space, identity, and embodiment in virtual reality. Minneapolis, MN: University of Minnesota Press; 1999.
17. Heim M. The metaphysics of virtual reality. New York: Oxford University Press; 1993.
18. Schroeder R. Virtual reality in the real world: history, applications and projections. Futures. 1993;25(9):963–73. https://doi.org/10.1016/0016-3287(93)90062-X.

19. Hudson S, Matson-Barkat S, Pallamin N, Jegou G. With or without you? Interaction and immersion in a virtual reality experience. J Bus Res. 2019;100:459–68. https://linkinghub.elsevier.com/retrieve/pii/S0148296318305514
20. Brown E, Cairns P. A grounded investigation of game immersion. In: Extended abstracts of the 2004 conference on Human factors and computing systems—CHI 04. Vienna: ACM Press; 2004. p. 1297. http://portal.acm.org/citation.cfm?doid=985921.986048.
21. Patrick E, Cosgrove D, Slavkovic A, Rode JA, Verratti T, Chiselko G. Using a large projection screen as an alternative to head-mounted displays for virtual environments. In: Proceedings of the SIGCHI conference on Human factors in computing systems—CHI '00. The Hague: ACM Press; 2000. p. 478–85. http://portal.acm.org/citation.cfm?doid=332040.332479
22. Zyda M. From visual simulation to virtual reality to games. Computer. 2005;38(9):25–32. http://ieeexplore.ieee.org/document/1510565/
23. Bollmer G, Suddarth A. Embodied parallelism and immersion in virtual reality gaming. Convergence. 2022;28(2):579–94. https://doi.org/10.1177/13548565211070691.
24. Bollmer G. From immersion to empathy: the legacy of Einfühlung in virtual reality and digital art. In: Shifting interfaces. Leuven: Leuven University Press; 2020. p. 17–30.
25. Bollmer G. Empathy machines. Media Int Austral. 2017;165(1):63–76. https://doi.org/10.1177/1329878X17726794.
26. Ross M. Virtual reality's new synesthetic possibilities. Television & New Media. 2018;21(3):297–314.
27. Friedman D, Leeb R, Guger C, Steed A, Pfurtscheller G, Slater M. Navigating virtual reality by thought: what is it like? Presence Teleoper Virtual Environ. 2007;16(1):100–10. https://direct.mit.edu/pvar/article/16/1/100-110/18658
28. Lécuyer A, Lotte F, Reilly RB, Leeb R, Hirose M, Slater M. Brain-computer interfaces, virtual reality, and videogames. Computer. 2008;41(10):66–72. http://ieeexplore.ieee.org/document/4640665/
29. Yeh SC, Hou CL, Peng WH, Wei ZZ, Huang S, Kung EYC, et al. A multiplayer online car racing virtual-reality game based on internet of brains. J Syst Archit. 2018;89:30–40. https://linkinghub.elsevier.com/retrieve/pii/S1383762118301619
30. Cattan G, Mendoza C, Andreev A, Congedo M. Recommendations for integrating a p300-based brain computer interface in virtual reality environments for gaming. Computers. 2018;7(2):34. http://www.mdpi.com/2073-431X/7/2/34
31. Salisbury DB, Dahdah M, Driver S, Parsons TD, Richter KM. Virtual reality and brain computer interface in neurorehabilitation. Baylor University Medical Center Proceedings. 2016;29(2):124–7. https://doi.org/10.1080/08998280.2016.11929386.
32. Miloff A, Lindner P, Hamilton W, Reuterskiöld L, Andersson G, Carlbring P. Single-session gamified virtual reality exposure therapy for spider phobia vs. traditional exposure therapy: study protocol for a randomized controlled non-inferiority trial. Trials. 2016;17(1):60. http://www.trialsjournal.com/content/17/1/60
33. Dunn J, Yeo E, Moghaddampour P, Chau B, Humbert S. Virtual and augmented reality in the treatment of phantom limb pain: a literature review. NRE. 2017;40(4):595–601. https://www.medra.org/servlet/aliasResolver?alias=iospress&doi=10.3233/NRE-171447
34. Vourvopoulos A, Ferreira A, Bermudez I, Badia S. Development and assessment of a self-paced BCI-VR paradigm using multimodal stimulation and adaptive performance. In: Holzinger A, Pope A, Plácido da Silva H, editors. Physiological computing systems. Cham: Springer International Publishing; 2019. p. 1–22. https://doi.org/10.1007/978-3-030-27950-9_1.
35. Hanakawa T. Organizing motor imageries. Neurosci Res. 2016;104:56–63. https://linkinghub.elsevier.com/retrieve/pii/S0168010215002837
36. Škola F, Liarokapis F. Embodied VR environment facilitates motor imagery brain–computer interface training. Comput Graph. 2018;75:59–71. https://linkinghub.elsevier.com/retrieve/pii/S009784931830089X
37. Coyle D, Garcia J, Satti AR, McGinnity TM. EEG-based continuous control of a game using a 3 channel motor imagery BCI: BCI Game. 2011; 1–7.

38. Ferrando F. From new materialisms to object-oriented ontology. In: Philosophical posthumanism. London/New York: Bloomsbury Academic; 2019. p. 158–65.
39. Stone AR. The war of desire and technology at the close of the mechanical age. Cambridge, MA: MIT Press; 1995. p. 212.
40. Latour B, Porter C. Crisis. In: We have never been modern. Cambridge, MA: Harvard University Press; 1993. p. 1–12.
41. Latour B. How to resume the task of tracing associations. In: Reassembling the social: an introduction to actor-network-theory. Oxford/New York: Oxford University Press; 2005. p. 1–20.
42. Latour B. Pandora's hope: essays on the reality of science studies. Cambridge, MA: Harvard University Press; 1999.
43. Barad KM. Meeting the universe halfway quantum physics and the entanglement of matter and meaning. Durham: Duke University Press; 2007.
44. Dolphijn R, Tuin IV. New materialism: interviews & cartographies. Ann Arbor: Open Humanities Press; 2012. (New metaphysics).
45. Wilde P. I, posthuman: a deliberately provocative title. Int Rev Qualitat Res. 2020;13(3):365–80.
46. Ma X, Yao Z, Wang Y, Pei W, Chen H. Combining brain-computer interface and eye tracking for high-speed text entry in virtual reality. In: 23rd International Conference on Intelligent User Interfaces [Internet]. Tokyo: ACM; 2018. p. 263–7. https://doi.org/10.1145/3172944.3172988.
47. Sundén J. Desires at play: on closeness and epistemological uncertainty. Games Cult. 2012;7(2):164–84. https://doi.org/10.1177/1555412012451124.

"The Trauma of Losing Your Own Identity Again": The Ethics of Explantation of Brain–Computer Interfaces

Paul Tubig and Frederic Gilbert

1 Introduction

Clinical trials are underway to investigate the effectiveness of implantable neurotechnologies to treat a range of serious and confounding medical conditions, such as epilepsy, treatment-resistant depression, paralysis, dementia, and severe enduring anorexia nervosa [1]. Such trials will be more frequent so long as implantable neurotechnologies are still held as promising modes of therapy and enhancement. Yet the involvement of research participants to test these experimental technologies raises a panoply of ethical quandaries. One prominent issue is identifying and weighing the moral risks of implanting a neural device in participants, which has inspired a robust neuroethics literature. [1] There are also broader ethical and societal concerns that arise from the development and use of invasive neurotechnologies for therapeutic and enhancement purposes, such as how they could exacerbate social

[1] The neuroethical literature focuses on various topics. We know, for example, brain implant technologies, such as Deep Brain Stimulation, raise a series of ethical issues, including (a) user safety and risk-benefit analysis [2], (b) implications on notions of identity and autonomy [3, 4], (c) research ethics and informed consent, (d) justice issues [5], (e) general placebo-controlled surgical trial concerns [6], and (f) the impact of enhancement via DBS [7], ethical consequences linked with increased life expectancy of patients [8].

P. Tubig
Department of Philosophy and Religious Studies, Georgia Southern University, Statesboro, GA, USA
e-mail: ptubig@georgiasouthern.edu

F. Gilbert (✉)
Philosophy Program, University of Tasmania, Hobart, TAS, Australia
e-mail: frederic.gilbert@utas.edu.au

V. Dubljević, A. Coin (eds.), *Policy, Identity, and Neurotechnology*, Advances in Neuroethics, https://doi.org/10.1007/978-3-031-26801-4_3

inequalities, intrude on mental privacy, and create novel forms of exploitation. [2] But one ethical question that is currently underexplored is the ethical implications of neural device explantation. By explantation, we mean the procedure of removing an implanted neural device from a user, which is an option that may be open, offered, or even required after a clinical trial is completed or discontinued. Some have already raised important ethical questions related to neural device explantation, such as who should pay for the costs of explantation when research participants want the device removed, or whether researchers and clinicians ought to honor the requests of patients for their devices to be explanted when doing so would threaten their life or health [14, 15]. With the increase of neural device implantations, ethical issues of neural device explantation become increasingly more pressing since a large number of implantations may call for their removal.

This chapter focuses on the ethics of neural device explantation [16, 17]. What are the possible moral harms that could come from removing a neural implant, and what are the post-trial responsibilities of researchers to prevent or mitigate such harms? We are particularly concerned about the effects of explantation to a participant's personality, identity, autonomy, authenticity, agency, and/or self (or PIAAAS for short). [3] [29] There are some empirical findings of participants perceiving the explantation of their neural device as a serious threat to important features of who they are. These testimonies call for ethicists and researchers to be more attentive to the PIAAAS-related harms that may result from explantation and develop practices that properly recognize and attend to these harms. Here, we argue that implanted persons have a strong moral claim to their devices when they support or constitute their PIAAAS. This should be considered in the overall individual assessments of whether a device ought to be explanted. If explanting a device is a live option after a clinical trial ends, then we argue that researchers have a post-trial obligation to provide ancillary care to participants to reduce the anticipated negative effects of explantation, including any serious PIAAAS-related harms.

---

[2] Many neuroethicists have raised important justice concerns regarding the distribution of benefits and burdens of neurotechnology in a socio-historical context of inequality and bias. For example, Sara Goering and Eran Klein raise a range of justice concerns for people with disabilities, including equitable access given how disabled people are historically marginalized while bearing the burdens of novel neurotechnology research [9]. Another major worry is that neurotechnologies may only be readily available to higher socioeconomic classes, which will likely exacerbate existing inequalities, especially when such technologies are used for capability and opportunity enhancement purposes [10–13].

[3] In this chapter, we will primarily use this acronym as an all-encompassing term for personality, identity, agency, authenticity, autonomy, and self. One reason is that these terms are often conflated with one another in the neuroethics discourse. The imprecision of how these terms is used and their close relations with one another lead to a certain lack of clarity regarding what critical aspect of the person is really threatened by their use of neurotechnologies. Many neuroethicists seek to disentangle the concepts to better explain the morally troubling changes to the person that are brought about by neurotechnologies [18–28]. Here, we will sidestep this issue and use the term PIAAAS to broadly refer to the important, intertwined aspects of our ways of being and acting in the world while recognizing the inexactness of the language used to refer to them.

This chapter will proceed as follows. In Sect. 2, we will explain what explantation is and the reasons for doing it on research participants who have been implanted with a neural device for research purposes. In Sect. 3, we will consider the perspective of a research participant whose neural device will be explanted and how it could lead to troubling PIAAAS-related changes. Then in Sect. 4, we will discuss how serious PIAAAS changes are widely appreciated as important moral considerations of whether to proceed with neural device implantation. We argue that such reasoning should also extend to neural device explantation, given that such interventions can also make participants vulnerable to troubling PIAAAS-related changes. We conclude in Sect. 5 by arguing that clinicians and researchers have responsibilities toward their patients and research participants to avoid or mitigate the serious negative effects of explantation, including any concerning PIAAAS-related changes. This includes recognizing that explantation can be experienced as a traumatic event and major disruption of their sense of self and the ethical imperative to provide support—like developing exclusion criteria for explantation and providing counseling to explantees—in response to it.

2 Explantations and Why They Are Done

Neural device explantation is the removal of a neural device that has been inserted and fixed to a person's brain or part of their central nervous system for therapeutic or investigational purposes. There are various reasons for explanting a device. One reason is that the device's continued presence may endanger the physical or psychological health of the user. Neural implantation involves introducing a foreign material into the body, which then brings risks of biocompatibility. Eran Klein maps out the various safety risks of implantable brain–computer interfaces (BCI), describing how BCI components—electrodes, power systems, and data processing systems—could cause tissue damage or adverse changes in the brain [30]. Furthermore, a device may bring about undesirable psychiatric after-effects [31]. These effects include dramatic alterations in a participant's mood, troubling emotional instability, depersonalization, and feelings of alienation. When a neural device proves to be unreasonably unsafe to the user, the principles of beneficence and nonexploitation call for its removal to protect the user's health.

Another reason for explantation is that a neural device proves to be inefficacious. If a device was designed to provide therapeutic benefit and it is not demonstrating this, then there is no therapeutic or exploratory reason for leaving these devices in the participants. Yet this may not in itself be a sufficient or weighty reason to explant a device, and other reasons may have to be coupled with it since explantation is an invasive procedure that brings its own risks to persons undergoing it. Without other confounding reasons, it may more preferrable to leave the implant in the body even if it may not be functioning properly or producing the desired result.

A third reason is that the clinical trial has ended. Investigational neural devices are only useable and manageable within the research trial. During the trial, participants are supported by interdisciplinary teams that monitor and maintain their devices, gather data, recalibrate treatment parameters, and observe participants' health. Beyond the temporary setting of the research trial, there is no established infrastructure to provide ongoing support and care. This conundrum is exemplified in the case of Rita Leggett, which was profiled in *Nature Medicine* and *The New Yorker* [32, 33]. Leggett, who struggled with epilepsy, participated in a research trial to explore the use of a neural implant to detect upcoming seizures. The device was effective in helping Leggett manage her epilepsy. But the trial abruptly ended because the researchers could not sustain funding and the company eventually folded. Leggett and her husband sought to purchase the device, but they were denied, in part because there was no infrastructure in place to handle the complications of the device, such as adjusting its settings and replacing its batteries. This contrasts with other implantable devices, such as cardiac pacemakers, where there are many institutions that can support its continued use. As Joseph Fins notes, any hospital with a cardiology service can provide technical support to people with implanted pacemakers. This is not the case for people with implanted neural devices since the technology is still novel, maturing, and not yet sustainable, so support is limited to highly specialized centers [34].

A fourth reason is the legal ownership of the device. As implied from the previous reason, industries that sponsor the neural device trials play some role in the ability of research participants to continue to have access to investigational neural devices. The level of control that a company has on when and how their devices are used is unclear and likely vary according to the terms agreed upon by all stakeholders in a research program. Neurolaw, a burgeoning field of law that seeks to address the legal implications of innovations in neuroscience and neural engineering, is still catching up to address difficult legal situations brought about by the practice of explantation. It raises a pressing ethical question of whether the private ownership of neural devices integrated in the bodies of research participants violate the bodily autonomy of participants who want to keep them. Although the physical removal of neural implants requires the consent and cooperation of implanted persons, the discretionary power of companies sponsoring the trials to deny providing support to research participants to continue using their investigational neural devices is a weighty consideration for explantation.

In summary, the major reasons for explanting a neural implant range from beneficence, futility, to proprietary rights of the device [14, 35, 36]. The principle of beneficence may call for the explantation of a device when it inflicts physical and psychiatric harms on the participant. Explanting a device may also be justified by concerns of futility, both in terms of the device not producing any kind of therapeutic or informational benefits and such devices lacking the background institutional support to use and maintain them properly. Lastly, there may be proprietary motivations for explanting a device. Here, we want to argue that the decision to explant a device from a participant should also take into account the adverse effects of explantation. One such effect is the PIAAAS-related harm that may result from explantation. In the next section, we will discuss some initial empirical findings of participants' views on the possibility of PIAAAS-related changes induced by explantation.

3 "You Are Experiencing That Trauma of Losing Your Own Identity Again": The Testimony of An Expected Explantee

The ethics of explantation is complicated by the potential harm of altering the PIAAAS states of participants and patients. As explained in the previous section, there may be positive reasons for removing an implanted device from a user. But this should be weighed against some of the moral risks associated with explantation. We would like to bring attention to a particular risk that has been underrepresented in the neuroethics discourse. It is the potential link between explantation and its effects on participants' PIAAAS states. In contrast to discussions about psychological disruptions linked with DBS implantation, which appeared as early as 2002 [37], discussions about psychological adversities related to explantation were not reported till more than 15 years later [38]. We turn to a particular testimony of a BCI user to underscore the point and to raise awareness of the ramifications of BCI explantation to a person's PIAAAS.

A patient with quadriplegia volunteered to be implanted with the first-in-human, experimental brain-computer devices [39]. He was implanted with a BCI device to send signals from his brain to his muscles, which would allow him to regain some movement in his right arm, hand, and wrist. We interviewed this patient as part of a project to gain novel insight in the phenomenological impacts of BCI on users' perceived sense of self through their first-person experiences. We used qualitative methodological tools grounded in phenomenology to conduct in-depth, open-ended, semi-structured individual interviews. Interviews were based on an adapted version of the qualitative instrument first developed and tested in earlier iterations [38, 40–42] and further elaborated in [43]. Interviews were transcribed verbatim. Here, we will focus on two extracts relevant to explantation.

Interviewer: With other technologies (Neurovista, Broaden trial, etc.) we observe patients refusing and resisting getting these devices out of their body or head. What do you think goes through these patients mind when refusing?

Patient #1: I understand, I think anything that is going to help a patient to experience a better quality of life, if that system is still working, they would be extremely hesitant to give that up. Because they understand what it is like without and they do not necessarily want to go back at it, because if it is a benefit to their life, they want to sustain these benefits. For me, it has been quite something different, because as you know, going into it [trial] this wasn't a forever device, I knew it wasn't going to be something I'd be able to always use... although I couldn't prepare for that, if I was at the other end thinking it is a forever device, repairing my lost abilities, and then someone telling me: "Oh no, we're going to have to take that away from you" it would be almost the same amount of trauma as if I had my spinal cord injury all over again. Losing this ability completely again.

Interviewer: Yes, you are right, it would be another trauma, clearly a psychological harm. Some of the patients we're talking about, with the device they find themselves, they find these new capacities [...].

Patient #1: You are experiencing that trauma of losing your own identity again and try to figure out who you are; because, you know, if you really do identify with the device it becomes a part of you, and when that change [...] it would be just very similar to the trauma and the adjustments I had to make after my initial spinal cord injury.

This patient's perspective brings attention to a serious moral cost of explantation. It may bring about a significant psychological harm, especially in circumstances where users think that they'll be able to have access to the device for a long period of time. This harm, in the words of the patient, would be a loss of identity, experienced as reliving the trauma of becoming disabled in the first place.

From the patient's testimony, one can then draw the source of the identity harm. First, the identity harm may be related to re-losing the valued abilities that may have sustained or been constitutive of that patient's identity. A person's sense of autonomy or self-conception may be intimately tied to certain roles, activities, or ways of living. Accomplishing these aspects about themselves may require the possession or exercise of certain capacities. Thus, it is understandable that losing or re-losing these capacities will likely lead to serious negative disruptions to a patient's PIAAAS states.

Second, the identity harm may be related to the integration of the implanted device to their sense of self or bodily integrity. The introduction of interactive prosthetics has blurred our bodily boundaries. Studies have shown that people can and do extend their bodily representations to include wheelchairs, exoskeletons, and protheses, where these devices aren't perceived as tools separable from their users, but as an integral part of themselves [44, 45]. Today, "cyborg" is a growing identity that is gaining wider recognition, or at least, increased calls for its recognition [46, 47]. Persons with neural implants can perceive their devices as integral to their sense of self. Removing the device can then be experienced as a dramatic alteration of their way of being to the extent that it requires a difficult readjustment or re-creation of the identity, like the kind of coping and readjustment period to a new-found embodiment resulting from spinal cord injury. This reaction was acutely felt by explantee Rita Leggett, stating, "The device and I were one. We were successful. It was like taking away that part of myself that made me complete" [32].

From the various ways in which identity harms induced by explantation could be articulated, it is at least clear that explantation can be experienced as deeply traumatizing, a dramatic rupture of a lived embodiment deeply entwined to an explantee's sense of being. When a neural implant enables a patient to regain some function, like the ability to walk, explantation is, in important ways, re-inducing a patient's disability, like quadriplegia. As such, explantation can involve the serious discontinuity of a lived embodiment that bounds a patient's identity and agency. When a person is in a symbiotic relationship with their neural device, explantation can be experienced as an intrusive, and even violent, way of taking away the patient's capacities or violation of their intimate sense of being.

These cases suggest that explantation can threaten the stability of a person's PIAAAS states. It is poignantly described by the patient as undergoing the trauma of losing their identity again. This kind of vulnerability should be recognized and factored in when assessing the benefits and risks around neural implant removal from research participants and the responsibilities of researchers to their participants when their devices are indeed removed. In the next section, we will elaborate how PIAAAS-related change has been widely regarded as a serious moral concern when considering the ethics of neural device implantation. We argue that if it is a serious moral consideration when it comes to whether, when, and how we proceed with neural device implantation, then it should also be a serious moral consideration when it comes to whether, when, and how we proceed with neural device explantation.

4 Implantation, Explantation, and PIAAAS Change

Putative PIAAAS changes from the application of neurotechnologies have garnered considerable attention in the neuroethics discourse. This moral concern follows from a growing body of empirical findings of implantees experiencing dramatic changes of their psychological characteristics while going through DBS treatment. Some of these cases are extreme, such as implantees undergoing total transformations of their psychological profile to the extent that they seem to be wholly different persons. One classic case, which is described by Walter Glannon, is of a patient with advanced Parkinson's disease (PD) who received DBS treatment to mitigate his motor disorder [48]. Although the stimulation helped restore his motor functions, it also made him manic and megalomaniacal, invoking the specter of Phineas Gage [4] and how intervening in the activities of the brain could lead to dramatic revisions of the self. In response, Karsten Witt and others argue that the risk of "becoming another person" is one of the most urgent ethical problems facing DBS treatment for conditions like PD. [50] Another set of cases involve implantees having difficulties adjusting to their newfound embodiment, feeling estranged or distant from the kind of being they've become from their neural implant. For example, PD patients have reported that they don't feel like themselves during DBS treatment, acquiring abilities, or psychological and motivational states that they can't identify with. Other patients reported felt experiences of inauthenticity and heteronomy, some

[4] Phineas Gage was a railroad foreman who in 1848 suffered a severe head injury from a construction accident. An errant explosion caused an iron rod to pierce through his brain. He miraculously survived the event, but his personality changed dramatically. Prior to the accident, Gage was known to be a reserved, even-tempered person. But after the accident, he was outgoing, impulsive, and profane. His personality transformed so dramatically that Gage's friends and acquaintances described Gage as "no longer Gage" [49].

describing themselves as robots since their neural devices seem to be major springs of their thoughts, desires, and action. [5]

These kinds of potential psychological changes and feelings of alienation and depersonalization from their neurotechnology-enabled embodiment are undoubtedly disturbing because they seem to amount to PIAAAS change, where fundamental components of the self that give persons their sense of individuality, psychological and narrative continuity, and autonomy are violated. Though the point of neurotechnological intervention is to alter the physical and psychological states of persons undergoing it, there may be certain accompanying changes that are perceived as serious affronts to their integrity, one being the revision or removal of important properties that are tied to persons' self-constitutions. Persons could incur a loss due to their neural implants, namely a loss of key aspects that may be unreasonable to accept. For this reason, neurotechnological intervention could be perceived as a serious harm even though it may fulfill its intended therapeutic purpose. Thus, the risk of PIAAAS change is a key question around the ethics of implantable neurotechnologies, generating a robust discourse on the nature of this harm, its normative significance, and what are the appropriate responses to this type of vulnerability.

Here, we argue that if PIAAAS change is an important consideration in the ethics of implanting a neural device in the embodiments of persons, then this consideration should also be extended to the ethics of explanting a neural device from the embodiments of persons. So far, the concern over PIAAAS change following the excision of a neural implant has not had the same kind of moral attention as it does in the context of neural device implantation. There could be a variety of reasons for the discrepancy. One reason is that PIAAAS change from neural device explantation is under-recognized. As noted earlier, psychological disruptions linked with DBS implantation appeared as early as 2002, and yet discussions about psychological adversities related to explantation were not reported till more than 15 years later [38]. Thus, neuroethicists may still be catching up to the empirical studies of the after-effects of explantation. When wider acknowledgment is achieved, then PIAAAS change would expectedly be taken into greater account when determining the morality of neural device explantation.

A second reason may be that certain presumptions are operating in the background, like that there is a crucial moral difference between implantation-induced PIAAAS change and explantation-induced PIAAAS change. One possible explanation is that PIAAAS change from implantation is due to the intrusion and artificial influence of a foreign device, whereas PIAAAS change from explantation stems from people returning to their original, biophysical state after the cessation and removal of a foreign device. One can then ground the normative significance in the idea that the former does direct harm, whereas the latter only allows harm to occur. [6]

[5] For further discussions on the putative postoperative impact of DBS on patients' PIAAAS, please refer to footnote 3.

[6] This plays on influential doing/allowing harm distinction that have shaped numerous ethical discourses, such as the euthanasia debate and whether there is a moral difference between active and passive euthanasia.

We do not think such lines of argumentation for the moral asymmetry of implantation and explantation will be successful or even relevant. What seems to matter ultimately is whether such procedures bring about distressing PIAAAS change. Though the class of psychological changes exemplified in cases of implantation may be qualitatively different from the class of psychological changes exemplified in the cases of explantation, the normative significance is whether the psychological changes can be reasonably characterized as serious disturbances to a person's PIAAAS states.

Since PIAAAS change is morally troubling, and explanting a neural device puts people undergoing the procedure at risk of experiencing these changes, it is important for researchers to be more considerate of this vulnerability when assessing whether, when, and how to proceed with neural device explantation. There is no reason to think that PIAAAS change is only morally relevant in the context of neural device implantation. The possibility of PIAAAS change generates post-trial responsibilities on researchers toward their patients and research participants to avoid or mitigate the serious negative effects of explantation, including any concerning PIAAS-related changes. In the next section, we will propose some recommendations on how researchers should proceed when it comes to explantation.

5 Recommendations

Given the prospects of patients and research participants experiencing troubling PIAAAS change if their neural implants were removed, we propose the following recommendations for clinicians and researchers for consideration to respond appropriately to this vulnerability:

The development of exclusion criteria. We believe that there are cases where explantation may not be a permissible option. Certain harms following the excision of a neural device, including PIAAAS-related harms, may be so severe that they outweigh other moral considerations, disqualifying certain persons with implants from becoming subjects for explantation. The degree of the harm will be dependent on a variety of factors, such as the nature of the illness that is being treated, the patient or participant's history with the device, and certain life circumstances. An adequate specification of the exclusion criteria will be sensitive to these features of the person being considered for explantation. One factor that should be weighed is the length of time that the participant had the device. The length of time that a neural implant is kept likely correlates with the degree in which a person's PIAAAS-related states are bounded to their neural implant. The longer a person lives with a neural device and is immersed in the physical and psychological life it enables, the more likely the person's self will be intertwined with their device. Therefore, removing the neural implant after this level of human-machine merger will likely lead to serious harms extended to the patient or participant.

Another factor is the nature of the condition or limitation that the BCI was ameliorating. If removing the device is going to dramatically diminish the quality of life the explantee, then this is a strong reason against explantation. Yet this also has to be weighed against the possibility that extended BCI use or treatment could also lead to a diminished quality of life. Additionally, we should be attentive to how long-term DBS treatment could also bring about advanced, novel stages of a disease that were never encountered before, a ramification of extending a patient's life through the implantation of DBS [8].

A third factor is the psychiatric history of the person being considered for explantation. It is common for researchers to disqualify prospective users from receiving a neural implant because they have a history of depression or other psychological conditions, making them unfit for implantation since these conditions could be exacerbated when undergoing invasive neurotechnological interventions. Similar exclusion criteria should also be extended to determinations for explantation. If a participant has a past clinical record of depression or suicidal ideations and excising their device would likely arouse or intensify these internal states, then the participant should not be eligible for explantation. What is to be avoided is imposing a range of serious traumas, including PIAAAS-altering traumas, on patients and participants. This moral consideration may require re-thinking certain values, like the strength of proprietary rights of neural devices when they are intertwined in people's physical and psychological being.

One could argue that the exclusion criteria for explantation should be involved in the process of selecting participants for neural implant research trials. If we can reliably predict in the recruitment phase which volunteers are susceptible to harmful PIAAAS change if they had to undergo explantation and exclude these volunteers from participation in clinical trials, then this would avoid the explantation-related dilemmas raised in this chapter. [7] We agree that susceptibility to harmful explantation-induced PIAAAS change should be part of the individual assessment of who is eligible as research subjects in neural implant research. Part of our argument is that exclusion criteria that appropriately recognize the vulnerabilities of serious PIAAAS-related harms from explantation should be involved in the decision-making of neural researchers, whether it is involved in the initial recruitment stage or at the end of a clinical trial so long as the well-being of participants from the threats of difficult PIAAAS change from explantation is considered. Also, whether exclusion criteria for explantation are unnecessary depending on whether it is always possible to reliably predict in the initiation stage who will experience the troubling effects of PIAAAS change if they undergo explantation. We cannot rule out the possibility that a participant could change in some new way that was not anticipated, given that implantable neurotechnologies can lead to transformative experiences, or experiences that are radically novel to the implanted person and may alter their identity in some fundamental way [51, 52]. Given the dynamic experience and changing relationship a participant may have with their neural device, there may be cases where prior risk assessments may become invalid, and researchers will have to re-evaluate whether an implanted person should be subjected to explantation.

Access to counseling for explantees. If neurotechnologically implanted persons consent or are required to undergo explantation, then we hold that clinicians

---

[7] We thank the two anonymous reviewers for raising this point.

and researchers, as well as funders and institutions, have a responsibility to provide post-trial or ancillary care to explantees to mitigate any PIAAAS-related harms that may follow from explantation. Currently, the lack of provision of care indicates a lack of acknowledgment and anticipation, if not lacking a sense of beneficence and justice, to the PIAAAS-related harms accompanying the removal of neural devices. Leggett, for example, described how she felt abandoned after the research trial ended. There was no expression of gratitude for her years-long participation or no offer of counseling to help her transition back to a life without an implant. As Liam Drew describes Leggett's experience in his article on explantation, "The day she travelled to the hospital to return the handheld device that had become an essential part of her life, she anticipated a poignant, reflective conversation with the trial coordinator who had accompanied her throughout the process. However, he was not there. Rita had to hand her device over to a stranger, who told her she could leave a note if she wanted" [32]. Researchers have an obligation to research participants, either generated from their relationship of trust and vulnerability or from the principle of reciprocity given the contributions of participants. Also, it is irresponsible not to provide explantees with support to reconceive themselves in ways that help them move on with life without an implant. Losing a cherished, meaningful identity and valued form of living is excruciating. Making peace with an estranged or unwelcoming embodiment and its attendant physical and psychological life is a very difficult, unsure process of acceptance, adjustment, and re-creation of a new identity. It is an unreasonable burden for explantees to face on their own without expert counseling support.

Further research on the PIAAAS-related effects of explantation. The PIAAAS-related effects of explantation are underexamined areas of neuroethical research. To develop clinical and research practices and therapies that are appropriately responsive to the risks of PIAAAS change from explantation, we need to have a better understanding of the phenomena. This investigation includes probing the phenomenological aspects of explantation. Elucidating the lived experience of loss of a neural device and the specific goods it provided will have valuable practical application in the clinical and research context. For instance, therapists can assist explantees in transitioning to embodiments they initially did not want, helping them to construct or repair their identity and meaningful pattern of living within that embodied context. This may require what Hilde Lindemann Nelson calls "narrative repair" when experiencing an injured identity, an approach that Marya Schechtman suggests for helping DBS users feel less alienated from themselves after the activation of their implanted device [53, 54]. This approach could also be extended to people having adjustment difficulties after going through neural device explantation.

6 Conclusion

The impetus of this chapter is to examine the ethics of neural device explantation from the normative lens of PIAAAS. PIAAAS change is an often-overlooked aspect of explantation, which then translates to how the option of explantation is perceived and judged. We push back against the characterization that explantation is a relatively benign resort with impacts only affecting physical health. The neuroethical discourse around PIAAAS change from neural device implantation has illuminated the normative significance of PIAAAS change, why it is morally concerning, and why this moral consideration should be integrated in how the development and use of implantable neurotechnologies are approached. Given empirical findings of troubling PIAAAS change following explantation, we argue that the moral analysis applied to neural device implantation should also be applied to neural device explantation. Thus, our approach to neural device explantation should be appropriately responsive to the vulnerability of PIAAAS change of explantees.

Although this chapter focuses on the difficult PIAAAS-related harms following neural device explantation, we do not think that this is a concern unique to implantable neurotechnologies. In some ways, the concern we highlight is distinctive to neural implants given the ways in which the brain is widely regarded as the principal seat of the self, the level of invasiveness of the device in the body, and how intervening in its activities can have wide-ranging effects on users' core aspects of their identity. But in many other ways, the concern of troubling PIAAAS change following explantation is also applicable to a broader range of implants, where people have integrated these devices to their sense of self, bodily integrity, and autonomy. Thus, the argument we present here can also extend to other types of explantation and speak to a larger set of concerns around the option of removing implants that have been supporting people's PIAAAS-constituting states after the end of research trials.

Acknowledgments Both authors would like to thank Sara Goering and the Neuroethics research group with the Center for Neurotechnology at the University of Washington. This work was supported by the EthicsLab. Funding from the Gwen Nettlefold Memorial Fellowship and the Goddard Sapin-Jaloustre Scholarship are gratefully acknowledged.

Ethics Approval University of Washington, Human Subject Division, IRB ID: MOD0000174 End-user experience of neural technologies for demand-driven management of symptoms.

References

1. Viaña J, Vickers JC, Cook MJ, Gilbert F. Currents of memory: recent progress, translational challenges, and ethical considerations in fornix deep brain stimulation trials for Alzheimer's disease. Neurobiol Aging. 2017;56:202–10. https://doi.org/10.1016/j.neurobiolaging.2017.03.001.
2. Klein E, Brown T, Sample M, Truitt AR, Goering S. Engineering the brain: ethical issues and the introduction of neural devices. Hast Cent Rep. 2015;45(6):26–35. https://doi.org/10.1002/hast.515.
3. de Haan S, Rietveld E, Stokhof M, Denys D. Effects of deep brain stimulation on the lived experience of obsessive-compulsive disorder patients: in-depth interviews with 18 patients. PLoS One. 2015;10(8):e0135524. https://doi.org/10.1371/journal.pone.0135524.
4. Gilbert F. The burden of normality: from 'chronically ill' to 'symptom free'. New ethical challenges for deep brain stimulation postoperative treatment. J Med Ethics. 2012;38:408–12. https://doi.org/10.1136/medethics-2011-100044.
5. Burwell S, Sample M, Racine E. Ethical aspects of brain computer interfaces: a scoping review. BMC Med Ethics. 2017;18:60. https://doi.org/10.1186/s12910-017-0220-y.

6. Rogers W, Hutchison K, Skea ZC, Campbell MK. Strengthening the ethical assessment of placebo-controlled surgical trials: three proposals. BMC Med Ethics. 2014;15(78) https://doi.org/10.1186/1472-6939-15-78.
7. Gilbert F, Tubig P. Cognitive enhancement with brain implants: the burden of abnormality. J Cogn Enhanc. 2018;2(4):364–8. https://doi.org/10.1007/s41465-018-0105-0.
8. Gilbert F, Lancelot M. Incoming ethical issues of deep brain stimulation: when long term treatment lead to a "new form of disease". J Med Ethics. 2021;47:20–5. https://doi.org/10.1136/medethics-2019-106052.
9. Goering S, Klein E. Neurotechnologies and justice by, with, and for disabled people. In: Cureton A, Wasserman DT, editors. The Oxford handbook of philosophy and disability. Oxford: Oxford University Press; 2019. p. 615–32. https://doi.org/10.1093/oxfordhb/9780190622879.013.33.
10. Goering S, Klein E, Specker-Sullivan L, Wexler A, Arcas B, Bi G, et al. Recommendations for responsible development and application of neurotechnologies. Neuroethics. 2021;14:365–86. https://doi.org/10.1007/s12152-021-09468-6.
11. Yuste R, Goering S, Arcas B, Bi G, Carmena JM, Carter A, et al. Four ethical priorities for neurotechnologies and AI. Nature. 2017;551:159–63. https://doi.org/10.1038/551159a.
12. Levy N. Neuroethics and social justice. The Neuroethics Blog. 2019; http://www.theneuroethicsblog.com/2019/04/neuroethics-and-social-justice.html
13. Buchman D, Wadhawan S. A global vision for neuroethics needs more social justice: brain imaging, chronic pain, and population health inequalities. AJOB Neurosci. 2019;10(3):130–2. https://doi.org/10.1080/21507740.2019.1632966.
14. Sierra-Mercado D, Zuk P, Beauchamp MS, Sheth SA, Yoshor D, Goodman WK, et al. Device removal following brain implant research. Neuron. 2019;103(5):P759–61. https://doi.org/10.1016/j.neuron.2019.08.024.
15. Hansson SO. The ethics of explantation. BMC Med Ethics. 2021;22(121) https://doi.org/10.1186/s12910-021-00690-8.
16. Sankary LR, Zelinsky M, Machado A, Rush T, White A, Ford PJ. Exit from brain device research: a modified grounded theory study of researcher obligations and participant experiences. AJOB Neurosci. 2022;13(4):215–26. https://doi.org/10.1080/21507740.2021.1938293.
17. Gilbert F, Tubig P, Harris A. Not-so-straightforward decisions to keep or explant a device: when does neural device removal become patient coercion? AJOB Neurosci. 2022;13(4):230–2. https://doi.org/10.1080/21507740.2022.2126544.
18. Bluhm R, Cabrera LY. It's not just counting that counts: a reply to Gilbert, Viaña, and Ineichen. Neuroethics. 2021;14:23–6. https://doi.org/10.1007/s12152-018-9391-6.
19. Pugh J, Pycroft L, Maslen H, Aziz T, Savulescu J. Evidence-based neuroethics, deep brain stimulation and personality—deflating, but not bursting, the bubble. Neuroethics. 2021;14:27–38. https://doi.org/10.1007/s12152-018-9392-5.
20. Gaillard M. Neuroessentialism, our technological future, and DBS bubbles. Neuroethics. 2021;14:39–45. https://doi.org/10.1007/s12152-019-09407-6.
21. Mosley PE, Robinson K, Coyne T, Silburn P, Breakspear M, Carter A. 'Woe betides anybody who tries to turn me down.' A qualitative analysis of neuropsychiatric symptoms following subthalamic deep brain stimulation for Parkinson's Disease. Neuroethics. 2019;14:47–63. https://doi.org/10.1007/s12152-019-09410-x.
22. Snoek A, de Haan S, Schermer M, Horstkötter D. On the significance of the identity debate in DBS and the need of an inclusive research agenda. A reply to Gilbert, Viaña, and Ineichen. Neuroethics. 2021;14:65–74. https://doi.org/10.1007/s12152-019-09411-w.
23. Erler A. Discussions of DBS in neuroethics: can we deflate the bubble without deflating ethics? Neuroethics. 2021;14:75–81. https://doi.org/10.1007/s12152-019-09412-9.
24. Zuk P, Lázaro-Muñoz G. DBS and autonomy: clarifying the role of theoretical neuroethics. Neuroethics. 2021;14:83–93. https://doi.org/10.1007/s12152-019-09417-4.
25. Kubu CS, Ford PJ, Wilt JA, Merner AR, Montpetite M, Zeigler J, et al. Pragmatism and the importance of interdisciplinary teams in investigating personality changes following DBS. Neuroethics. 2021;14:95–105. https://doi.org/10.1007/s12152-019-09418-3.

26. Thomson CJ, Segrave RA, Carter A. Changes in personality associated with deep brain stimulation: a qualitative evaluation of clinician perspectives. Neuroethics. 2021;14:109–24. https://doi.org/10.1007/s12152-019-09419-2.
27. Gilbert F, Viaña JM, Ineichen C. Deflating the deep brain stimulation causes personality changes bubble: the authors reply. Neuroethics. 2021;14:125–36. https://doi.org/10.1007/s12152-020-09437-5.
28. Mackenzie C, Walker M. Neurotechnologies, personal identity and the ethics of authenticity. In: Clausen J, Levy N, editors. Springer handbook of neuroethics. Springer: Dordrecht; 2014. p. 373–92.
29. Gilbert F, Viaña JM, Ineichen C. Deflating the 'DBS causes personality changes' bubble. Neuroethics. 2021;14:1–17. https://doi.org/10.1007/s12152-018-9373-8.
30. Klein E. Informed consent in implantable BCI research: identifying risks and exploring meaning. Sci Eng Ethics. 2016;22(5):1299–317. https://doi.org/10.1007/s11948-015-9712-7.
31. Pinsker M, Amtage F, Berger M, Nikkhah G, van Elst LT. Psychiatric side-effects of bilateral deep brain stimulation for movement disorders. Acta Neurochir Suppl. 2013;117:47–51. https://doi.org/10.1007/978-3-7091-1482-7_8.
32. Drew L. "Like taking away a part of myself"—life after a neural implant trial. Nat Med. 2020;26(8):1154–6. https://doi.org/10.1038/d41591-020-00028-8.
33. Kenneally C. Do brain implants change your identity? New Yorker. https://www.newyorker.com/magazine/2021/04/26/do-brain-implants-change-your-identity.
34. Fins JJ. Deep brain stimulation, deontology, and duty: the moral obligation of non-abandonment at the neural interface. J Neural Eng. 2009;6:1–4. https://doi.org/10.1088/1741-2552/6/5/050201.
35. Lawton J, Blackburn M, Rankin D, Werner C, Farrington C, Hovorka R, et al. Broadening the debate about post-trial access to medical interventions: a qualitative study of participant experiences at the end of a trial investigating a medical device to support type 1 diabetes self-management. AJOB Empir Bioeth. 2019;10(2):100–12. https://doi.org/10.1080/23294515.2019.1592264.
36. Lázaro-Muñoz G, Yoshor D, Beauchamp MS, Goodman WK, McGuire AL. Continued access to investigational brain implants. Nat Rev Neurosci. 2018;19:317–8. https://doi.org/10.1038/s41583-018-0004-5.
37. Houeto JL, Mesnage V, Mallet L, Pillon B, Gargiulo M, Tezenas du Moncel M, et al. Behavioural disorders, Parkinson's disease and subthalamic stimulation. J Neurol Neurosurg Psychiatry. 2002;72(6):701–7. https://doi.org/10.1136/jnnp.72.6.701.
38. Gilbert F. Self-estrangement & deep brain stimulation: ethical issues related to forced explantation. Neuroethics. 2015;8(2):107–14. https://doi.org/10.1007/s12152-014-9224-1.
39. Bouton CE, Shaikhouni A, Annetta NV, Bockbrader MA, Friedenberg DA, Nielson DM, et al. Restoring cortical control of functional movement in a human with quadriplegia. Nature. 2016;533:247–50. https://doi.org/10.1038/nature17435.
40. Gilbert F. A threat to autonomy? The intrusion of predictive brain devices. AJOB Neurosci. 2015;6(4):4–11. https://doi.org/10.1080/21507740.2015.1076087.
41. Gilbert F, Cook M, O'Brien T, Illes J. Embodiment and estrangement: results from a first-in-human "intelligent BCI" trial. Sci Eng Ethics. 2019;25:83–96. https://doi.org/10.1007/s11948-017-0001-5.
42. Gilbert F, Goddard E, Viaña JM, Carter A, Horne M. "I miss being me": phenomenological effects of deep brain stimulation. AJOB Neurosci. 2017;8(2):96–109. https://doi.org/10.1080/21507740.2017.1320319.
43. Gilbert F, Brown T, Dasgupta I, Martens H, Klein E, Goering S. An instrument to capture the phenomenology of implantable brain device use. Neuroethics. 2019;14:333–40. https://doi.org/10.1007/s12152-019-09422-7.
44. Soekadar S, Chandler J, Ienca M, Bublitz C. On the verge of the hybrid mind. Morals Machines. 2021;1(1):30–43. https://doi.org/10.5771/2747-5174-2021-1-30.

45. Miletić T, Gilbert F. Does AI brain implant compromise agency? Examining potential harms of brain-computer interfaces. In: Gouveia SS, editor. The age of artificial intelligence: an exploration. Wilmington: Vernon Press; 2020. p. 253–72.
46. Weise J. The dawn of the Tryborg. The New York Times December. 2016;1. https://www.nytimes.com/2016/11/30/opinion/the-dawn-of-the-tryborg.html
47. Weise J. Common cyborg. Granta September. 2018;24. https://granta.com/common-cyborg/
48. Glannon W. Stimulating brains, altering minds. J Med Ethics. 2009;35:289–92. https://doi.org/10.1136/jme.2008.027789.
49. O'Discoll K, Leach JP. "No longer gage": an iron bar through the head: early observations of personality change after injury to the prefrontal cortex. BMJ. 1998;317(7174):1673–4. https://doi.org/10.1136/bmj.317.7174.1673a.
50. Witt K, Kuhn J, Timmerman L, Zurowski M, Woopen C. Deep brain stimulation and the search for identity. Neuroethics. 2013;6:499–511. https://doi.org/10.1007/s12152-011-9100-1.
51. Pugh J. Clarifying the normative significance of 'personality changes' following deep brain stimulation. Sci Eng Ethics. 2020;26(3):1655–80. https://doi.org/10.1007/s11948-020-00207-3.
52. Paul LA. Transformative experience. Oxford: Oxford University Press; 2014.
53. Nelson HL. Damaged identities, narrative repair. New York: Cornell University Press; 2001.
54. Schechtman M. Philosophical reflections on narrative and deep brain stimulation. J Clin Ethics. 2010;21(2):133–9.

Ethical Considerations of Endovascular Brain–Computer Interfaces

Adam Fry, Erica Breyman, Edward LaGrassa, Thomas Oxley, and David Putrino

1 Introduction

A recent early feasibility trial by Oxley and colleagues [1] demonstrated effective control of a personal computer using a brain–computer interface (BCI) in two individuals with paralysis due to amyotrophic lateral sclerosis. The trial was noteworthy as it provided the first demonstration of a working BCI using brain signals recorded endovascularly in humans. This approach involved placement of a stent into the superior sagittal sinus adjacent to the motor cortices via a catheter accessing the jugular vein. The stent housed sixteen recording electrodes, thus providing access to electrocorticographic recordings from within the brain without the need for craniotomy. This endovascular BCI arrangement is illustrated in Fig. 1.

A. Fry
Department of Rehabilitation and Human Performance, Abilities Research Center, Icahn School of Medicine at Mount Sinai, New York, NY, USA

Synchron, Inc., Campbell, CA, USA
e-mail: afry@synchron.com

E. Breyman · D. Putrino (✉)
Department of Rehabilitation and Human Performance, Abilities Research Center, Icahn School of Medicine at Mount Sinai, New York, NY, USA
e-mail: erica.breyman@mountsinai.org; david.putrino@mountsinai.org

E. LaGrassa
Stevens Institute of Technology, Hoboken, NJ, USA
e-mail: elagrass@stevens.edu

T. Oxley
Synchron, Inc., Campbell, CA, USA

Department of Neurosurgery, Icahn School of Medicine at Mount Sinai, New York, NY, USA
e-mail: toxley@synchron.com

V. Dubljević, A. Coin (eds.), *Policy, Identity, and Neurotechnology*, Advances in Neuroethics, https://doi.org/10.1007/978-3-031-26801-4_4

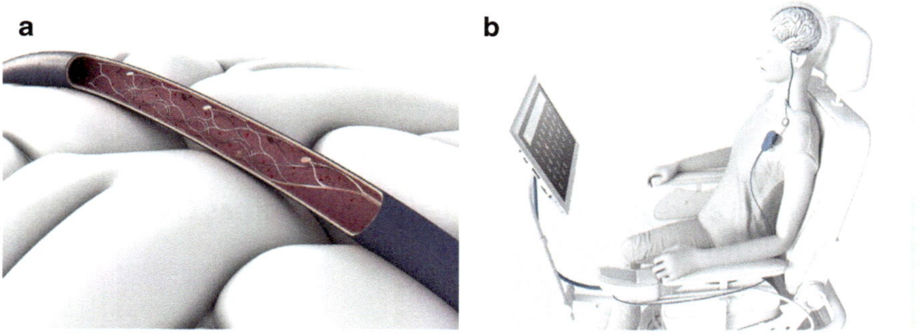

Fig. 1 **Panel a** illustrates a stent-based endovascular electrode array deployed within a cortical vein. Once placed, the stent may be incorporated into the wall of the blood vessel via endothelialization. **Panel b** depicts a patient using an endovascular BCI system. Such systems may be fully implanted with recorded signals transmitted wirelessly out of the body to control various devices such as a personal computer

While the use of endovascular electrode arrays within BCI systems is a recent development, the first demonstration of endovascular electrophysiological recordings took place almost four decades ago [2]. Development of endovascular recording technologies has since progressed from wire electrodes to nanowire and catheter electrodes and finally stent-based electrode arrays [3]. Coupled with an unmet clinical need, these technological advances have attracted significant commercial investment, putting fully implantable BCIs such as the Stentrode™ system used by Oxley et al. [1] on the path to market entry and clinical adoption [4].

In this chapter, we explore some of the bioethical considerations specific to the use of endovascular electrode arrays for BCI. We first examine the safety of endovascular electrode arrays including how they compare to previous approaches to BCI and what unique risks they might pose. As risks should be weighed against the potential for benefit, we also present the efficacy of endovascular recording devices. A unique consideration of endovascular electrode arrays among approaches to BCI is that implantation may be permanent. We therefore explore how this might affect bioethical considerations of BCIs that rely upon this method of recording from the brain. The consequences of endovascular approaches to BCI on the informed consent process are briefly considered. Finally, we also reflect on the impact the recent publicity surrounding BCIs may have on the immediate expectations of newly developed endovascular BCIs and how commercial ventures in this technology may differ from academic pursuits. Our discussion focuses on the bioethical impact of using endovascular recording electrodes within BCIs from a hardware perspective, but does not offer extensive consideration of the applications of endovascular BCI as the initial applications may be similar to BCIs using other sensor modalities.

2 Safety of Endovascular BCIs

Endovascular BCIs are novel among BCIs as they are implanted via a minimally invasive neurointerventional procedure and might, therefore, pose a substantially different risk profile to existing BCI procedural approaches. Rather than requiring burr hole craniotomies for implantation, endovascular BCIs require that a small incision be made to gain entry into a blood vessel, such as the jugular vein. A guide catheter is then advanced to the implant location where the stent is deployed (Fig. 1a, and the trailing lead is tunneled and connected to an internal telemetry unit (ITU) within a subcutaneous pocket in the chest [1] (Fig. 1b)). This approach to BCI placement involves unique safety challenges, and the potential for benefits compared to other BCI approaches should be assessed under a bioethical lens.

All invasive BCIs carry risks for infection, hemorrhage, and other hardware related complications. Although there is limited literature discussing the safety profile of endovascular BCIs [5], endovascular stenting is considered standard of care for idiopathic intracranial hypertension (IIH), thrombectomy, vascular malformation, and aneurysms [6, 7]. As such, we can make inferences regarding the potential for adverse events associated with stent based endovascular BCI implantation and compare them to that of other invasive systems.

Infection can be a serious complication following BCI implantation, as it can result in meningitis, brain abscesses, septic emboli, or other life-threatening consequences. The infection risk for subdural electrode arrays has been reported to be 2.3% [8]. Similarly, infections have been found in 1–5% of patients after implantation of depth electrodes [9]. In a systematic review including 174 deep brain stimulation (DBS) articles that reported data on infection, the incidence rate of infection was identified as 3.79%. Of the 104 papers that included the location of the infection, 44.2% occurred at the site of the internal pulse generator, 17.8% occurred at the scalp or burr hole, 13.6% occurred at the connector and extension cable, and 11.1% occurred in the brain along the electrode lead [10]. The procedural-related risk of infection for the Neuropace RNS system, a BCI with both ECoG and depth electrode components, has been reported to be 4.1%, with an overall incidence rate of infection of 12.1% over the cumulative 1895 patient-implantation years [11]. Alternatively, infections associated with venous sinus stenting for IIH are rare, with only one case of infection identified in several systematic reviews [12–14], which was a urinary tract infection [15]. However, the only existing report of a stent based BCI technology used in humans [1] included a hardware piece, an ITU, that is comparable to that of the internal pulse generator in DBS systems and may carry a similar infection rate when placed in the chest. It is difficult to make a direct comparison to the infection rate of Utah arrays due to limited literature on complications and adverse events in chronically implanted Utah arrays [10].

Hemorrhaging after a BCI implant procedure can also result in serious life-threatening complications including neurological damage. Non-seizure-related hemorrhage was reported in 2.7% of patients implanted with the Neuropace RNS system [11]. Bullard et al. [10] reported a similar incidence rate of 2.9% for DBS systems with 86.9% being intracerebral hemorrhages. A meta-analysis and systematic review reporting on 21 studies and 2542 patients with subdural electrode arrays noted a slightly higher incidence rate of 4.0% [8]. Rates of hemorrhages may be slightly lower for endovascular stenting procedures in clinical settings. For the endovascular treatment of venous sinus stenosis in IIH, a systematic review comprising of 17 studies with a total of 185 patients reported hemorrhages in 3 patients (1.6%) [14]. It is worth noting that humans implanted with the Stentrode™ endovascular BCI underwent dual antiplatelet therapy [1] which might increase the risk of systemic hemorrhage for those not already taking this medication [16, 17].

Other hardware related complications have been reported including device migration. In the endovascular treatment of intracranial aneurysms, migration of a coil or stent occurs in 2–6% of cases [18, 19]. For IIH, Teleb et al. [20] reported a single intraprocedural event of stent migration out of 25 procedures. These rates are comparable to that of DBS hardware complications; Bullard et al. [10] reported that lead migration occurred 3.49% of the time, followed by lead fractures (2.53%), internal pulse generator malfunctions (2.33%), and extension cable malfunctions (1.95%). Events such as these may require additional surgical procedures or removal which may pose additional surgical risks [21].

Endovascular BCIs may also carry additional risks including perforation or occlusion of the vein/artery, stenosis, and catheter related complications. For instance, intraprocedural vessel perforation during stent retriever thrombectomy occurred in 16 of 1599 cases [22]. Occlusion or perforation of the vessel can lead to hemorrhage or stroke [23]. At the time of writing this, endovascular BCI technologies have used venous implantation routes, and it is important to consider the risk profile of venous versus arterial implantation routes to understand why this is the case. Alawieh et al. [24] noted the advantage of the venous route by highlighting its lack of smooth muscle layers and lower likelihood of provoking vasospasm. As recording technology changes over time, there may be a functional advantage of using an arterial route if it allows an endovascular recording technology to achieve a location that is closer to the target brain region than the venous route. Should this occur, a separate risk-benefit analysis will need to be conducted. Also, the additional procedural risks associated with stenosis must be considered, as in IIH. Starke et al. [14] reported in-stent stenosis in 6 patients (3.4%) and stent-adjacent stenosis in 19 patients (11.4%), with 10 patients requiring restenting. Finally, the diameter of the catheter and vessels accessed may also influence the rates of adverse events. For example, venous dissection, subdural hemorrhage, and acute thrombus formation occurred when using a 5-Fr or 6-Fr catheter in animal testing, whereas no complications occurred when using a 4-Fr or 2-Fr catheter. In some cases, these complications led to non-recovery after anesthesia [25].

The first in-human study of the Stentrode™ found no instances of infection, device migration, stenosis, or device related adverse events. They did however find a post-procedural-related adverse event of syncope associated with two sinus pauses which was attributed to post-procedural vagal tone and required no intervention. It is important to note that there were only two participants involved in this study [1].

A clinical trial with a larger sample size needs to be performed before the safety profile of this stent based endovascular BCI can be established.

Endovascular stent electrodes may also have several advantages. For example, they circumvent the need for craniotomy which can open additional avenues of care for individuals with stroke or traumatic brain injury [24]. Also, considering the relatively smaller scale and shorter duration of surgery, implantation of endovascular BCIs would require smaller total doses of anesthesia. This would likely limit surgical risk for individuals with respiratory compromise. Additionally, the implantation of other invasive BCI systems is still a niche skill, whereas the advent of thrombectomy in stroke has made stent placement a ubiquitous skillset for interventional neurosurgeons [26–28]. As such, the surgical risk should be adjusted accordingly.

Finally, non-invasive BCI systems including scalp electroencephalography, magnetoencephalography, and functional near-infrared spectroscopy-based devices pose lower procedural risks than invasive BCIs as they do not penetrate the skin [23]. However, non-procedural risks should also be considered. Unlike endovascular BCIs, some users have found headsets used for non-invasive BCI to be uncomfortable [29, 30]. Application of these systems also requires assistance which can be burdensome and may increase feelings of dependency [31]. Their visible machinery and medical connotation may also contribute to stigma [32], thereby affecting patients' self-consciousness, willingness to socialize, and social status. Conversely, endovascular BCIs may be fully implantable and may add less to stigma, burden, and feelings of frustration.

A summary of our safety considerations is provided in Table 1. While the short- and long-term safety profile of endovascular BCIs in humans has yet to be established, endovascular stenting for a range of indications has shown slightly lower rates of infection and hemorrhage compared to that of other invasive hardware. Circumventing the need for craniotomy during implantation of an endovascular BCI may also play a role in lowering the rate of these complications and avoiding other procedural risks. Larger human clinical trials investigating the rate of new endovascular BCI-specific safety concerns and advantages are required to develop pre- and post-implant mitigation procedures and enable comprehensive risk-benefit analyses.

Table 1 Safety considerations of endovascular electrode arrays for brain–computer interfaces (BCIs)

Pros	Cons
• Stent placement associated with slightly lower rate of infection vs. subdural and depth electrodes • Stent placement associated with slightly lower rate of hemorrhage vs. subdural and depth electrodes • Minimally invasive surgery avoids craniotomy and associated medical management • Minimally invasive surgery requires shorter duration under anesthesia • Compared to non-invasive BCIs, fully implantable endovascular systems may have lower non-procedural risks (discomfort, stigma, increased feelings of dependency)	• Higher procedural risks than non-invasive BCIs • Placement involves risk of perforation of the blood vessel • Placement involves risk of venous stenosis or occlusion of the blood vessel • Risk of secondary surgery following device migration or lead fracture

3 Efficacy of Endovascular BCIs

The overall effectiveness and benefit derived from a BCI may depend on several factors including the signal decoding algorithms, device software, system usability, ergonomics, and the design/engineering of any effector. Additionally, the type of sensor used to record the brain signal may also impact BCI effectiveness. The reliability, the number of outputs, and customizability of a BCI may all be affected by inadequate recording quality. Therefore, the efficacy of endovascular electrodes for recording neural signals is worth considering within our bioethics discussion as it will impact the risk-benefit profile of any BCI that uses this technology.

3.1 Comparisons with Other Sensor Types

Comparisons between endovascular and scalp EEG were included in the earliest demonstrations of endovascular recordings. Penn and colleagues [2] demonstrated that endovascular EEG recordings were able to detect localized activity in intracerebral structures that were not accessible to routine scalp EEG. Their report concluded that the endovascular technique offered advantages over non-invasive scalp EEG recordings for the detection of clinically relevant, paroxysmal EEG activity. Comparing endovascular and scalp EEG signal quality, Nakase et al. [33] reported a 2–5 times stronger EEG voltage potential in endovascular EEG compared to scalp EEG. Similarly, Stoeter et al. [34] found higher peak-to-peak amplitudes of both somatosensory and auditory evoked potentials when comparing concurrent endovascular and scalp EEG recordings in patients undergoing interventional angiographies. These results indicated that endovascular EEG signals were subject to less attenuation than extracranial recordings. More recently, He et al. [35] compared visual and auditory evoked potentials recorded using both endovascular and scalp EEG in rabbits. Their results confirmed that higher amplitudes of evoked responses can be observed in endovascular recordings and indicated that their endovascular recordings were of a higher quality than their scalp recordings with up to 100 times better signal-to-noise ratio. These results were shown to be unrelated to non-neural biological signals or differences in electrode materials.

Endovascular EEG recordings have also been compared to surface ECoG. Nakase and colleagues [33] found that interictal spike discharges were simultaneously visible in concurrent endovascular and subdural strip electrode recordings from the medial temporal lobe in three patients with epilepsy. Likewise, Bower et al. [36] found that epileptiform spikes recorded by endovascular and subdural electrodes were similar in both amplitude and waveshape. Here, recordings were collected from pigs using endovascular electrodes within the superior sagittal sinus and subdural electrode arrays placed in parallel, and epileptic activity was induced via direct cortical injection of penicillin. Using low-amplitude sinusoidal currents passed between opposing corners of the subdural grid, these authors reported that the frequency response of the intravascular recordings were reduced by 11.0% and 24.2% at 30 Hz and 100 Hz, respectively, relative to the recording electrodes of the subdural array [36]. Nevertheless, the endovascular microelectrodes were able to record the low-amplitude, high-frequency signals arising from spatially localized seizure activity that were only observed by the microelectrodes within the subdural grid and missed by the standard clinical macroelectrodes of the subdural array.

Oxley et al. [37] compared the spectral content of recordings from their stent-based endovascular electrode array (Stentrode™) with contemporaneous recordings collected using commercially available subdural and epidural arrays in sheep. No difference in absolute power was found in mu, beta or low gamma bands between the three array types, whereas higher power was seen in the higher gamma bands in the subdural recordings compared to both the endovascular and epidural recordings, which were similar. Maximum bandwidth was also highest in the subdural array and similar between the endovascular and epidural devices.

John et al. [38] also compared the signal quality of the Stentrode™ array with both subdural and epidural ECoG interfaces. Again, sheep were used, and measurements were recorded 3–4 weeks after electrode placement to allow for stabilization of signal quality following endothelialization of the stent-based electrode array. The authors found no significant effect of recording location or electrode sizes on the bandwidth of the recorded signal, indicating the quantity of information that can be obtained from endovascular, subdural, and epidural electrode arrays is similar. There was also no significant effect of electrode location or size on the signal-to-noise ratio indicating similar signal quality. However, the variability between subdural electrodes was higher, meaning select subdural electrodes achieved a higher signal-to-noise ratio than the endovascular electrodes. Finally, the difference in spatial resolution between the arrays was frequency dependent between 8 and 180 Hz, differing only in the lowest frequency band investigated (8–24 Hz), where subdural electrode arrays produced the best spatial resolution.

In the same study, John et al. [38] demonstrated that the similar signal qualities between the epidural, subdural, and endovascular arrays meant that there was no appreciable difference in single-trial decoding performance (classification of the presence or absence of an evoked potential). In addition, Forsyth et al. [39] compared decoding performance between Stentrode™, subdural, and epidural arrays implanted in a sheep trained to perform left and right head movements in response to an external stimulus. Classification of these movements was computed offline. Using 50% of the trials for training data and 50% for test data, results demonstrated that the epidural array had a slightly lower classification accuracy (80%) than the endovascular and subdural arrays (both 85%) when comparing movement vs. rest. All were higher than the minimum accuracy required to exceed chance, which was dependent on the number of trials used to test the classification. When comparing left vs. right movements, classification accuracies from the endovascular (58%), subdural (55%), and epidural (51%) arrays were again similar, but only the endovascular array distinguished between left and right movements at a higher rate than chance.

To the authors knowledge, no direct comparisons in signal quality or classification performance between endovascular and penetrative electrode arrays have been made. This may be more complicated as these sensor types might record different neural signals for classification. For example, upon implantation, penetrative microelectrode arrays offer spike recordings from individual neurons. This resolution of neural activity is not available to endovascular electrode arrays, which record synchronized post-synaptic potentials from within the wall of the blood vessel. However, over time, encapsulation of microelectrodes may occur in response to the damage caused to the surrounding tissues during implantation. Thereafter, microelectrode arrays may be limited to recording local field potentials [40] making comparisons of endovascular and microelectrode arrays more valid.

3.2 Longitudinal Assessments

Oxley et al. [37] performed a series of experiments over a 28-day period following implantation of their Stentrode™ into a superficial cortical vein overlying the motor cortex in sheep. This time window allowed for examination of changes in recording signal quality as the stent was incorporated into the blood vessel, its permanent recording location, via endothelialization. Direct median nerve stimulation was used to elicit somatosensory evoked potentials, which were detected in 98% of all functional channels. This increased from approximately 50% on day 1 post-implant to 92% by day 4. The peak-to-peak amplitudes were unchanged over the 28-day study period indicating stability of the recorded signals over this time. Anesthesia-induced theta burst suppression (see [41]) was used to examine changes in recording sensitivity. Higher burst-suppression ratios at 1 month versus baseline indicated an increase in sensitivity over this time. In 10 sheep, neural recordings were continued for up to 190 days. Maximum bandwidth was recorded and found to be stable throughout, indicating long-term viability of the Stentrode™. Maximum bandwidth was lower than contemporaneous subdural ECoG recordings but similar to epidural recordings.

In another study, Opie et al. [42] examined the electrical signals recorded by the same Stentrode™ device following implantation within the superior sagittal sinus of 15 sheep. Using electrochemical impedance spectroscopy, this study found that 1 kHz impedances were consistent across the 91-day examination period, and capacitance stabilized after approximately 8 days after initial increases accompanying the early stages of endothelialization. Opie and colleagues [43] also reported that the bandwidth of the recordings (~193 Hz) was consistent both between sheep and over time across their 6-month study period.

More recently, a study with human subjects using the Stentrode™ for BCI demonstrated effective control of a computer up to 238 days following implantation [1]. Signal decoder settings were fixed as early as the third training session (53 days post-implantation) and, thereafter, required only calibration of the feature normalization constants, which took 30 seconds to complete.

These results demonstrate the potential for chronically stable endovascular recordings from within the superior sagittal sinus, adjacent to the motor cortices. Signal quality may even improve with biological activity at the electrode–tissue interface as the stent carrying the electrodes is integrated into the wall of the blood vessel via endothelialization. This contrasts with other invasive electrode arrays, where numerous biological reactions to the implanted device may occur often resulting in a decrease in signal quality [44]. These reactions are commonly cited as a limitation to the use of these BCIs. However, it should be noted that the length of study of recent endovascular array based BCIs does not yet exceed previous investigations that have demonstrated high utility of both surface ECoG and microelectrode based BCIs after 3–5 years despite degradation of the underlying signals [45, 46].

3.3 Endovascular Neurostimulation

The potential to use endovascular electrodes for brain stimulation has also been explored. Gerboni et al. [47] demonstrated cortical activation in response to monopolar, endovascular electrical stimulation in sheep. Clear evoked potentials were demonstrated that were graded in relation to the current applied. No observable responses were recorded after culling the animal, verifying the neural origin of the responses. In another proof-of-concept study, Opie et al. [43] elicited motor responses in the lip, face, jaw, neck, and limbs of 25 sheep using cortical stimulation delivered via endovascular electrodes. The observed responses were similar to those elicited using invasive penetrating and subdural arrays. These results demonstrated the ability of endovascular electrodes to deliver focal stimulation to neural tissue without open brain surgery.

Endovascular electrode arrays offer superior signal quality compared to noninvasive EEG and comparable recordings to existing subdural ECoG arrays. Early evidence indicates that these high-quality signals may be stable over time when using stent-based endovascular arrays that are incorporated into the wall of the blood vessel. Thus, with some potential safety advantages compared to other invasive approaches that require craniotomy, endovascular electrodes may offer an advantageous risk-benefit profile for certain uses. Moreover, in the future, access to smaller blood vessels may enable recordings from regions of the brain not readily accessible using surface ECoG or microelectrode arrays, and without the neural damage caused by depth electrodes. The potential for neurostimulation in these brain regions may also increase the utility of endovascular electrode arrays.

4 Endovascular BCI Permanency

The implantation of endovascular BCIs is intended to be permanent. Stents carrying the electrode arrays are incorporated into the wall of the blood vessel via endothelialization. Partial endothelialization may be observed within days and may be complete within 4 weeks of placement [37, 48]. Thereafter, removal from the brain would require major vascular surgery. The permanency of endovascular BCIs raises important bioethical considerations for their development and use.

First-in-human trials of endovascular BCIs have begun [1]. This research required participants' consent to the permanent implantation of the electrode array in their brains. This is a departure from previous invasive BCI research that has mostly relied on the temporary implantation of electrode arrays for clinical purposes as a window of opportunity. For example, the use of ECoG arrays in perioperative observation for epilepsy surgery has enabled much of the opportunity for invasive BCI research [49–51]. Whereas, to the authors' knowledge, only one case of a chronically implanted ECoG-based BCI has been reported in the literature [45, 52]. Similarly, most microelectrode array implants have also been temporarily implanted to study epilepsy, anesthesia, cognition, memory, or language [10]. Relatively few have been chronically implanted. In their review, Bullard and colleagues reported that only 18 of 48 people implanted with the Utah array were implanted for longer than 30 days [10]. Accordingly, bioethical considerations regarding the permanent implantation of investigational BCI devices has received a paucity of attention in the literature. Here we consider the ethical impact of BCI permanency using the four principles of bioethics as a framework: non-maleficence, beneficence, autonomy, and justice.

4.1 Nonmaleficence

At the early feasibility phase of research, the risks of any investigational BCI are uncertain. Various safety and biocompatibility data need first be collected in bench top and animal testing. This process will be mostly unchanged by the permanency of a BCI, although longer durations of monitoring during animal testing may be warranted prior to human studies. The risk of a malfunction or toxicity may increase with the duration of implantation, which would become the lifetime of the patient when using endovascular BCIs. Moreover, the consequences of adverse reactions to an endovascular BCI may be greater if they necessitate a major neurovascular surgery to remove the device. However, the biggest difference in the risk of harm between permanent and temporary implants may be the potential for indirect harm to permanently implanted patients. For example, if the endovascular BCI is not MRI compatible, then implantation would preclude the patient from undergoing an MRI scan for any future medical need.

In addition to the risks being uncertain, the efficacy of investigational BCIs for individuals with different neurological deficits will be unknown. Research subjects who fail to learn how to operate their BCI as intended (e.g., control a computer cursor or prosthetic arm) may suffer psychological harm due to their failure to meet perceived expectations or achieve an expected therapeutic benefit [53]. This psychological harm may be amplified or prolonged if the user is unable to have the ineffective device removed from their person. Thus, carefully considered selection criteria may become even more essential when admitting study subjects to a trial of an investigational endovascular BCI.

Permanent implantation of a BCI could also lead to reductions in harm. Invasive BCIs intended to be temporary would necessitate a second surgery for explantation.

This would have associated risks that might exceed the risks of not removing the device. This exposure to potential harm may be unwarranted if the patient does not wish the device to be removed and there is low anticipated risk from the device remaining in place.

Finally, the permanency of endovascular BCIs may further blur the murky distinction between BCI research and treatment [54]. Research has a finite timeframe. Outcome measures must be collected at preset times following implantation. However, permanently implanted BCIs will remain in place beyond any scheduled data collection. Researchers may need to incorporate a plan for continued care beyond the collection of primary outcome measures.

4.2 Beneficence

BCIs hold the potential for myriad benefits for individuals with a wide variety of neurological conditions. For example, BCIs may bestow a sense of agency, increase the opportunity for social participation, and have a positive effect on a user's self-image [55]. Temporary implantation of investigational BCIs may offer benefits to the patient, only for these to be taken away when the device is explanted. This practice could arguably be seen as unethical. It may at least be unattractive to potential study participants who may not want to end their participation [55]. Although research with an indefinite timeline raises concerns of exploitation, in exploratory studies where the goal of the research has not yet been fully realized, the subject does not wish to end their participation, and the research continues to progress, it may be appropriate to continue participation in the research until these criteria are met [56]. Similarly, restricting potential research subjects to temporary use of a BCI may be a barrier to participation in BCI research. This may obstruct the potential for benefit to the individual and hinder the progress of BCI development toward a clinically available and beneficial product for the wider patient population.

Permanent implantation of an endovascular BCI may offer benefits to the user that extend beyond the time span of a research study. This possibility requires considerations regarding the need for continued technical and medical support [57]. Commitment to life-long follow-up should be necessitated by funders of endovascular BCI research. However, guarantees made by entrepreneurial health technology companies could be unreliable given the high business risks involved in entering this emerging market. If the manufacturer of a BCI goes out of business and the device is no longer available, then any benefit the user received could discontinue. Replacing or maintaining components of the device may not be possible, which could transform a beneficial situation into a potentially harmful one [53].

The permanent implantation of endovascular BCIs could also make recipients ineligible for future opportunities including subsequent generations of devices or different upgrades that may offer greater benefit. This should be made clear within the informed consent process. As much as possible, manufacturers should consider the compatibility of first-generation endovascular BCIs with anticipated future device generations to ensure the maximization of benefit from the implantation of investigational endovascular BCIs.

From a research and development perspective, the long-term safety outcomes of a BCI are not clear from transient implantation of devices. Permanent implantation during typically low *n* feasibility studies early in the development process would provide the most information on safety prior to larger pivotal trials and post-market surveillance involving larger cohorts. Similarly, some EEG features are more robust over time ("permanent") than others [58]. Therefore, the efficacy of BCI systems over extended periods may remain unknown without longitudinal monitoring across the lifespan of a device, even if biological interactions with the implanted materials can be demonstrated as stable or mitigated. These benefits must be weighed against the risk of harm to the first recipients of permanent devices.

4.3 Justice

The permanency of endovascular BCIs engenders deliberation on which patient populations might be appropriate for feasibility studies of investigational devices. Some exclusions should certainly be applied. Pediatric populations with still developing vascular anatomy would be inappropriate, at least for the current generation of devices. Other decisions may be less clear. For example, individuals with life-limiting disorders such as ALS may be less impacted by the inability to receive future upgrades (see *Beneficence*). As such, researchers might seek to target these patient populations. However, access to these opportunities should not be restricted from eligible individuals with other neurological conditions, providing they are fully informed prior to giving their consent. Equally, those with limited expected life spans may be viewed as being more willing to take on higher levels of risk or less impacted by severe adverse events; however, the potential for harm should not be concentrated in this population.

4.4 Autonomy

Individuals with severe communication difficulties are a target population of some BCIs. For these individuals, providing informed consent may be difficult or even impossible, especially if simple yes/no responses are considered insufficient and amounting only to assent at best [59]. In these cases, legally authorized representatives may issue consent on the patient's behalf [57]. However, the BCI may restore the capacity for higher levels of communication and subsequently the ability to exercise greater autonomy [60] including the ability to consent for themselves. Should the recipient not wish to continue the use of their BCI they may be unable to have the device removed. This would leave only the potential for harm from the unused implanted device. Therefore, the use of endovascular BCIs might be precluded for individuals unable to provide autonomous informed consent. Alternatively, the use of non-invasive BCIs might enable the ability to communicate informed consent for more invasive BCIs including endovascular devices.

While it is not beyond the realms of possibility that temporary endovascular BCIs will be developed, the endothelialization of the materials in the current generation of devices dictates that, for the near future at least, their placement within the brains of humans will be permanent. This will necessitate an adjustment to existing bioethical considerations surrounding the development and use of invasive BCIs. This chapter section outlined some initial considerations; however, with early feasibility testing of investigational endovascular BCIs underway, further attention is urgently required.

5 Informed Consent for Endovascular BCIs

The informed consent process serves to provide the information pertinent to participation in a research trial, as well as comprehension of that information, to enable an informed, voluntary decision on participation to be made by the potential subject or their legally authorized representative. The information provided on the informed consent form, along with the opportunity to ask questions about the research trial, helps subjects to weigh the unique risks and benefits for them as individuals. The uncertain risk profile and permanency of endovascular BCIs creates additional challenges for researchers and participants to convey and comprehend any potential risks and benefits. The goal of this chapter section is to highlight some of these challenges.

Endovascular BCI is still an investigational treatment, and its therapeutic viability has not been established [1]. Caution should be applied when participants are determining the clinical benefit from clinical trials [61, 62]. Unrealistically high expectations pose a risk of psychological harm to subjects [63] and should be screened for before and during the informed consent process. Comprehension of the information provided during consent should help to provide a realistic level of expectation of benefit. For example, researchers consenting for feasibility trials should emphasize that the primary goal of the study is to investigate safety outcomes and that no expectation of benefit can be assured. Researchers should also be careful not to over-compensate for a lack of supporting evidence for BCIs by overstressing the technological limitations [64, 65]. As mentioned previously, the permanency of endovascular BCIs introduces potential risks including MRI incompatibility, ineligibility for future devices or upgrades, and the possibility for tech support to become unavailable. These risks should also be covered during informed consent to guide realistic expectations of the potential for harm.

Prior literature has identified some of the psychosocial challenges of high importance to BCI users and several target populations. Users regard the ability to feel like the author of one's actions have a social life, self-esteem, freedom, and empowerment as critical aspects in the use of BCI [66]. Additionally, the aesthetics and medical connotations of BCI systems have been considered a significant barrier to adoption by many users [32, 67–69]. In the context of endovascular BCI, the electrodes of the device might be hidden, but wires or devices connected to the patient still pose a risk for users to feel stigma. These and other considerations reported as important by BCI user groups should be addressed during the informed consent process to adequately convey the potential for both risks and benefits. However, simply disclosing psychosocial risks to potential participants may not be sufficient for an informed decision to be made. Risks such as challenges to personal identity, loss of agency, and stigma may require a deeper exploration [63]. Moreover, the meaning of a shift in identity, loss of agency, or the experience of stigma can only be determined on an individual basis [65, 70]. These experiences related to BCI have been evaluated over the course of an intervention, but their exploration during the informed consent process is rarely reported. Moreover, assessing these risks requires high self-awareness, and many subjects may not be accustomed to health-related self-reflection [71]. Newly disabled patients may not possess this self-awareness due to insufficient experience with their condition [72]. The process of self-reflection prior to consenting to participation in an endovascular BCI trial may be additionally complicated due to both the novelty and permanency of this technology. For procedures as significant as the implantation of a permanent endovascular BCI, the study team might require a trained psychologist to explore a potential subject's suitability beyond meeting a set of rigid inclusion and exclusion criteria.

In outlining their frameworks for informed consent best practices for implantable BCIs, Klein and Ojemann [72] emphasized the importance of self-reflection and exploration of participant's preferences and values. They expressed that the subjects are required to understand their values, preferences, and goals to project how their life will change in the years following implantation of a BCI. Qualitative information on current and potential BCI end-users are available and could be used to create a method or tool(s) to help participants explore their needs, values, and goals with respect to BCI [55, 68, 73, 74]. However, much of the current BCI research has not incorporated these existing insights or suggestions [66]. Researchers and clinicians may also need to enhance their understanding of common BCI user perspectives to develop effective informed consent processes [75].

Frameworks for guiding value-based decision-making in healthcare may be applicable to informed consent for undergoing clinical procedures within a BCI research trial. For example, the Patient Priorities Care prototype is a clinically feasible approach to identify and link what matters most to a patient to the health care they are willing and able to participate in [76, 77]. This framework was developed for patients with multiple chronic conditions, which may have similar values to many BCI candidates with chronic conditions [55, 78]. Clinicians on the research team can be trained in identifying patient priorities [79]. Although this training might require additional time and resources, failure to accommodate for a subject's values ahead of and during the informed consent process could lead to rejection of a potentially efficacious BCI technology.

Due to the permanency of endovascular BCI, greater attention to the needs and values of the potential subjects is warranted during the consent process. If researchers do not attend to end users' perspectives during the informed consent process informed consent might be provided for the implantation of a permanent device that may not align with their needs, goals, and values. Effectively evaluating patient suitability on these criteria may require additional resources and expertise, which should be considered during the design stage of research trials and encouraged by funding agencies.

6 Ethical Considerations of Industry Funded Endovascular BCI Research

Traditionally, access to invasive BCI technologies by people living with severe paralysis has occurred under the auspices of government-funded research initiatives, with a primary focus on scientific discovery. However, the field of invasive BCI technology is currently at crossroads with the advent of multiple private and for-profit corporations that have the goal of developing consumer oriented BCI products. This produces a variety of novel and unprecedented ethical challenges that have not been previously navigated in the field of BCI technology. Some of these ethical issues may be common to all invasive BCI technologies, but at the time of writing, there is a particular relevance to endovascular BCIs as the first and, to date, only published human trial of an endovascular BCI [1] investigated the safety and efficacy of the Stentrode™, an endovascular BCI developed by a for-profit corporation: Synchron Inc. (Brooklyn, New York). This section of this chapter will largely focus on how industry funded, invasive BCI research influences both the *pace* and the *publicity* of the research and the ethical implications of both.

Industry-sponsored BCI development differs starkly from government sponsored initiatives when it comes to the pace of the research that is to be performed. The field of health technology has frequently cited the concept of the "valley of death," which is the idea that novel health technologies take, on average, almost two decades to successfully traverse the path from "bench to bedside" [80]. Traditionally, technology developed by industry moves faster than technology that is developed by federal research funding simply due to the differences in timelines [81]. This is due to the disparity in risk tolerance in the two funding sources. By its nature, government-sponsored research has an apparent responsibility to the public to be low-risk and incremental, meaning that the intuitive and innovative leaps forward necessary to keep a health technology startup alive cannot be easily supported within the natural lifespan of a startup company [82]. Thus, industry-sponsored research is required to move at a much faster pace, with a higher risk of failure. However, while the pace of both government- and industry-sponsored research must be tempered by ethics and the well-being of research participants, lowering the overall risk of the implantation procedure for the BCI technology will allow research to move forward at a faster pace. If research involving endovascular BCI technologies do, in fact, pose a lower risk to research participants, they may have greater potential for successful rapid development when compared with intracranial approaches that require penetration of the skull. This difference in research pace, driven by ethics, may be dramatic enough to influence the number of invasive BCI companies that choose an endovascular route in the future. This is a powerful example of how the ethics of BCI technology can influence the focus of research and technological development in the field.

In addition to an accelerated research pace, industry-sponsored research is often much more public-facing, with marketing and media relations professionals who are dedicated to sharing product progress and presenting research findings in the best possible light. Across the field of BCI research, this creates ethical tension, because it can lead to people with disabilities having unrealistic expectations of a BCI product or being led toward a scenario that creates false hope in that product [83]. Notably, at the time of writing this chapter, Elon Musk's BCI company Neuralink is frequently making headlines in mainstream media about seemingly audacious goals for their novel, implantable BCI device [84]. To be clear, these are not claims that are related to an existing product (because such claims would be subject to the scrutiny of regulatory agencies), these are hopeful statements about the long-term vision of the company. Traditionally trained scientists tend to avoid statements of this style, because their scientific reputation relies upon restricting their public statements to what can be empirically proven by their science. Technology startup CEOs tend to enjoy more latitude when it comes to making such claims, and, as the technology startup ecosystem is increasingly entering the healthcare sector, ethical concerns are certain to emerge. Although it may be tempting to dismiss these concerns as a standard occurrence in consumer product marketing, the ethical implications of these actions could be far-reaching. For instance, a large virtual footprint paired with attractive virtual advertising will significantly increase the number of individuals who express interest in ongoing clinical trials [85, 86]. This strategy is known to be particularly effective for identifying research participants who have rare conditions and may be marginalized or difficult to otherwise reach, which is highly relevant to those with a severe disability [87, 88]. However, research participants who are recruited via social media typically display lower rates of retention than those recruited through more conventional means, especially when participation in clinical trial activities require more than a single interaction with researchers [89–91]. This concern is particularly relevant to endovascular BCIs due to the permanency of the technology and the need for long-term follow-up in participants who engage with such a novel, implanted technology. For a technology in its early stages of development with few active users due to regulatory restrictions, the potential loss of data from a premature dropout could be catastrophic to the development of that technology. As such, researchers investigating endovascular technologies must engage in a detailed study recruitment process and mitigation strategies in order to avoid poor retention that can occur as a result of social media recruitment [91]. The disparities that exist between the real-world efficacy and the research needs of endovascular BCI technology and the way that industry sponsored BCI technologies may be advertised must be considered carefully to ensure that false hope in potential research participants does not affect the quality of the research that is conducted.

A final ethical consideration, given the novelty of endovascular BCI technology, relates to the protection of privacy of research participants. Highly publicized research (especially as it pertains to industry-led publicity) that involves very few participants, or participants that are a part of small communities, can run the risk of identifying research participants who may not wish to disclose their involvement in a clinical trial. In addition, researchers who are leading clinical trials involving public–private partnerships, such as government-sponsored research of a privately owned technology, must be particularly vigilant to ensure that recruitment of research participants is not predicated upon the participant's willingness to be involved in publicity activities, only the research itself.

7 Conclusion

The advent of endovascular electrode arrays in BCI has brought new bioethical considerations and challenges that should be considered as the use of this technology progresses. In particular, the permanency of stent-based electrode arrays has wide-ranging bioethical implications. Endovascular electrode arrays can be placed using minimally invasive surgical procedures, which might attenuate some of the safety concerns in comparison to other types of invasive BCI that require craniotomy. This might contribute to an improved risk-benefit ratio. However, substantially more investigation is required before either the full risks or benefits of endovascular BCIs will become apparent. This considerable uncertainty poses challenges to the informed consent process for clinical trials of BCI technologies. Additionally, subjects entering the consent process may have unrealistic expectations of novel BCIs due to the publicity surrounding some for-profit BCI companies. Overall, this chapter outlined some initial considerations of the bioethics of endovascular approaches to BCI, with further discussion needed.

References

1. Oxley TJ, Yoo PE, Rind GS, et al. Motor neuroprosthesis implanted with neurointerventional surgery improves capacity for activities of daily living tasks in severe paralysis: first in-human experience. J NeuroInterventional Surg. 2021;13:102–8.
2. Penn R. Intravascular intracranial EEG recording Technical note. J Neurosurg. 1973;38:239–43.
3. Sefcik RK, Opie NL, John SE, Kellner CP, Mocco J, Oxley TJ. The evolution of endovascular electroencephalography: historical perspective and future applications. Neurosurg Focus. 2016;40:E7.
4. Rapeaux AB, Constandinou TG. Implantable brain machine interfaces: first-in-human studies, technology challenges and trends. Curr Opin Biotechnol. 2021;72:102–11.
5. Raza SA, Opie NL, Morokoff A, Sharma RP, Mitchell PJ, Oxley TJ. Endovascular neuromodulation: safety profile and future directions. Front Neurol. 2020;11:351.
6. John SE, Grayden DB, Yanagisawa T. The future potential of the Stentrode. Expert Rev Med Devices. 2019;16:841–3.
7. Rajah G, Saber H, Singh R, Rangel-Castilla L. Endovascular delivery of leads and Stentrodes and their applications to deep brain stimulation and neuromodulation: a review. Neurosurg Focus. 2018;45:E19.
8. Arya R, Mangano FT, Horn PS, Holland KD, Rose DF, Glauser TA. Adverse events related to extraoperative invasive EEG monitoring with subdural grid electrodes: a systematic review and meta-analysis. Epilepsia. 2013;54:828–39.
9. Zentner J. Complications. In: Surgical treatment of Epilepsies. Cham: Springer; 2020. p. 331–70.

10. Bullard AJ, Hutchison BC, Lee J, Chestek CA, Patil PG. Estimating risk for future intracranial, fully implanted, modular neuroprosthetic systems: a systematic review of hardware complications in clinical deep brain stimulation and experimental human intracortical arrays. Neuromodulation. 2020;23:411–26.
11. Nair DR, Laxer KD, Weber PB, et al. Nine-year prospective efficacy and safety of brain-responsive neurostimulation for focal epilepsy. Neurology. 2020;95:e1244–56.
12. Nicholson P, Brinjikji W, Radovanovic I, Hilditch CA, Tsang ACO, Krings T, Mendes Pereira V, Lenck S. Venous sinus stenting for idiopathic intracranial hypertension: a systematic review and meta-analysis. J NeuroInterventional Surg. 2019;11:380–5.
13. Puffer RC, Mustafa W, Lanzino G. Venous sinus stenting for idiopathic intracranial hypertension: a review of the literature: table 1. J NeuroInterventional Surg. 2013;5:483–6.
14. Starke RM, Wang T, Ding D, Durst CR, Crowley RW, Chalouhi N, Hasan DM, Dumont AS, Jabbour P, Liu KC. Endovascular treatment of venous sinus stenosis in idiopathic intracranial hypertension: complications, neurological outcomes, and radiographic results. Sci World J. 2015;2015:1–8.
15. Kumpe DA, Bennett JL, Seinfeld J, Pelak VS, Chawla A, Tierney M. Dural sinus stent placement for idiopathic intracranial hypertension: clinical article. J Neurosurg. 2012;116:538–48
16. Fan JZ, Lopez-Rivera V, Sheth SA. Over the horizon: the present and future of endovascular neural recording and stimulation. Front Neurosci. 2020;14:432.
17. Saber H, Lewis W, Sadeghi M, Rajah G, Narayanan S. Stent survival and stent-adjacent stenosis rates following venous sinus stenting for idiopathic intracranial hypertension: a systematic review and meta-analysis. Interv Neurol. 2018;7:490–500.
18. Ding D, Liu KC. Management strategies for intraprocedural coil migration during endovascular treatment of intracranial aneurysms: table 1. J NeuroInterventional Surg. 2014;6:428–31.
19. Turek G, Kochanowicz J, Lewszuk A, Lyson T, Zielinska-Turek J, Chwiesko J, Mariak Z. Early surgical removal of migrated coil/stent after failed embolization of intracranial aneurysm. J Neurosurg. 2015;123:841–7.
20. Teleb MS, Cziep ME, Issa M, Lazzaro M, Asif K, Hun Hong S, Lynch JR, Fitzsimmons B-FM, Remler BF, Zaidat OO. Stenting and angioplasty for idiopathic intracranial hypertension: a case series with clinical, angiographic, ophthalmological, complication, and pressure reporting: an IIH case series with clinical, angiographic, and pressure reporting. J Neuroimaging. 2015;25:72–80.
21. Boviatsis EJ, Stavrinou LC, Themistocleous M, Kouyialis AT, Sakas DE. Surgical and hardware complications of deep brain stimulation. A seven-year experience and review of the literature. Acta Neurochir. 2010;152:2053–62.
22. Mokin M, Fargen KM, Primiani CT, et al. Vessel perforation during stent retriever thrombectomy for acute ischemic stroke: technical details and clinical outcomes. J NeuroInterventional Surg. 2017;9:922–8.
23. Leuthardt EC, Moran DW, Mullen TR. Defining surgical terminology and risk for brain computer Interface technologies. Front Neurosci. 2021;15:599549.
24. Alawieh A, Fernando Gonzalez L, Feng W. Barriers and opportunities of cortical stimulation via cerebral venous approach. Brain Stimulat. 2020;13:401–2.
25. Oxley TJ, Opie NL, Rind GS, et al. An ovine model of cerebral catheter venography for implantation of an endovascular neural interface. J Neurosurg. 2018;128:1020–7.
26. Campbell BCV, Mitchell PJ, Kleinig TJ, et al. Endovascular therapy for ischemic stroke with perfusion-imaging selection. N Engl J Med. 2015;372:1009–18.
27. Jovin TG, Chamorro A, Cobo E, et al. Thrombectomy within 8 hours after symptom onset in ischemic stroke. N Engl J Med. 2015;372:2296–306.
28. Saver JL, Goyal M, Bonafe A, et al. Stent-retriever thrombectomy after intravenous t-PA vs. t-PA alone in stroke. N Engl J Med. 2015;372:2285–95.
29. Peters B, Bieker G, Heckman SM, Huggins JE, Wolf C, Zeitlin D, Fried-Oken M. Brain computer Interface users speak up: the virtual users' forum at the 2013 international brain computer Interface meeting. Arch Phys Med Rehabil. 2015;96:S33–7.

30. Rashid M, Sulaiman N, Majeed APPA, Musa RM, Ab. Nasir AF, Bari BS, Khatun S. Current status, challenges, and possible solutions of EEG-based brain-computer Interface: a comprehensive review. Front Neurorobot. 2020;14:25.
31. Holz EM, Botrel L, Kaufmann T, Kübler A. Long-term independent brain-computer Interface home use improves quality of life of a patient in the locked-in state: a case study. Arch Phys Med Rehabil. 2015;96:S16–26.
32. Carmichael C, Carmichael P. BNCI systems as a potential assistive technology: ethical issues and participatory research in the BrainAble project. Disabil Rehabil Assist Technol. 2014;9:41–7.
33. Nakase H, Ohnishi H, Touho H, Karasawa J, Yamamoto S, Shimizu K. An intra-arterial electrode for intracranial electro-encephalogram recordings. Acta Neurochir. 1995;136:103–5.
34. Stoeter P, Dieterle L, Meyer A, Prey N. Intracranial electroencephalographic and evoked-potential recording from intravascular guide wires. AJNR Am J Neuroradiol. 1995;16(6):1214–7.
35. He BD, Ebrahimi M, Palafox L, Srinivasan L. Signal quality of endovascular electroencephalography. J Neural Eng. 2016;13:016016.
36. Bower MR, Stead M, Van Gompel JJ, Bower RS, Sulc V, Asirvatham SJ, Worrell GA. Intravenous recording of intracranial, broadband EEG. J Neurosci Methods. 2013;214:21–6.
37. Oxley TJ, Opie NL, John SE, et al. Minimally invasive endovascular stent-electrode array for high-fidelity, chronic recordings of cortical neural activity. Nat Biotechnol. 2016;34:320–7.
38. John SE, Opie NL, Wong YT, et al. Signal quality of simultaneously recorded endovascular, subdural and epidural signals are comparable. Sci Rep. 2018;8:8427.
39. Forsyth IA, Dunston M, Lombardi G, et al. Evaluation of a minimally invasive endovascular neural interface for decoding motor activity. In: 2019 9th International IEEE/EMBS Conference on Neural Engineering (NER). IEEE, San Francisco, CA, pp. 2019. 750–753.
40. Moran D. Evolution of brain–computer interface: action potentials, local field potentials and electrocorticograms. Curr Opin Neurobiol. 2010;20:741–5.
41. Lukatch HS, Kiddoo CE, MacIver MB. Anesthetic-induced burst suppression EEG activity requires glutamate-mediated excitatory synaptic transmission. Cereb Cortex. 2005;15:1322–31.
42. Opie NL, John SE, Rind GS, Ronayne SM, Grayden DB, Burkitt AN, May CN, O'Brien TJ, Oxley TJ. Chronic impedance spectroscopy of an endovascular stent-electrode array. J Neural Eng. 2016;13:046020.
43. Opie NL, John SE, Rind GS, et al. Focal stimulation of the sheep motor cortex with a chronically implanted minimally invasive electrode array mounted on an endovascular stent. Nat Biomed Eng. 2018;2:907–14.
44. Polikov VS, Tresco PA, Reichert WM. Response of brain tissue to chronically implanted neural electrodes. J Neurosci Methods. 2005;148:1–18.
45. Pels EGM, Aarnoutse EJ, Leinders S, Freudenburg ZV, Branco MP, van der Vijgh BH, Snijders TJ, Denison T, Vansteensel MJ, Ramsey NF. Stability of a chronic implanted brain-computer interface in late-stage amyotrophic lateral sclerosis. Clin Neurophysiol. 2019;130:1798–803.
46. Colachis SC, Dunlap CF, Annetta NV, Tamrakar SM, Bockbrader MA, Friedenberg DA. Long-term intracortical microelectrode array performance in a human: a 5 year retrospective analysis. J Neural Eng. 2021;18:0460d7.
47. Gerboni G, John SE, Ronayne SM, Rind GS, May CN, Oxley TJ, Grayden DB, Opie NL, Wong YT. Cortical brain stimulation with endovascular electrodes. In: 2018 40th Annual International Conference of the IEEE Engineering in Medicine and Biology Society (EMBC). IEEE, Honolulu, HI, 2018. pp. 3088–3091.
48. Opie NL, van der Nagel NR, John SE, Vessey K, Rind GS, Ronayne SM, Fletcher EL, May CN, O'Brien TJ, Oxley TJ. Micro-CT and histological evaluation of an neural Interface implanted within a blood vessel. IEEE Trans Biomed Eng. 2017;64:928–34.
49. Felton EA, Wilson JA, Williams JC, Garell PC. Electrocorticographically controlled brain–computer interfaces using motor and sensory imagery in patients with temporary subdural electrode implants: report of four cases. J Neurosurg. 2007;106:495–500.

50. Vansteensel MJ, Hermes D, Aarnoutse EJ, Bleichner MG, Schalk G, van Rijen PC, Leijten FSS, Ramsey NF. Brain-computer interfacing based on cognitive control. Ann Neurol. 2010;67:809–16.
51. Levine SP, Huggins JE, BeMent SL, Kushwaha RK, Schuh LA, Passaro EA, Rohde MM, Ross DA. Identification of electrocorticogram patterns as the basis for a direct brain Interface. J Clin Neurophysiol. 1999;16:439.
52. Vansteensel MJ, Pels EGM, Bleichner MG, et al. Fully implanted brain–computer interface in a locked-in patient with ALS. N Engl J Med. 2016;375:2060–6.
53. Glannon W. Ethical issues in neuroprosthetics. J Neural Eng. 2016;13:021002.
54. Desmoulin-Canselier S. Patient's lived experience with DBS between medical research and care: some legal implications. Med Health Care Philos. 2019;22:375–86.
55. Kögel J, Jox RJ, Friedrich O. What is it like to use a BCI?—Insights from an interview study with brain-computer interface users. BMC Med Ethics. 2020;21:2.
56. Klein E, Peters B, Higger M. Ethical considerations in ending exploratory brain–computer interface research studies in locked-in syndrome. Camb Q Healthc Ethics. 2018;27:660–74.
57. Hendriks S, Grady C, Ramos KM, et al. Ethical challenges of risk, informed consent, and posttrial responsibilities in human research with neural devices: a review. JAMA Neurol. 2019;76:1506.
58. Sadeghi K, Lee J, Banerjee A, Sohankar J, Gupta SKS. Permanency analysis on human electroencephalogram signals for pervasive brain-computer Interface systems. In: 39th Annual International Conference of the IEEE Engineering in Medicine and Biology Society (EMBC). IEEE, Seogwipo. 2017. pp. 767–770.
59. Fins JJ, Schiff ND. In the blink of the mind's eye. Hast Cent Rep. 2010;40:21–3.
60. Friedrich O, Racine E, Steinert S, Pömsl J, Jox RJ. An analysis of the impact of brain-computer interfaces on autonomy. Neuroethics. 2021;14:17–29.
61. Sim J. Distinctive aspects of consent in pilot and feasibility studies. J Eval Clin Pract. 2021;27:657–64.
62. Fisher CE, Dunn LB, Christopher PP, Holtzheimer PE, Leykin Y, Mayberg HS, Lisanby SH, Appelbaum PS. The ethics of research on deep brain stimulation for depression: decisional capacity and therapeutic misconception: ethics of DBS for depression research. Ann N Y Acad Sci. 2012;1265:69–79.
63. Klein E. Informed consent in implantable BCI research: identifying risks and exploring meaning. Sci Eng Ethics. 2016;22:1299–317.
64. Grübler G, Al-Khodairy A, Leeb R, Pisotta I, Riccio A, Rohm M, Hildt E. Psychosocial and ethical aspects in non-invasive EEG-based BCI research—a survey among BCI users and BCI professionals. Neuroethics. 2014;7:29–41.
65. Specker Sullivan L, Illes J. Ethics in published brain–computer interface research. J Neural Eng. 2018;15:013001.
66. Kögel J, Schmid JR, Jox RJ, Friedrich O. Using brain-computer interfaces: a scoping review of studies employing social research methods. BMC Med Ethics. 2019;20:18.
67. Holz EM, Höhne J, Staiger-Sälzer P, Tangermann M, Kübler A. Brain–computer interface controlled gaming: evaluation of usability by severely motor restricted end-users. Artif Intell Med. 2013;59:111–20.
68. Blain-Moraes S, Schaff R, Gruis KL, Huggins JE, Wren PA. Barriers to and mediators of brain–computer interface user acceptance: focus group findings. Ergonomics. 2012;55:516–25.
69. Zickler C, Halder S, Kleih SC, Herbert C, Kübler A. Brain painting: usability testing according to the user-centered design in end users with severe motor paralysis. Artif Intell Med. 2013;59:99–110.
70. Richmond F, Loeb G. Dissemination: getting BCIs to the people who need them. In: Wolpaw JR, editor. Brain–computer interfaces: principles and practice; 2012.
71. Lim CY, Berry ABL, Hartzler AL, Hirsch T, Carrell DS, Bermet ZA, Ralston JD. Facilitating self-reflection about values and self-care among individuals with chronic conditions. In: Proceedings of the 2019 CHI Conference on Human Factors in Computing Systems. ACM, Glasgow Scotland UK, 2019. pp. 1–12.

72. Klein E, Ojemann J. Informed consent in implantable BCI research: identification of research risks and recommendations for development of best practices. J Neural Eng. 2016;13:043001.
73. Huggins JE, Wren PA, Gruis KL. What would brain-computer interface users want? Opinions and priorities of potential users with amyotrophic lateral sclerosis. Amyotroph Lateral Scler. 2011;12:318–24.
74. Liberati G, Pizzimenti A, Simione L, Riccio A, Schettini F, Inghilleri M, Mattia D, Cincotti F. Developing brain-computer interfaces from a user-centered perspective: assessing the needs of persons with amyotrophic lateral sclerosis, caregivers, and professionals. Appl Ergon. 2015;50:139–46.
75. Gilbert F, Cook M, O'Brien T, Illes J. Embodiment and estrangement: results from a first-in-human "intelligent BCI" trial. Sci Eng Ethics. 2019;25:83–96.
76. Naik AD, Dindo LN, Liew JR, et al. Development of a clinically feasible process for identifying individual health priorities. J Am Geriatr Soc. 2018;66:1872–9.
77. Tinetti ME, Esterson J, Ferris R, Posner P, Blaum CS. Patient priority–directed decision making and care for older adults with multiple chronic conditions. Clin Geriatr Med. 2016;32:261–75.
78. Uhlig K, Leff B, Kent D, et al. A framework for crafting clinical practice guidelines that are relevant to the care and Management of People with multimorbidity. J Gen Intern Med. 2014;29:670–9.
79. Freytag J, Street RL, Barnes DE, Shi Y, Volow AM, Shim JK, Alexander SC, Sudore RL. Empowering older adults to discuss advance care planning during clinical visits: the PREPARE randomized trial. J Am Geriatr Soc. 2020;68:1210–7.
80. Putrino D. Introduction—why me? In: Putrino D, editor. Hacking health make money save lives heal. World. Cham: Springer; 2018. p. 5–11.
81. Simon GE, Richesson RL, Hernandez AF. Disseminating trial results: we can have both faster and better. Healthcare. 2020;8:100474.
82. Putrino D. Identifying opportunities to innovate and create your niche. In: Evidence-based leadership, innovation, and entrepreneurship in nursing and healthcare: a practical guide for success. 1st ed. New York: Springer; 2019. p. 297–312.
83. Nijboer F, Clausen J, Allison BZ, Haselager P. The Asilomar survey: stakeholders' opinions on ethical issues related to brain-computer interfacing. Neuroethics. 2013;6:541–78.
84. Rogers A. Neuralink is impressive tech, Wrapped in Musk Hype. Wired.
85. Bisset CN, Carter B, Law J, et al. The influence of social media on recruitment to surgical trials. BMC Med Res Methodol. 2020;20:201.
86. Christensen T, Riis AH, Hatch EE, Wise LA, Nielsen MG, Rothman KJ, Sørensen HT, Mikkelsen EM. Costs and efficiency of online and offline recruitment methods: a web-based cohort study. J Med Internet Res. 2017;19:e6716.
87. Russomanno J, Patterson JG, Tree JMJ. Social media recruitment of marginalized, hard-to-reach populations: development of recruitment and monitoring guidelines. JMIR Public Health Surveill. 2019;5:e14886.
88. Topolovec-Vranic J, Natarajan K. The use of social Media in Recruitment for medical research studies: a scoping review. J Med Internet Res. 2016;18:e5698.
89. Leonard A, Hutchesson M, Patterson A, Chalmers K, Collins C. Recruitment and retention of young women into nutrition research studies: practical considerations. Trials. 2014;15:23.
90. Murray E, Khadjesari Z, White I, Kalaitzaki E, Godfrey C, McCambridge J, Thompson S, Wallace P. Methodological challenges in online trials. J Med Internet Res. 2009;11:e1052.
91. Pedersen ER, Kurz J. Using Facebook for health-related research study recruitment and program delivery. Curr Opin Psychol. 2016;9:38–43.

Future Developments in Brain/Neural–Computer Interface Technology

Surjo R. Soekadar, Mareike Vermehren, Annalisa Colucci, David Haslacher, Christoph Bublitz, Marcello Ienca, Jennifer A. Chandler, and Benjamin Blankertz

1 Introduction

Mind-reading devices were already conceptualized in the late nineteenth century and have, since then, inspired numerous science fiction authors and filmmakers. After the invention of devices that turned the invisible, e.g., sound, into something measurable, it seemed plausible that thoughts too could be measured and translated into something visible. Although the biological substrates of thought remained elusive, effects of direct electric stimulation of the brain suggested that thought, memories, and emotions have an electric manifestation in the brain that could influence electrical properties of the body. Using a galvanometer, the first attempts to measure

S. R. Soekadar (✉) · M. Vermehren · A. Colucci · D. Haslacher
Clinical Neurotechnology Laboratory, Department of Psychiatry and Neurosciences, Charité Campus Mitte (CCM), Charité—Universitätsmedizin Berlin, Berlin, Germany
e-mail: surjo.soekadar@charite.de; mareike.vermehren@charite.de; annalisa.colucci@charite.de; david.haslacher@charite.de

C. Bublitz
Faculty of Law, University of Hamburg, Hamburg, Germany
e-mail: christoph.bublitz@uni-hamburg.de

M. Ienca
School of Medicine, Technical University of Munich (TUM), Munich, Germany

College of Humanities, Swiss Federal Institute of Technology (EPFL), Lausanne, Switzerland
e-mail: marcello.ienca@tum.de

J. A. Chandler
Faculty of Law, University of Ottawa, Ottawa, ON, Canada
e-mail: chandler@uottawa.ca

B. Blankertz
Faculty of Electrical Engineering and Computer Science, Technical University Berlin, Berlin, Germany
e-mail: benjamin.blankertz@tu-berlin.de

V. Dubljević, A. Coin (eds.), *Policy, Identity, and Neurotechnology*, Advances in Neuroethics, https://doi.org/10.1007/978-3-031-26801-4_5

these influences, e.g., for lie detection, seemed promising [1] but remained rather unspecific and inaccurate.

In the early twentieth century, Hans Berger, who self-reportedly searched for the substrate of telepathy, successfully invented electroencephalography (EEG), turning the brain's invisible electric oscillations directly into curves and patterns painted on long tapes of paper [2]. Using such an apparatus, it was found that brain responses evoked by external stimuli, e.g., a flash, tone, or touch, could allow inferences about mental states or cognitive processes. The finding that modulations of EEG activity e.g., alpha oscillations (9–15 Hz), were state- or task-dependent and apparently confined to circumscribed brain regions (e.g., occipital alpha oscillations related to visual function or central alpha oscillations related to sensory and motor function) raised the expectation that it would be possible to decipher the mental processes underlying perception and action. After initial enthusiasm, it turned out, however that mapping these EEG patterns to specific thoughts was exceptionally difficult. Nonetheless and despite its variability and non-stationarity, EEG provided useful information about a person's level of alertness and proved very helpful in the characterization of brain disorders, such as epilepsy [3].

With the advent of computers in the 1950s, it was hoped that the automatic interpretation of electric brain signals would lead to new insights into the workings of the brain. Thus, in 1961, the National Institutes of Health (NIH) funded the first computer facility for this purpose at the Brain Research Institute of the University of California in Los Angeles (UCLA) [4]. Here, Thelma Estrin, a pioneer in biomedical engineering, successfully established the first analog-to-digital converter to realize an online EEG digital computing system. Previously, Grey Walter developed the first automatic EEG frequency analyzer [5], which was later used to provide online feedback of brain activity. In this context, it was soon discovered that operant conditioning, i.e., increasing or decreasing the probability of a specific behavior depending on the presence or absence of a contingent reward or punishment, was also valid for the behavior of brain signals [6]. Such operant conditioning of neural events (OCNE) soon became the subject of broad interest. It was not only demonstrated for widespread synchronized activity of neuronal cell ensembles in the sensorimotor cortex of cats [7] but also for single cells in the motor cortex of macaques [8].

Building on these developments, the first brain–computer interface (BCI) project that aimed at direct brain–computer communication using EEG signals was started at UCLA in the early 1970s. This project was born of the conviction that EEG does not only consist of random noise, but contains "concomitances of conscious and unconscious experiences" resulting in a complex mixture of neural events [9]. Building on well-characterized evoked responses, e.g., those in the visual domain and other phenomena such as OCNE or the contingent negative variation (CNV), the proposed BCI system was geared toward the use of both spontaneous EEG and specific evoked responses to establish a direct "brain-computer dialogue."

Today, 50 years later, we may still not have achieved precise and accurate mind-reading as envisioned in the nineteenth century, but the feasibility, utility, and limits of BCI-enabled human–computer interaction have been increasingly well charted. Besides active or directed control of external devices [10], additional generally unanticipated forms of BCI applications were established, e.g., BCIs designed to trigger neural recovery (also termed *restorative* BCIs) [11] or BCIs conveying information regarding the user state to improve the ergonomics of human–computer interaction (also termed *passive* BCIs) [12, 13].

Increasing digitalization of processes and interactions, ubiquitous computing, and wireless connectivity have paved the way for so-called immersive technologies [14] that are now increasingly blurring the boundaries between the human mind and body and digital tools [15]. The ambitions and collective efforts to merge the human mind and machines have never been higher. But how could state-of-the-art BCIs contribute to this endeavor? Where are the limits on this path? And where should these limits be?

In this chapter, we aim to provide an overview of the most well-established BCI concepts and will introduce current state-of-the-art medical and non-medical use cases that involve BCI technology. We will then describe current efforts to extend the scope of BCI applications and map out the prospects of merging BCIs with artificial intelligence (AI) and other emerging technologies, such as quantum sensors. At the end of this chapter, we will shortly outline the possible impact of future BCI technology on society and human self-conception.

2 Modes of Operation and Applications of Brain/Neural–Computer Interfaces

The original ambition of the first BCI project at UCLA was to establish electrical brain signals as carriers of information in human–computer interaction or for the purpose of controlling external devices, such as prosthetics or even spaceships [9]. To implement such BCI, three distinct components are required: (1) a brain signal recording unit translating analogue electric, magnetic, or metabolic measures into digital data streams, (2) a real-time signal processing unit for interpretation of this incoming data, and (3) an actuator, output generator, or basically any device that uses this information in a purposeful way [16].

While first conceptions of BCIs were mainly designed to convey information (specifically control commands) from the brain to the computer, e.g., to restore movement or communication in paralysis [17], subsequent concepts extended from unidirectional toward bidirectional BCIs, i.e., systems that not only read-out brain activity but also write into it, e.g., via brain stimulation of the central nervous system (CNS) [18, 19]. Here, the stimulator acts as a BCI output creating a direct feedback loop between brain states and brain stimulation that does not involve or depend on the user's sensory system.

However, since it is not entirely clear how information is represented and stored in the brain, the ability to transfer information into the brain using brain stimulation is still very limited. Thus, the notion of "read and write" may be misleading in the context of current bidirectional BCIs, to the extent that it implies an analogy between the brain and a hard disk or magnetic tape. In contrast to conventional computers that use a two-symbol system to represent information (i.e., zeros and ones), the symbol system underlying information processing and its physiological representation in the brain is not well understood [20, 21]. As long as this relationship remains an enigma, information transfer from the computer to the brain will be very limited.

Before outlining more specific BCI applications, the following section introduces the established modes of operation and categories of BCIs that influence the range and scope of applications across various possible use cases.

From the user's perspective, three different modes of operation can be distinguished: *active, reactive, and passive*. Active BCIs are designed for the voluntary control of external devices, and typically translate brain activity linked to goal-directed behavior into control commands directed at tools that assist in achieving the intended goal [22, 23]. Reactive BCIs, in contrast, infer their output from the brain's reaction to external sensory or direct stimulation [24–26]. Due to the relatively high signal-to-noise ratio of evoked brain responses, such paradigms achieved the highest information transfer rate (ITR) in noninvasive BCIs, e.g., in a speller task [27]. Finally, passive BCIs derive their output from automatic or spontaneous brain activity serving as implicit input to support an ongoing task [12]. Here, depending on the level of interactivity, four categories can be distinguished: (1) Mental state assessment to replace or support other data (e.g., from questionnaires or behavioral observations) without direct feedback to the user [28], (2) online state assessment (e.g., of fatigue or mental workload) with feedback to the user (e.g., indicated by a warning light), (3) state assessment with the purpose of directly influencing the assessed state in a closed control loop manner (e.g., reducing workload of an air traffic controller by limiting irrelevant sensory input in critical situations) [29, 30], and (4) automated adaptation, where the BCI system continuously learns and adapts according to the user state [31]. Here, the system builds a model to represent aspects of the user's affective or cognitive responses that serve as a basis for the system's autonomous behavior. For example, in case of the air traffic controller the system would have learned which situations will most likely provoke a mental overload and automatically call for assistance ahead of time.

Besides these different modes of operation, BCI systems are generally categorized according to the invasiveness or location of the brain signal recordings, as well as the purpose of use. When brain signals are recorded from inside the skull, a system is referred to as an *invasive* or *implantable* BCI. Here, a variety of signals have been established for BCI applications ranging from synaptic or local field potentials (LFP) using electrocorticography (ECoG) [32] to action potential spike trains [33]. *Noninvasive* BCIs typically use brain signals recorded from the surface of or at some distance from the scalp. Here, mainly six types of brain signals are used for BCI applications so far. Four were established primarily for use in active BCIs: (1) Slow cortical potentials (SCP) [34], (2) sensorimotor or mu rhythms (9–15 Hz) [35, 36], (3) blood-oxygen-level-dependent (BOLD) signals in functional magnetic resonance imaging (fMRI) [37, 38], and (4) concentration changes of oxy/deoxy hemoglobin using near-infrared spectroscopy (NIRS) [39, 40]. Two other types were established primarily in reactive BCIs: (5) event-related potentials (ERPs) and (6) steady-state visual or auditory evoked potentials (SSVEP/SSAEP) [41, 42]. SCPs, i.e., slow potential drifts with a duration of 500 ms to several seconds, were among the first signals used for noninvasive control of a BCI [43]. When related to motor activity, SCPs were specified as motor-related cortical potentials (MRCP) and can be further sub-categorized into the self-initiated Bereitschaftspotential (BP) [44], also termed the readiness potential (RP), and the externally triggered contingent negative variation (CNV) [45]. Since the BP builds up 2 s before voluntary movements, it was mainly explored in the context of controlling motor prosthetics or exoskeletons [46, 47]. BP-based BCIs were also used in ways beyond such medical applications, e.g., as a research tool [48] or in the context of detecting emergency braking intention [49].

To increase classification accuracy, BCI control paradigms were designed that use pre-defined time windows during which the presence or absence of a particular signal feature is evaluated. Reactive BCIs commonly use this approach by design because the external stimulus determines the relevant time window in which brain responses are evaluated. Active BCIs can be either operated with such pre-defined time windows, which is called a *synchronous* mode of operation, or in a self-paced or self-initiated mode, which is then called an *asynchronous* mode of operation [50]. The synchronous mode of operation vastly limits the applicability and practicality of active BCI control in daily life contexts since users must wait for the pre-defined time windows before active control is possible. Moreover, since time windows are indicated by external stimuli or triggers, such a paradigm requires an intact (afferent) sensory domain as well as the user's attention (like reactive BCIs). Thus, this type of BCI has been mainly established in the auditory [51] or visual [52] domain since these two sensory domains are typically less affected in many neurological disorders (e.g., stroke, spinal cord injury or motor neuron diseases). In contrast, the *asynchronous* mode of operation allows for self-initiated or self-paced BCI control providing more autonomy, flexibility, and intuitiveness to the user. However, classification accuracies of self-paced BCIs are typically lower compared to BCIs operating in a synchronous mode.

To increase control accuracy, BCI paradigms were established that combine different types of brain signals, e.g., MRCP or ERP with mu rhythms [53, 54]. Such hybrid BCIs also include error-related potentials (ErrPs), i.e., stimulus-triggered negative and positive EEG deflections that are related to error processing, reward prediction, and conscious error perception [55]. Such signals can be used to correct false classifications or to improve classification algorithms. In addition, other biosignals, e.g., related to eye movements or peripheral muscle activity, were implemented to improve the applicability of brain-controlled devices in real-life scenarios. When brain-controlled systems also infer mental states, including intentions, from such peripheral signals, these systems are typically referred to as brain/neural–computer or –machine interfaces [10, 56].

Given that the physiological representations of the symbol system underlying information processing in the brain is largely unknown, how can classification accuracies of BCIs be further improved?

A very important strategy to increase the classification accuracies of active BCIs is operant conditioning of neural events (OCNE) which was mentioned before [57]. These neural events are typically measures of electrical, magnetic, or metabolic brain activity and their derivatives.

Behavior of single cells or larger neuronal cell ensembles (e.g., increased firing rates or desynchronization/synchronization of sensorimotor EEG activity) are contingently rewarded (e.g., by delivery of a banana-flavored pellet in the case of a monkey, or monetary incentive in the case of humans) to increase the likelihood of the behavior's recurrence. It was found that the intactness of the cortical-basal ganglia-thalamic feedback loop plays an important role for OCNE [58] and might explain why people with brain lesions or neurodegenerative disorders struggle in acquiring BCI control. Still, depending on the brain signal and signal-to-noise ratio, BCI learning based on operant conditioning can require days to weeks in healthy volunteers, or even months of training in patients. Another strategy to increase BCI control accuracy is feedback learning. Being independent of external reward, feedback learning depends, however, on the user's intrinsic motivation and the involvement of internal reward mechanisms ascribed to intact cortico-striatal circuitry including the ventromedial prefrontal cortex (VMPFC) [59]. Using such an approach, more complex measures of brain activity, e.g., cortico-thalamic BOLD connectivity, could be trained [38]. With the advent of novel machine learning approaches, including convolutional neural networks (CNN), new and more complex derivatives of brain activity have been proposed for BCI applications [60, 61]. The main challenge with these more complex derivatives is to ensure that meaningful brain activity and not (systematic) signal artifacts are decoded, and that the underlying algorithms are real-time compatible.

In contrast to active BCIs, reactive BCIs based on evoked responses do not require any learning. This increases their applicability, e.g., in BCI naïve users, but comes with other disadvantages mentioned before (e.g., dependence on the sensory system and user attention, or possible distraction from and interference with other tasks).

An aspect of OCNE and feedback learning that was rather neglected in the early history of BCIs is the impact of BCI learning on brain physiological processes and, thus, brain function. After automatic EEG frequency analyzers became available in the early 1960s, M. Barry Sterman discovered that operant conditioning of SMR can reduce seizure frequency in cats exposed to seizure-inducing agents or persons diagnosed with epilepsy [7]. Later it was found that such conditioning of brain activity, commonly termed *neurofeedback*, can also improve symptoms of attention deficit and hyperactivity disorder (ADHD) in children [62]. These first successes raised great hopes that EEG neurofeedback could improve symptoms of a multitude of brain disorders. While some positive effects, e.g., in ADHD, could be replicated in larger controlled and double-blinded studies [63, 64], others could not be confirmed. Although it became widely accepted that OCNE can influence brain function and behavior, the underlying mechanisms remained incompletely understood, and the specificity and effect size of neurofeedback seemed to depend on a variety of factors [65]. Thus, the initial enthusiasm gradually abated over the ensuing decades but became later revived in the context of BCI research.

While the first BCI paradigms used OCNE to improve BCI control, e.g., to operate a prosthesis, it was found that repeated use of such BCI can also have restorative effects on the brain. For instance, it was shown that repeated use of a BCI-controlled exoskeleton can trigger motor recovery, even in chronic paralysis after stroke [35, 66] or cervical spinal cord injury [67]. Thus, *assistive* and *restorative* BCIs came to be distinguished on the purpose of the application [11, 68]. In this sense, restorative BCIs build on the old research tradition of neurofeedback and may now—with the technological advances implemented in the BCI field—elucidate some of the unknown mechanisms underlying neurofeedback. Although first conceptions of BCIs mainly focused on the external effects (i.e., a BCI conceptualized as a tool to act on the environment), it becomes increasingly clear that any form of BCI interaction also impacts the brain itself, to a larger (restorative BCIs) or lesser extent (reactive or passive BCIs). This aspect is even more apparent in bidirectional BCIs that involve direct stimulation of the CNS.

Building on the described modes of operation, various medical and non-medical BCI applications have been realized so far. These include versatile control of multi-joint prosthetics, robotic arms, or functional electric stimulation (FES) that assist in grasping and manipulating objects of daily living [69, 70]. By using a bidirectional interface to restore sensory capacity during prosthesis control, manipulation could be substantially improved [71]. Here, a simple linear relationship between external force sensors and applied stimulation intensities could induce the feeling of touch. Using a combination of noninvasive EEG/EOG signals enabled patients with complete hand paralysis after a high spinal cord injury to independently eat and drink in a restaurant [10]. In this study, closing motions were initiated using modulations in EEG, and hand opening motions were controlled by EOG. Such a paradigm was also successfully implemented for whole-arm exoskeleton control [72]. In the context of robot or exoskeleton control, also error-related potentials (ErrP) were used to optimize brain control of robotic devices [73].

Another important medical application aims at the restoration of communication. Here, early demonstrations of SCP-based systems allowed patients who were diagnosed with locked-in syndrome (LIS), i.e., the inability to speak or move, to spell full sentences by selecting letters on a display [34]. Different modalities were applied, including electrocorticography (ECoG) and functional NIRS, with variable success [74, 75]. The main limitation of these approaches was that once patients entered a complete locked-in state (CLIS), BCI-enabled communication failed. The reason for this is still not well understood, but fragmentation of sleep patterns [76] and progressive neurodegenerative processes affecting cell metabolism may play a role. Since, by definition, LIS patients can still communicate with eye movements or subtle muscle twitches, the necessity of developing BCIs for this clinical population was, thus, quite controversially discussed [77]. By providing an alternative communication channel, BCIs in LIS were nonetheless positively received and regularly used by the patients [78]. Only recently, successful communication in CLIS was reported using two microelectrode arrays. By modulating neural firing rates based on auditory feedback, a 35yo patient diagnosed with amyotrophic lateral sclerosis (ALS), an often rapidly progressing neurodegenerative disorder, could sustain communication despite CLIS [79]. Interestingly, the communication rates the patient could achieve with the implanted device were comparable to those of noninvasive BCIs used in healthy volunteers. Communication often dealt with requests related to body position, health status, food, personal care, and social activities. Since all attempts to restore communication using noninvasive BCIs in CLIS have failed before, this first successful use case may now increase the willingness of ALS patients to undergo implantation. However, this implantation is costly and not broadly available, yet. Furthermore, its use still requires a team of experts to maintain the functionality of the system. It is unclear whether health insurance will cover the associated costs. Nevertheless, the prospect of successfully overcoming the profound communication impairment of CLIS marks an important milestone in the history of BCIs and provides important reasons to use this technology in the medical field.

Beyond these examples, the use of active BCIs in non-medical applications is still confined to research environments. While it has been shown that active BCIs can be also used to steer an airplane or drone [80, 81], or to play video games [82], acquiring the necessary BCI control is cumbersome and time consuming. Although BCI learning itself could be part of a game, this would necessitate that BCI hardware be inexpensive and accessible and BCI control be of sufficient reliability. To date, despite extensive investment into technology development, this has not been achieved [83]. It can be anticipated, though, that inexpensive and accessible hardware will become available, and this will facilitate the implementation of BCI technology as part of video games or other forms of entertainment. Especially reactive and passive BCIs may find future applications beyond entertainment. For instance, reactive BCIs, e.g., based on SSVEP, are currently explored in autonomous driving [84, 85]. Passive BCIs have been mainly tested in the context of the ergonomics of human–computer or human–machine interaction, but they may incorporate other bio- or neuroadaptive approaches that take peripheral measures such as muscle tone, dynamic posture, pupillometry, or skin conductance into account [86].

While state assessment with feedback in passive BCIs resembles classical neurofeedback paradigms, the aim of such systems is mainly to improve human–computer interaction and not to normalize or alter brain states [12]. However, it can be argued that passive BCIs may also exert such an effect when used over longer periods of time. More systematic studies are needed to address this question.

3 Next-Generation Brain/Neural–Computer and Machine Interfaces

Current BCI technology mainly covers the motor domain (e.g., to re-establish movement and communication) and seeks to improve human–computer or human–machine interaction by integrating measures related to error processing, attention, cognitive workload, or fatigue. The next-generation B/NCIs will extend toward other domains, e.g., emotion regulation, memory formation, cognitive control, and perception. These brain functions are often affected across various brain disorders, such as depression, ADHD, obsessive-compulsive or anxiety disorder, addiction, or dementia [87]. With the development and clinical application of these next-generation interfaces, it is hoped that it will be possible to alleviate the burden of brain disorders and to better understand the relationship between brain functions and clinical symptoms.

To establish such BCI system, the so-called brain–behavior relationship, i.e., the link between specific individual neurophysiological measures and domain-specific brain function or behaviors and symptoms, must be revealed [88]. Despite tremendous progress in neuroscience over the last decades, this has not been achieved yet. Although various correlations between certain brain physiological measures and brain functions have been found, their causal relationships have remained largely unclear. Moreover, attempts to reveal these causal relationships often included the averaging of brain physiological measures over tens to hundreds of trials (i.e., repetitions of experimental tests) to reduce variability and noise. Such averaging may, however, also reduce relevant information for precise mapping to brain functions or behavioral outcome measures.

A possible approach to overcome this challenge is to use brain stimulation targeting specific physiological measures, e.g., brain oscillations at certain frequencies, and assess the impact on brain function and behavior at millisecond-to-millisecond precision [89–91]. To achieve this, several technical challenges must be mastered: (1) Brain physiological measures must be recorded and analyzed in real-time, (2) stimulation must be delivered at high temporal and spatial precision, (3) stimulation artifacts must be sufficiently eliminated to assess the online stimulation effects.

The first two points are equally necessary to establish bidirectional BCIs, and thus, share the same technological framework. Typically, implantable bidirectional BCIs that deliver electric stimulation to the brain do so at some distance from the recording electrodes to reduce interference of stimulation with BCI classification [71]. Currently, as previously mentioned, it is unclear which stimulation parameters are most effective to interact with the human brain. In this context, novel approaches have been introduced that use deep learning to derive effective stimulation patterns, e.g., for optic nerve stimulation to restore normal vision in disorders affecting the retina [92].

In the noninvasive field, a few important milestones toward bidirectional BCIs have been reached, e.g., in vivo assessment of brain oscillations during transcranial direct current stimulation using magnetoencephalography (MEG) [89] and, recently, recovery of targeted brain oscillations during transcranial alternating current stimulation (tACS) using EEG [90]. The MEG has the advantage that neuromagnetic activity passes through transcranial stimulation electrodes undistorted, so that brain activity immediately underneath the stimulation electrodes can be reconstructed. This is not the case in EEG, and due to the variable path of the electric currents through the skull, precise and focal stimulation using transcranial electric stimulation (tES) remains a challenge. Brain oscillations can be also targeted with transcranial magnetic stimulation (TMS) [93], but due to the magnitude of artifacts that are associated with magnetic fields at 2–3 Tesla, sufficient artifact elimination is difficult to achieve. Moreover, TMS commonly uses short magnetic pulses of 160–250 μs duration that do not resemble the targeted neural activity. Nevertheless, successful implementation of the first EEG-based closed-loop TMS targeting brain oscillations at millisecond-precision represented an important milestone on the path toward effective neurostimulation.

Recent advances in brain stimulation methods using temporal interference of electric [94] or magnetic [95] fields or ultrasound [96] may overcome some of the current limitations in noninvasive brain stimulation and neuromodulation by offering higher focality and penetration depth. Both temporal interference and ultrasound stimulation can reach deeper brain areas and provide higher focality than other established methods like TMS or tES. Currently, stimulation effects of these new methods and their underlying mechanisms have not yet been well explored, and it is unclear whether closed-loop operation is feasible, since stimulation artifacts have not been well characterized yet. Consequently, these new stimulation methods have not been implemented in the context of bidirectional BCI but promise to further advance the field.

Another frontier in the development of next-generation B/NCIs relates to the precise recording of neural activity at high temporal and spatial resolution. In this context, new high-density microelectrodes were developed [97]. However, these necessitate implantation with the risk of bleedings or infections [98]. Moreover, replacement or repair of implanted device components requires another surgical procedure. Other approaches include semi-invasive methods such as intravascular [99] or sub-scalp recordings to assess LFP or EEG. It is most likely that all these approaches will remain in the experimental, medical domain for the foreseeable future because the cost-benefit-ratio across users and applications has yet to be determined. It is unclear whether recording from more neurons will automatically result in higher decoding accuracy, precision in the assessment of brain states, or, using OCNE, in more degrees of freedom for active BCI control, or whether there are some inherent boundaries for implantable BCIs [100]. To further advance our understanding of brain–behavior relationships and to possibly elucidate the symbol system underlying information processing in the brain, these methods are, however, of critical importance.

In analogy to BCI-triggered motor recovery after stroke, it is conceivable that next-generation B/NCIs may be only used for a defined time or intermittently to facilitate or maintain recovery of brain function, e.g., to recover from depression or to maintain memory function in neurodegenerative disorders. This is another reason for the assumption that rather noninvasive BCIs will be more prevalent than implantable solutions. Nevertheless, there might be cases in which recovery of brain functions is not possible so that continuous use of such a neuroprosthesis is necessary. The main challenge with noninvasive means to record brain activity is their susceptibility to noise and their lack of spatial resolution compared to implanted electrodes or sensors [101]. Moreover, when using EEG, brain signals are dampened and distorted by various tissues. This reduces signal quality, particularly in the upper frequency bands above 25 Hz. Recently, quantum sensors were introduced that promise to increase precision of brain recordings. These quantum sensors, e.g., optically pumped magnetometers (OPM), can measure neuromagnetic fields passing the skull undistorted [102]. Provided proper calibration, OPMs could reach much higher spatial resolution than any other established noninvasive neuroimaging tool, e.g., conventional helium-cooled magnetometry using super-conducting quantum interference devices (SQUID). Moreover, they allow for the assessment of brain signal frequencies up to the kilohertz range, i.e., they might make physiological information in higher frequencies, e.g., the gamma range, more accessible. The main drawback with these highly sensitive magnetometers is that they require substantial magnetic shielding from environmental fields. Magnetometers based on synthetic diamonds or polymers that can be operated in the earth's magnetic field may overcome this limitation, however [103].

Taken together, important advances in sensor technology, stimulation techniques, real-time signal processing, and machine learning have been made that now allow the study of brain–behavior relationships with unprecedented precision. Although many unsolved questions remain (e.g., the link between brain functions and specific clinical symptoms, the cause for reoccurring episodes of mental illness), the prospects for extending the scope of B/NCIs toward domains such as emotion regulation, memory formation, cognitive control or perception have never been better. Particularly, the combination of BCI technology with other emerging technologies, such as AI and quantum computing, may catalyze the feasibility of new applications outlined in the next section.

4 Merging Brain/Neural-Computer Interfaces with Artificial Intelligence

For many decades, machine learning has been used as part of BCI technology to improve the accuracy of brain signal classification [104]. In recent years, new machine learning-enabled tools and applications have been developed outside of the BCI field, e.g., image or speech recognition using neural networks, subsumed under the term artificial intelligence (AI) technology, and they are now increasingly being merged with BCI technology [105]. Overall, there are three main areas in which the implementation of AI components is being explored in the development of new BCI applications: 1. improving classification of brain/neural activity patterns, 2. identifying effective stimulation parameters for bidirectional BCIs, 3. implementing shared control of brain/neural-controlled external devices, e.g., robots or exoskeletons.

While the most established and broadly used BCI algorithms use linear classification because of their robustness (in the presence of noise or signal outliers and a comparably small amount of training data) and low computational cost, kernel-based, e.g., support vector machines (SVM), and other non-linear methods have also been implemented [106]. These methods reach slightly higher classification accuracies but at the cost of higher computation time and memory. Since this is nowadays less of a problem due to broader availability of computational power and memory, non-linear classifiers, e.g., convolutional neural networks (CNN), are being increasingly explored in the context of BCI feature classification [107]. Here, it should be underlined that the use of non-linear methods requires good understanding of the data, because several parameters and network design decisions must be chosen in an informed way. For example, signal artifacts (e.g., related to heartbeat, pulse waves, breathing, voluntary movements, eye blinks, increased muscle tone, eye movements, or other artifacts from the environment) may unintentionally influence classification. Consequently, with non-linear methods it is more difficult to verify what kind of information (e.g., neural vs. artifactual sources) is being exploited by a particular model although concepts for explainable AI exist [108, 109]. In the case of implantable BCIs, this might be less of a challenge, but caution needs to be taken. Nevertheless, neural networks may be superior under certain conditions and paradigms compared to other approaches. With the advent of quantum sensors, non-linear classifiers might also prove particularly useful in noninvasive or minimally invasive BCIs.

The second area where the implementation of AI may play an increasing role for future BCI technology is the development of effective stimulation parameters to interact with the human brain. This would be a critical prerequisite to establish bidirectional BCIs and sensory neuroprostheses, i.e., systems that substitute for motor, sensory, or cognitive functions. For instance, by using a convolutional neural network (CNNs) as a model of the ventral visual stream, optic nerve stimulation patterns could be derived to elicit static and dynamic visual scenes [92]. In other words, specific stimulation patterns were identified that could convey a certain visual image directly to the brain. While this has been only achieved in silico so far and not in real time, these results indicate that neural networks may play an important role for computer-to-brain communication, and may also help to gain better understanding of how information is represented in the brain [110]. Once it becomes feasible to convey rich sensory information to the brain, neuroadaptive algorithms could be implemented, e.g., to reduce or augment sensory information depending on the user's state. It is very likely, though, that—in analogy to active BCIs used in paralysis—this technology will first be used to restore compromised or lost sensory function in blind or deaf individuals. Besides improving function of a sensory system, such technology could be also used to provide information beyond accurate representations of the environment, but it is unclear to what degree this would be possible due to top-down regulation of sensory input [111] and whether this would negatively interfere with normal perception of the environment. Since the brain can also generate quasi-sensory experience in the absence of input from the sensory system (e.g., during dreaming), it is not entirely inconceivable that sensory experience could be influenced or induced in such state with a bidirectional BCI.

The third area where merging B/NCI technology and AI has been explored and successfully demonstrated is the so-called shared control of external devices, e.g., robots or exoskeletons [72, 112]. Controlling a multi-joint robotic arm or any other device or manipulator with many degrees of freedom requires high ITRs that cannot be achieved with noninvasive means. Currently, the boundaries for OCNE-based BCI control in terms of precision, reliability, and robustness are still unclear and are the subject of further exploration. Even implantable BCIs using several hundreds of electrodes are limited in their capacity to control such devices.

Thus, concepts of human-machine collaboration have been implemented in which the intended goal is inferred from brain activity, and the best solution to achieve the goal is computed and executed by the machine. To compute such a solution, the availability of a precise model of the environment is an important prerequisite. Various AI-enabled tools, such as 3D-object recognition, have been successfully implemented to create such a model allowing for, e.g., brain/neural whole-arm exoskeleton control [72]. Using a shared-control paradigm, grasping and manipulating a bottle, then moving it to the mouth and drinking was feasible, for example [113].

To increase applicability and practicality, it is essential to implement a veto function in shared-control paradigms that include an AI-enabled autonomous machine [114]. Such a veto function is also critical for questions of accountability and liability. Since a BCI-triggered veto is subject to the same inaccuracies inherent to all active BCI applications, other bio-signals providing higher control accuracy, e.g., related to eye movements, are currently used to trigger a veto [10, 115]. Independent of the machine's precision or capabilities, it will be always the user's responsibility to decide the contexts and situations in which to use shared-control systems.

Beyond these three main areas, machine learning methods are also contributing to the advancement of neuroadaptive technologies. For instance, real-time analysis of brain responses was used to create a continuously updated user model of expectations [116]. With such a model, a computer could learn to adapt to the user's mindset to optimize goal congruency. Here, it is important to note that the model of expectation can be derived without user awareness, i.e., implicitly [110]. Whereas increasing applicability, this subconsciously informed brain–computer interaction raises important neuroethical questions, however.

5 Neuroethical Perspective

The development of bidirectional BCIs relying on AI methods and coupled to the human brain marks an interesting new step in human–machine or –computer interaction. It generates a hybrid cognitive system that runs on, or is fed by inputs from, the organic hardware of the brain as well as the AI implementing BCI. This creates, as some say, hybrid minds [117]. Surely, many technologies have deeply influenced the workings and perhaps even the evolution of the human mind. Philosophers subscribing to the Extended Mind Thesis may even go as far as saying that cognitive tools such as iPads or even pen and paper sometimes become part of the mind. Nonetheless, in those cases the interaction between the human brain and the cognitive system is less direct; the input of the extended cognitive systems proceeds via external sensory perception. BCIs afford a direct coupling at the physiological level. Also, BCIs may be more deeply integrated in the operation of the brain, and future applications may well restore or replace mental functions today carried out by the brain. The novelty thus lies in the direct, internal coupling of computers, AI, and the human brain and their resulting functional integration [118]. First cases of such *hybridminds* are on the verge, e.g., in people engaging with adaptive DBS or closed-loop BCIs. In the near future, closed-loop applications that affect and regulate emotions or other mental functioning are to be expected raising a range of unanswered neuroethical questions.

Apart from safety and side-effects, a central question concerns the user perspective, the experience of having-or being-a hybrid mind. Will users realize if part of their mental functioning is executed or influenced by a BCI, and how will that feel? A related question concerns the attitude people take toward their hybrid mind. There are indications that some users of DBS, for instance, feel alienated from themselves and ascribe this to the influence of the DBS on their mental life [119]. Others experience substantial changes in emotions and behavior, apparently resulting from the DBS, but welcome those changes [120]. More broadly, there is an open debate about whether brain stimulation with DBS causes personality changes, how frequently this might occur, and whether there are other explanations for the apparent changes [121]. Thus, the subjective experience of users of BCIs is an intriguing aspect to be explored, e.g., via phenomenological methods.

The blending of minds and machines raises further questions: Do BCIs become part of the person, in a strong sense; or at least, do the BCIs become part of their bodies—or do they remain tools, despite their functional and sometimes inseparable integration with the brain? Answers to these conceptual questions lead to a range of further moral and legal questions, from responsibility for negative outcomes of the actions of hybrid minds to manufacturer liability; from intellectual property in the code that might become part of the mind to issues of privacy. The range of practical problems that arise is broad and is already confronting society. Patients using the retinal implant Second Sight have encountered trouble accessing information about the device, patient support, and replacement parts following the company's financial collapse [122]. Given the reliance of users, not just for sensory or motor functions but perhaps also for mental functions on these advanced devices, should we be approaching them as we would any normal medical device, or are different social and legal arrangements warranted?

The issues are complex and fascinating, and they will vary according to the application in question. The tendency with these technologies is to focus on the immediate replacement, support, or enhancement of a human function. For example, the success of a motor BCI is largely a matter of how usable it is for the individual. However, some human functions are inherently relational, i.e., they require two human beings. The example of communication neuroprostheses offers a good example. Routine communications about their basic daily needs and wishes are a first and important objective for paralyzed people. However, if these devices are to support high-stakes communications (e.g., requesting or refusing medical care with potentially fatal consequences, or testifying in court), the listener's ability to judge the voluntariness and accuracy of communications will be critical [123]. The incorporation of layers of non-transparent machine learning within decoders of neural signals to produce communication will greatly complicate this effort, and the roles and needs of listeners must be built into the technology to maximize the utility of the technology.

Perhaps the most important question is how far the blending of minds and machines can and should go. The possibility of writing specific information to the brain—creating sensations, memories, and other mental states—while still far-off, is in view as an eventual possibility, as discussed above. These technologies need not be sophisticated and fully successful to be ethically important. Even a relatively modest or partial restoration of lost sensory functions is evidently exciting and valuable, but the abilities to do so would open up a host of other possibilities from the non-therapeutic creation of desired sensations (e.g., possible risk of developing dependencies that interfere with daily functioning) to the infliction of undesired sensations or states (e.g., in interrogations). Together all of this raises the perennial question with all human technological invention—that of how to capture the benefits while limiting the possible risks and unethical applications. Are there ethical lines that should not be crossed—or should technologies be developed but used only for alleviating severe disorders? These questions become pressing as BCI technology advances; in the best case, following careful consideration of the ethical implications for individuals and society.

6 Summary and Conclusions

Innovative neurotechnologies are rapidly evolving, rendering new medical and non-medical applications possible that were previously not anticipated. However, implementation of brain-controlled technology into everyday life environments is challenging. Due to the limited reliability and robustness of control, the most promising areas of application for active BCI systems remain in the medical field, e.g., for restoration of movement and communication. Here, however, the availability of versatile, robust, and certified actuators, e.g., individually tailored prostheses or exoskeletons, represent an important bottleneck for further adoption. In contrast, reactive BCIs can achieve higher classification accuracies but seem primarily attractive in scenarios in which providing feedback to the computer by other means, e.g., voice, gestures, or touch, is either undesirable or not feasible. Since high classification accuracy of passive BCI systems is less critical, such technology may become adopted faster in non-medical applications, e.g., to augment learning or to optimize ergonomics of human–computer interaction.

The implementation of brain state-informed or closed-loop stimulation of the brain marks an important milestone in BCI technology because it allows for interacting with ongoing brain activity independently of the sensory system. Besides providing direct feedback to the brain during prosthesis control, it could also sustain communication in neurodegenerative brain disorders affecting the motor and sensory system. Moreover, it could be used to suppress pathological brain activity to improve brain function. Importantly, combining neuromodulation with in vivo assessment of brain physiology could contribute to elucidating brain–behavior relationships, e.g., in the domains of cognitive, emotional, and memory function. Here, machine learning could contribute to the development of artificial models of the brain that inform new and effective stimulation patterns. Combining BCIs with AI-enabled tools, e.g., 3D-object recognition or tracking, does not only enhance versatility and performance of brain-controlled systems, but may catalyze the emergence of new applications beyond the medical field. The possibility to merge biological and artificial cognitive system gives rise to new entities that are referred to as *hybridminds*. Feasibility of such an entity depends, however, on understanding and direct manipulation of the mind's symbol system and its manifestation in the brain.

Acknowledgments This chapter and the presented studies were supported by the ERA-NET NEURON project HYBRIDMIND (BMBF, 01GP2121A and -B), the European Research Council (ERC) under the project NGBMI (759370) and TIMS (101081905), the Federal Ministry of Research and Education (BMBF) under the projects SSMART (01DR21025A), NEO (13GW0483C), QHMI (03ZU1110DD) and QSHIFT (01UX2211), and the Einstein Foundation Berlin (A-2019-558).

References

1. Peterson F, Jung CG. Psycho-physical investigations with the galvanometer and pneumograph in normal and insane individuals. Brain. 1907;30(2):153–218.
2. Berger H. Über das Elektrenkephalogramm des Menschen. Arch Psychiatr Nervenkr. 1929;87(1):527–70.
3. Gibbs FA, Davis H, Lennox WG. The electro-encephalogram in epilepsy and in conditions of impaired consciousness. Arch Neurol Psychiatr. 1935;34(6):1133–48.
4. Estrin T. The UCLA Brain Research Institute data processing laboratory: In Proceedings of ACM conference on History of medical informatics. Bethesda, MD: Association for Computing Machinery; 1987. p. 75–83.
5. Walter WG. An automatic low frequency analyser. Electron Eng. 1943;16:9–13.
6. Kamiya J. Conditioned discrimination of the EEG alpha rhythm in humans. San Francisco, CA: Western Psychological Association; 1962.
7. Sterman MB, Wyrwicka W, Howe R. Behavioral and neurophysiological studies of the sensorimotor rhythm in the cat. Electroencephalogr Clin Neurophysiol. 1969;27(7):678–9.
8. Fetz EE. Operant conditioning of cortical unit activity. Science. 1969;163(3870):955–8.
9. Vidal JJ. Toward direct brain-computer communication. Annu Rev Biophys Bioeng. 1973;2(1):157–80.
10. Soekadar SR, et al. Hybrid EEG/EOG-based brain/neural hand exoskeleton restores fully independent daily living activities after quadriplegia. Sci Robot. 2016;1(1):eaag3296.
11. Soekadar S, et al. Brain–machine interfaces in neurorehabilitation of stroke. Neurobiol Dis. 2015;83:172–9.
12. Zander TO, Kothe C. Towards passive brain-computer interfaces: applying brain-computer interface technology to human-machine systems in general. J Neural Eng. 2011;8(2):025005.
13. Blankertz B, et al. The berlin brain-computer interface: non-medical uses of BCI technology. Front Neurosci. 2010;4:198.
14. Suh A, Prophet J. The state of immersive technology research: a literature analysis. Comput Hum Behav. 2018;86:77–90.
15. Lee LH, et al. All one needs to know about metaverse: a complete survey on technological singularity, virtual ecosystem, and research agenda. 2021.
16. Wolpaw JR. Brain-computer interfaces as new brain output pathways. J Physiol. 2007;579(3):613–9.
17. Soekadar SR, Haagen K, Birbaumer N. Brain-computer interfaces (BCI): restoration of movement and thought from neuroelectric and metabolic brain activity. Coordination: neural, behavioral and social dynamics, 2008: 229.
18. Jackson A, Zimmermann JB. Neural interfaces for the brain and spinal cord—restoring motor function. Nat Rev Neurol. 2012;8(12):690.
19. Esmaeilpour Z, et al. Temporal interference stimulation targets deep brain regions by modulating neural oscillations. Brain Stimul. 2021;14(1):55–65.

20. Barsalou LW. Perceptual symbol systems. Behav Brain Sci. 1999;22(4):577–609; discussion 610–60
21. Reilly J, et al. Linking somatic and symbolic representation in semantic memory: the dynamic multilevel reactivation framework. Psychon Bull Rev. 2016;23(4):1002–14.
22. Wolpaw JR. Chapter 6—Brain–computer interfaces. In: Barnes MP, Good DC, editors. Handbook of clinical neurology. Amsterdam: Elsevier; 2013. p. 67–74.
23. Steinert S, et al. Doing things with thoughts: brain-computer interfaces and disembodied agency. Philosophy Technol. 2019;32(3):457–82.
24. Farwell LA, Donchin E. Talking off the top of your head: toward a mental prosthesis utilizing event-related brain potentials. Electroencephalogr Clin Neurophysiol. 1988;70(6):510–23.
25. Sellers EW, Donchin E. A P300-based brain-computer interface: initial tests by ALS patients. Clin Neurophysiol. 2006;117(3):538–48.
26. Muller-Putz GR, et al. Steady-state visual evoked potential (SSVEP)-based communication: impact of harmonic frequency components. J Neural Eng. 2005;2(4):123–30.
27. Chen X, et al. High-speed spelling with a noninvasive brain-computer interface. Proc Natl Acad Sci U S A. 2015;112(44):E6058–67.
28. Miklody D, Blankertz B. Cognitive workload of tugboat captains in realistic scenarios: adaptive spatial filtering for transfer between conditions. Front Hum Neurosci. 2022;16:818770.
29. Venthur B, et al. Novel applications of BCI technology: Psychophysiological optimization of working conditions in industry. In 2010 IEEE International Conference on Systems, Man and Cybernetics. 2010.
30. Borghini G, et al. A multimodal and signals fusion approach for assessing the impact of stressful events on air traffic controllers. Sci Rep. 2020;10(1):8600.
31. Krol LR, Andreessen LM, Zander TO. Passive brain–computer interfaces: a perspective on increased interactivity. In: Brain–computer interfaces handbook: technological and theoretical advances. Boca Raton, FL: CRC Press; 2018. p. 69–86.
32. Leuthardt EC, et al. A brain-computer interface using electrocorticographic signals in humans. J Neural Eng. 2004;1(2):63–71.
33. Stavisky SD, et al. A high performing brain-machine interface driven by low-frequency local field potentials alone and together with spikes. J Neural Eng. 2015;12(3):036009.
34. Birbaumer N, et al. A spelling device for the paralysed. Nature. 1999;398(6725):297–8.
35. Ramos-Murguialday A, et al. Brain-machine interface in chronic stroke rehabilitation: a controlled study. Ann Neurol. 2013;74(1):100–8.
36. Soekadar SR, et al. ERD-based online brain-machine interfaces (BMI) in the context of neurorehabilitation: optimizing BMI learning and performance. IEEE Trans Neural Syst Rehabil Eng. 2011;19(5):542–9.
37. Sitaram R, et al. fMRI brain-computer interfaces. IEEE Signal Process Mag. 2007;25(1):95–106.
38. Liew SL, et al. Improving motor corticothalamic communication after stroke using real-time fMRI connectivity-based neurofeedback. Neurorehabil Neural Repair. 2016;30(7):671–5.
39. Naseer N, Hong KS. Classification of functional near-infrared spectroscopy signals corresponding to the right- and left-wrist motor imagery for development of a brain-computer interface. Neurosci Lett. 2013;553:84–9.
40. Soekadar SR, et al. Optical brain imaging and its application to neurofeedback. Neuroimage Clin. 2021;30:102577.
41. Wolpaw JR, et al. Brain-computer interfaces for communication and control. Clin Neurophysiol. 2002;113(6):767–91.
42. Chen X, et al. High-speed spelling with a noninvasive brain-computer interface. Proc Natl Acad Sci. 2015;112(44):E6058–67.
43. Kübler A, et al. The thought translation device: a neurophysiological approach to communication in total motor paralysis. Exp Brain Res. 1999;124(2):223–32.
44. Kornhuber HH, Deecke L. Hirnpotentialänderungen beim Menschen vor und nach Willkürbewegungen, dargestellt mit Magnetbandspeicherung und Rückwärtsanalyse. Pflügers Arch. 1964;281(1):52.

45. Walter WG, et al. Contingent negative variation : an electric sign of sensori-motor association and expectancy in the human brain. Nature. 1964;203(4943):380–4.
46. Pereira J, et al. EEG neural correlates of goal-directed movement intention. NeuroImage. 2017;149:129–40.
47. Savić AM, et al. Online control of an assistive active glove by slow cortical signals in patients with amyotrophic lateral sclerosis. J Neural Eng. 2021;18(4):046085.
48. Schultze-Kraft M, et al. The point of no return in vetoing self-initiated movements. Proc Natl Acad Sci U S A. 2016;113(4):1080–5.
49. Haufe S, et al. Electrophysiology-based detection of emergency braking intention in real-world driving. J Neural Eng. 2014;11(5):056011.
50. de Almeida Ribeiro PR, et al. Controlling assistive machines in paralysis using brain waves and other biosignals. Adv Hum Comput Interact. 2013;2013:1–9.
51. Simon N, et al. An auditory multiclass brain-computer interface with natural stimuli: usability evaluation with healthy participants and a motor impaired end user. Front Hum Neurosci. 2014;8:1039.
52. Treder MS, Blankertz B. (C)overt attention and visual speller design in an ERP-based brain-computer interface. Behav Brain Funct. 2010;6(1):28.
53. Hong K-S, Khan MJ. Hybrid brain–computer interface techniques for improved classification accuracy and increased number of commands: a review. Front Neurorobot. 2017;11:35.
54. Muller-Putz G, et al. Towards noninvasive hybrid brain-computer interfaces: framework, practice, clinical application, and beyond. Proc IEEE. 2015;103(6):926–43.
55. Schmidt NM, Blankertz B, Treder MS. Online detection of error-related potentials boosts the performance of mental typewriters. BMC Neurosci. 2012;13(1):19.
56. Soekadar SR, et al. An EEG/EOG-based hybrid brain-neural computer interaction (BNCI) system to control an exoskeleton for the paralyzed hand. Biomed Tech (Berl). 2015;60(3):199–205.
57. Moritz CT, Perlmutter SI, Fetz EE. Direct control of paralysed muscles by cortical neurons. Nature. 2008;456(7222):639–42.
58. Hinterberger T, et al. Neuronal mechanisms underlying control of a brain-computer interface. Eur J Neurosci. 2005;21(11):3169–81.
59. Haber SN. Corticostriatal circuitry. Dialogues Clin Neurosci. 2016;18(1):7–21.
60. Schirrmeister RT, et al. Deep learning with convolutional neural networks for EEG decoding and visualization. Hum Brain Mapp. 2017;38(11):5391–420.
61. Zhang J, Yan C, Gong X. Deep convolutional neural network for decoding motor imagery based brain computer interface. in 2017 IEEE International Conference on Signal Processing, Communications and Computing (ICSPCC). 2017.
62. Lubar JF, Shouse MN. EEG and behavioral changes in a hyperkinetic child concurrent with training of the sensorimotor rhythm (SMR): a preliminary report. Biofeedback Self Regul. 1976;1(3):293–306.
63. Strehl U, et al. Neurofeedback of slow cortical potentials in children with attention-deficit/hyperactivity disorder: a multicenter randomized trial controlling for unspecific effects. Front Hum Neurosci. 2017;11:135.
64. Enriquez-Geppert S, et al. Neurofeedback as a treatment intervention in ADHD: current evidence and practice. Curr Psychiatry Rep. 2019;21(6):46.
65. Ros T, et al. Consensus on the reporting and experimental design of clinical and cognitive-behavioural neurofeedback studies (CRED-nf checklist). Brain. 2020;143(6):1674–85.
66. Cervera MA, et al. Brain-computer interfaces for post-stroke motor rehabilitation: a meta-analysis. Ann Clin Transl Neurol. 2018;5(5):651–63.
67. Donati AR, et al. Long-term training with a brain-machine INTERFACE-based gait protocol induces partial neurological recovery in paraplegic patients. Sci Rep. 2016;6:30383.
68. Soekadar SR, Birbaumer N, Cohen LG. Brain-computer-interfaces in the rehabilitation of stroke and neurotrauma. 2011.
69. Hochberg LR, et al. Reach and grasp by people with tetraplegia using a neurally controlled robotic arm. Nature. 2012;485(7398):372–5.

70. Ajiboye AB, et al. Restoration of reaching and grasping movements through brain-controlled muscle stimulation in a person with tetraplegia: a proof-of-concept demonstration. Lancet. 2017;389(10081):1821–30.
71. Flesher SN, et al. A brain-computer interface that evokes tactile sensations improves robotic arm control. Science. 2021;372(6544):831–6.
72. Crea S, et al. Feasibility and safety of shared EEG/EOG and vision-guided autonomous whole-arm exoskeleton control to perform activities of daily living. Sci Rep. 2018;8(1):10823.
73. Dornhege G, et al. Error-related EEG potentials in brain-computer interfaces. In: Toward brain-computer interfacing. Cambridge, MA: MIT Press; 2007. p. 291–301.
74. Birbaumer N, et al. Direct brain control and communication in paralysis. Brain Topogr. 2014;27(1):4–11.
75. Miller KJ, Hermes D, N.P. Staff. The current state of electrocorticography-based brain-computer interfaces. Neurosurg Focus. 2020;49(1):E2.
76. Soekadar SR, et al. Fragmentation of slow wave sleep after onset of complete locked-in state. J Clin Sleep Med. 2013;9(9):951–3.
77. Birbaumer N, et al. Ideomotor silence: the case of complete paralysis and brain-computer interfaces (BCI). Psychol Res. 2012;76(2):183–91.
78. Kubler A, Birbaumer N. Brain-computer interfaces and communication in paralysis: extinction of goal directed thinking in completely paralysed patients? Clin Neurophysiol. 2008;119(11):2658–66.
79. Chaudary U, et al. Spelling interface using intracortical signals in a completely locked-in patient enabled via auditory neurofeedback training. Nat Commun. 2022;13(1):1236.
80. Kryger M, et al. Flight simulation using a brain-computer interface: a pilot, pilot study. Exp Neurol. 2017;287(Pt 4):473–8.
81. LaFleur K, et al. Quadcopter control in three-dimensional space using a noninvasive motor imagery-based brain-computer interface. J Neural Eng. 2013;10(4):046003.
82. Congedo M, et al. "Brain invaders": a prototype of an open-source P300- based video game working with the OpenViBE platform. In BCI 2011—5th International Brain-Computer Interface Conference. Graz, Austria. 2011.
83. Cattan G. The use of brain–computer interfaces in games is not ready for the general public. Front Computer Sci. 2021:20.
84. Wang M, et al. A wearable SSVEP-based BCI system for quadcopter control using head-mounted device. Ieee Access. 2018;6:26789–98.
85. Stawicki P, Gembler F, Volosyak I. Driving a semiautonomous Mobile robotic car controlled by an SSVEP-based BCI. Comput Intell Neurosci. 2016;2016
86. Arico P, et al. Passive BCI beyond the lab: current trends and future directions. Physiol Meas. 2018;39(8):08TR02.
87. Menon V. Large-scale brain networks and psychopathology: a unifying triple network model. Trends Cogn Sci. 2011;15(10):483–506.
88. Buzsáki G. The brain–cognitive behavior problem: a retrospective. eneuro. 2020;7(4):ENEURO.0069-20.2020.
89. Soekadar SR, et al. In vivo assessment of human brain oscillations during application of transcranial electric currents. Nat Commun. 2013;4:2032.
90. Haslacher D, et al. Stimulation artifact source separation (SASS) for assessing electric brain oscillations during transcranial alternating current stimulation (tACS). NeuroImage. 2021;228:117571.
91. Garcia-Cossio E, et al. Simultaneous transcranial direct current stimulation (tDCS) and whole-head magnetoencephalography (MEG): assessing the impact of tDCS on slow cortical magnetic fields. NeuroImage. 2016;140:33–40.
92. Romeni S, Zoccolan D, Micera S. A machine learning framework to optimize optic nerve electrical stimulation for vision restoration. Patterns. 2021;2(7):100286. (this issue)
93. Zrenner C, et al. Real-time EEG-defined excitability states determine efficacy of TMS-induced plasticity in human motor cortex. Brain Stimul. 2018;11(2):374–89.

94. Grossman P, Taylor EW. Toward understanding respiratory sinus arrhythmia: relations to cardiac vagal tone, evolution and biobehavioral functions. Biol Psychol. 2007;74(2):263–85.
95. Xin Z, et al. Magnetically induced temporal interference for focal and deep-brain stimulation. Front Hum Neurosci. 2021;15:693207.
96. Darmani G, et al. Non-invasive transcranial ultrasound stimulation for neuromodulation. Clin Neurophysiol. 2022;135:51–73.
97. Musk E, Neuralink. An integrated brain-machine interface platform with thousands of channels. J Med Internet Res. 2019;21(10):e16194.
98. Leuthardt EC, Moran DW, Mullen TR. Defining surgical terminology and risk for brain computer interface technologies. Front Neurosci. 2021;15:599549.
99. Watanabe H, et al. Intravascular neural interface with nanowire electrode. Electron Commun Jpn. 2009;92(7):29–37.
100. Baranauskas G. What limits the performance of current invasive brain machine interfaces? Front Syst Neurosci. 2014;8:68.
101. Zhang X, et al. Tiny noise, big mistakes: adversarial perturbations induce errors in brain–computer interface spellers. Natl Sci Rev. 2020;8(4):nwaa233.
102. Boto E, et al. Moving magnetoencephalography towards real-world applications with a wearable system. Nature. 2018;555(7698):657–61.
103. Webb JL, et al. Nanotesla sensitivity magnetic field sensing using a compact diamond nitrogen-vacancy magnetometer. Appl Phys Lett. 2019;114(23):231103.
104. Blankertz B, et al. The Berlin Brain-Computer Interface: Progress beyond communication and control. Front Neurosci. 2016;10:530.
105. Cao Z. A review of artificial intelligence for EEG-based brain−computer interfaces and applications. Brain Sci Adv. 2020;6(3):162–70.
106. Müller K-R, et al. Machine learning and applications for brain-computer interfacing. Berlin, Heidelberg: Springer; 2007.
107. Dai G, et al. HS-CNN: a CNN with hybrid convolution scale for EEG motor imagery classification. J Neural Eng. 2020;17(1):016025.
108. Samek W, et al. Explainable AI: Interpreting, Explaining and Visualizing Deep Learning. Explainable AI: interpreting, explaining and visualizing deep learning. 2019.
109. Goebel R, et al. Explainable AI: the new 42? In machine learning and knowledge extraction. Cham: Springer; 2018.
110. Nasr K, Haslacher D, Soekadar S. Advancing sensory neuroprosthetics using artificial brain networks. Patterns. 2021;2(7):100304.
111. Gilbert CD, Li W. Top-down influences on visual processing. Nat Rev Neurosci. 2013;14(5):350–63.
112. Downey JE, et al. Blending of brain-machine interface and vision-guided autonomous robotics improves neuroprosthetic arm performance during grasping. J Neuroeng Rehabil. 2016;13(1):28.
113. Nann M, et al. Restoring activities of daily living using an EEG/EOG-controlled semi-autonomous and mobile whole-arm exoskeleton in chronic stroke. IEEE Syst J. 2020;15(2):2314–21.
114. Clausen J, et al. Help, hope, and hype: ethical dimensions of neuroprosthetics. Science. 2017;356(6345):1338–9.
115. Nann M, et al. Feasibility and safety of bilateral hybrid EEG/EOG brain/neural-machine interaction. Front Hum Neurosci. 2020;14:580105.
116. Zander TO, et al. Neuroadaptive technology enables implicit cursor control based on medial prefrontal cortex activity. Proc Natl Acad Sci U S A. 2016;113(52):14898–903.
117. Soekadar SR, et al. On the verge of the hybrid mind. Morals Machines. 2021;1(1):30–43.
118. Bublitz C, Chandler J, Ienca M. Human–machine symbiosis and the hybrid mind: implications for ethics, law and human rights. In: Ienca M, et al., editors. Cambridge handbook of information technology, life sciences and human rights. Cambridge: Cambridge University Press; 2022.

119. Gilbert F, et al. I miss being me: phenomenological effects of deep Brain stimulation. AJOB Neurosci. 2017;8(2):96–109.
120. Mosley PE, et al. Woe betides anybody who tries to turn me down.' A qualitative analysis of neuropsychiatric symptoms following subthalamic deep brain stimulation for Parkinson's disease. Neuroethics. 2021;14(1):47–63.
121. Gilbert F, Viaña JNM, Ineichen C. Deflating the "DBS causes personality changes" bubble. Neuroethics. 2021;14(1):1–17.
122. Strickland E, Harris M. Their bionic eyes are now obsolete and unsupported. IEEE Spectrum. 2022 [cited 2022 March 14th 2022]; https://spectrum.ieee.org/bionic-eye-obsolete
123. Chandler JA, et al. Brain Computer interfaces and communication disabilities: ethical, legal, and social aspects of decoding speech from the Brain. Front Hum Neurosci. 2022;16:841035.

A Path to Science Fiction Style Technology Applications? The Example of Brain-to-Brain Interfaces

Elisabeth Hildt

1 Introduction

Brain-to-brain interfaces (BBIs, B2BI, or BTB) allow direct communication between brains. The highly complex and currently entirely experimental technology involves both a brain–computer interface and a computer–brain interface. Several studies have used non-invasive BBIs for conscious information transmission between humans [1–3]. Linxing Jiang and colleagues even presented a multi-person non-invasive BBI network that enabled three persons to collaborate to solve a task resembling a Tetris game [4].

The recent developments in brain-to-brain interfaces build on the broader and more established field of brain–computer interface (BCI) technology [5] and other neurotechnologies. That's why, as an introduction to the topic and before focusing on BBIS, it is worthwhile to consider the broader technological context of BBIs, which is BCIs and brain–machine interfaces.

Jerry J. Shih and colleagues ([6], p. 268) define a BCI as "a computer-based system that acquires brain signals, analyzes them, and translates them into commands that are relayed to an output device to carry out a desired action." A BCI consists of three components: a component that detects and records brain signals; a system that decodes and processes the recorded brain signals; and a device that uses the transformed brain signals to control or navigate an output device.

Focusing on the output side, Rutger J. Vlek and colleagues ([7], p. 94) define a BCI as "a system that allows its user to control a machine (e.g., a computer, an automated wheelchair, or an artificial limb) solely with brain activity rather than the

E. Hildt (✉)
Center for the Study of Ethics in the Professions, Illinois Institute of Technology, Chicago, IL, USA
e-mail: ehildt@iit.edu

V. Dubljević, A. Coin (eds.), *Policy, Identity, and Neurotechnology*, Advances in Neuroethics, https://doi.org/10.1007/978-3-031-26801-4_6

peripheral nervous system." Others, like Mark A. Attiah and Martha J. Farah ([8], p. 1), characterize BCI technologies as "systems that enable the brain to send and receive information to and from a computer, bypassing the body's own efferent and afferent pathways." This broader definition includes cortical neural prostheses, cochlear implants, and retinal implants.

Also, more comprehensive terms such as "brain-machine interface" or "brain-hardware interface" have been used to categorize neurotechnologies, i.e., technical devices that directly connect to the nervous system. According to a characterization by Mikhail A. Lebedev and Miguel A.L. Nicolelis, "Brain-machine interfaces (BMIs) combine methods, approaches, and concepts derived from neurophysiology, computer science, and engineering in an effort to establish real-time bidirectional links between living brains and artificial actuators." ([9], p. 767). According to Andreas K. Demetriades and colleagues, brain–machine interfaces allow direct communication between a brain and an external device, involving the use of transducing or stimulating electrodes. They consider BMIs to include BCIs, direct neural interfaces, brain-machine applications, or deep brain stimulating electrodes [10].

Others have used brain–hardware interfaces as an umbrella term covering a broad spectrum of devices, such as BCIs, brain–machine interfaces (BMI), and brain stimulators. Accordingly, brain–hardware interfaces enable a more or less direct contact with the human brain that allows the exchange of electrical signals [11].

While these seem like minor differences in terminology, they point to the fact that different definitions exist and that BCIs are part of a larger group of neurotechnologies. A survey published in 2013 revealed that among BCI researchers, there are variations in what they consider the term BCI can refer to [12].

It seems that articles that reflect on the ethical and social implications of BCIs tend to use broader conceptions of the technology or refer to neurotechnology in general. This may point to authors considering the various forms of BCIs, BMIs, and neurotechnologies as technologies that raise similar issues, making it plausible to discuss the various neurotechnologies together and reflect on their ethical and social implications under a broader umbrella term.

In this chapter, I focus on brain-to-brain interfaces (BBIs), in that I introduce the technology and reflect on possible future uses and factors driving potential BBI technology development. Part of what I am depicting here is specific to BBIs; part of it holds for a relatively broad spectrum of similar technologies. Most of the driving factors discussed here are not specific to BBI technology in the narrow sense but are drivers of neurotechnology in general. However, BBIs certainly do raise specific issues.

After delineating the current state of BBI technology, I will reflect on factors that have been influencing and driving the development of neurotechnology and BCIs in general and BBIs in particular. I then discuss ethical aspects of the technology and reflect on possible ways to shape the potential future development.

2 Brain-to-Brain Interfaces: State of the Art

Brain-to-brain interfaces allow direct communication between two (or more) brains without using the peripheral nervous system. Andrea Stocco and colleagues ([13], p. 1) define BBIs as "technologies that combine neuroimaging and neurostimulation methods to exchange information between brains directly in neural code." Brain-to-brain interfaces involve two components: a brain–computer interface (BCI) that reads a sender's brain signals and sends them to a computer where they are decoded and processed, and a computer–brain interface (CBI) that transmits information to a receiver's brain. BBIs allow the exchange of information between two brains, in that the BCI part reads and decodes a sender's brain activity, and the CBI part transfers the decoded neural information to a receiver's brain. Studies involving human subjects have relied on non-invasive methods: primarily electroencephalography (EEG) for the BCI part and transcranial magnetic stimulation (TMS) or, though less often, transcranial focused ultrasonic stimulation (tFUS) for the CBI part [3, 13, 14].

BBI research is a very new and fledgling experimental field. So far, only a few research studies have been conducted. A recent review identifies 15 publications of brain-to-brain interface research [14]. In this, Chang S. Nam and colleagues stress the diversity of BBI systems. They categorize BBIs according to directionality (the flow of information) and directness (the use of brain stimulation to send information). Accordingly, most studies conducted so far are of a direct unidirectional collaboration style: information is transferred directly in one direction, from the brain of a sender directly to the brain of a receiver, by using direct neuromodulation technology. Some studies use an indirect bidirectional approach in that information is transferred in both directions, but only in one direction is it transferred directly by neuromodulation, whereas in the other direction, the transfer is done indirectly, e.g., by using visual feedback on a computer screen. Direct bidirectional BBIs, which have not been built yet, would directly transfer information between two brains in two directions, i.e., both from sender to receiver and backwards [14].

So far, only very few BBI studies that involve more than two participants, i.e., more than one sender and one receiver, have been conducted. Miguel Pais-Vieira and colleagues [15] presented a N:N collaboration model in which four rats collaborated in a system called Brainet. With this, synchronization of behavior in rats was achieved.

In 2019, Linxing Jiang and colleagues [4] published an article on what they called BrainNet, the first multi-person non-invasive brain-to-brain interface that allows collaborative problem solving. It consists of three persons, two senders and one receiver, whose brains interact directly by means of the BBI. Relying on BrainNet, three participants collaborated successfully to solve a task that resembles a Tetris game. Using electroencephalography (EEG), brain signals of the senders were recorded and, after having undergone a decoding and translation process, submitted through TMS to the receiver's occipital cortex. The receiver then made game decisions based on the stimulation input perceived as phosphenes, which functioned as indicators of yes-no-responses. For the study, the idea of a social network played a central role. The study setting involved a feedback loop that allowed the senders to give feedback to the receiver about whether they agreed with the receiver's decisions. Also, the setting enabled the receiver to find out about the senders' information reliability.

Potential future applications of BBIs could include assisting patient rehabilitation. This prospect was discussed in a study by M. Ebrahim M. Mashat and colleagues [16] in which a direct BBI and a muscle-to-muscle interface (MMI) were combined in a closed-loop pattern. The EEG-based BCI component transferred one person's motor intention via TMS to the receiver person's brain, where it induced hand motion. The hand motion in the receiver's arm was recorded by electromyography (EMG) and triggered functional electrical stimulation (FES) applied to the sender's arm to generate hand motion. The authors see potential future applications of this system in the rehabilitation or co-rehabilitation of stroke patients [16]. They write that the approach "might potentially enhance the rehabilitation of stroke-related motor impairments by promoting neuroplasticity and altering motor cortical areas. Therefore, rehabilitation is likely to be remarkably improved" ([16], p. 7). The authors see the potential to realize co-rehabilitation in two patients who engage as two players in the closed-loop system. Similarly, others have suggested that relying on a BBI system, a physical therapist could directly send commands to a patient's brain, which could support rehabilitation [14].

Andrea Stocco and colleagues [13] published a study with a noninvasive BBI in humans (EEG and TMS) in which pairs of participants successfully collaborated to solve a problem by completing a series of question and answer rounds. The authors suggest that the BBI paradigm could allow conversations with a non-verbal participant, e.g., a person with Broca's aphasia, or could be used with pairs of participants that do not speak the same language.

As Vladimir A. Maksimenko and colleagues [17] suggest, BBIs could serve to share the cognitive load among co-workers accomplishing a task together, such as office workers, pilots, or power plant operators. Accordingly, these interfaces could distribute a common task between the group members, which could improve the team's performance, e.g., by increasing sustained attention or alertness. However, what Maksimenko and colleagues describe is not an example of a direct BBI, but instead two persons each use a BCI, and their output is collected and analyzed by a joint computer.

3 Factors Driving BBI Technology Development

BCIs and BBIs are being developed within the broader context of neurotechnologies. Generally speaking, this development is enabled by an enormous increase in neuroscience knowledge and technological know-how.

Right now, it is unclear whether there will ever be applications of BBI technology. In medical contexts, it is conceivable that in the future, BBIs or BBI networks could allow direct brain-to-brain communication for paralyzed patients and patients with locked-in syndrome or support rehabilitation in stroke patients. Beyond these very narrow potential clinical applications, potential non-clinical uses are less clear. For except in clinical contexts, where patients could benefit from the technology in rehabilitation and where BBIs could serve as assistive devices that allow communication for people with severe motor restrictions, there is no immediate need for technologies like BBIs in non-clinical contexts. To be attractive to healthy users, the technology would have to offer something that cannot be easily achieved otherwise. As John B. Trimper and colleagues discuss, BBIs could potentially be used to enhance human cognition, for example, by coupling human brains to facilitate knowledge or skill acquisition [18].

If the technology was to be used more broadly, considerable challenges would have to be overcome. Future BBIs would have to be non-invasive and of minimal risk. They would have to be relatively small, wearable devices that are easy to handle and not too expensive. And they would have to offer some attractive functions to potential future users. It is far from evident whether it is possible to overcome these challenges. It is certainly not clear at this moment how the technology will develop, and whether there will be a path to any future BBI application beyond research and experimentation. In addition to the practical and technological challenges, there are unresolved ethical questions (see below).

However, several factors can be identified that drive the potential future development of neurotechnology in general and BBIs in particular. These include more general tendencies or trends such as scientific curiosity, a widespread technology fascination, and people interested in new technologies and more specific factors including financial interests, science fiction fascination, and theoretical underpinnings. In the following section, I identify and discuss several drivers: potential future markets, science fiction scenarios, enhancement and transhumanism, and social media analogies.

3.1 Potential Future Markets

It is unclear whether there will ever be any market for the complex BBI technology. Overall, however, an increasing trend toward augmented reality and virtual reality in gaming and other contexts seems to favor the BCI technology approach in general. This may positively influence BBI development and potential future markets.

Various companies, such as Neuralink or Kernel, invest in the development of neurotechnological devices in general and brain–computer interfaces in particular ([19]; https://www.kernel.com/#products; https://neuralink.com/). While technology companies invest considerably in this field, there is no identified need or specified potential future application yet, especially not with regard to BBIs. The development in the field can be seen as driven by technology fascination, vague business ideas, and the expectation of or hope for potential future markets.

While the computer-to-brain component currently is the more experimental BBI component requiring bulky technology, researchers and companies are more experienced with the BCI part of BBIs. For the BCI part, wearable headset solutions have been developed, and several EEG-based consumer neuromonitoring devices are being marketed already. Their performance is clearly limited, however [20].

Technology companies can certainly be considered drivers of neurotechnology development. Also, the military has been showing interest in BCIs and potential BCI applications in combat situations. The US Defense Advanced Research Projects Agency (DARPA), for example, has funded BCI research and research into technology to enhance brain function. This includes EEG-based silent user-to-user communication [19, 21–23]. The military can be seen as driving the development through funding research and as a potential future market for BCI and BBI technology. Both aspects are clearly connected, though. Potential future technological applications in the military may shape the development insofar as research funding goes in this direction.

3.2 Science Fiction Scenarios

Science fiction literature, movies, and scenarios have for a long time reflected the fascination of human–machine interaction, humans becoming part of technology, or humans fusing with technology. Musings over this science fiction merging of man and machine and ideas of immortality incite both fascination and fright. While the merging of man and machine is not totally fiction, these scenarios go considerably beyond what is realistically feasible.

Also, in the interdisciplinary literature on neurotechnology and BCIs, allusions to science fiction scenarios have played a role. For example, in the editorial titled "Brain-Machine Interface: The future is now," Neeraj Jain alludes to science fiction fantasies about "fusing the power of the human brain with the strength of machines" ([24], p. 321) and makes the point that some of the science fiction-based concepts have moved from science fiction to science journals.

The concept of a cyborg, i.e., a *cyb*ernetic *org*anism which consist of both human and technological body parts, or the fusion of man and machine, or brains and silicon, not only paved the way to increase interest in neurotechnology, but also stressed ethical and societal implications of neurotechnology and the need for policy and regulation to shape the development [22, 25–29]. This is reflected, for example, by the title of an article published in 2007—"Becoming Borg to Become Immortal: Regulating Brain Implant Technologies" [29]. Also, the concept has been adopted by cyborg artists like Stelarc or Neil Harbisson.[1]

While science fiction depictions allow speculation on what may lie ahead in neurotechnology development and what the social and ethical implications might be, there are also downsides to this reference or allusion to science fiction. Science fiction scenarios suggest that anything is possible, that a technological future with applications that go well beyond current technology lies ahead. Science fiction familiarizes all of us with the idea of direct human–machine interaction and

---

[1] http://stelarc.org/_.php; https://www.theguardian.com/artanddesign/2014/may/06/neil-harbisson-worlds-first-cyborg-artist; https://www.cyborgart.org/

merging with technology. It allows us become accustomed to technical scenarios that, in reality, are far-off and would otherwise be considered out of reach. This may come with inadequate hopes or aspirations for potential technological futures and may hinder a realistic perspective on ongoing technological developments.

Mark A. Attiah and Martha J. Farah [8] characterize science fiction scenarios as problematic for the ethical analysis of BCIs. Under the heading "The problem with science fiction" (p. 1), they point out that, in their view, addressing potential future ethical challenges of BCI technology such as giving people superhuman abilities or controlling their thoughts and desires is not helpful. Accordingly, these speculative and vivid images come with two challenges for ethical analysis: first, we have strong emotional reactions toward the futuristic technology scenarios and, second, given our limited real-world experience with BCIs, our ethical analysis lacks a foundation of practical empirical knowledge. They explain that both factors lead to our views on the ethical and social impacts of these technologies based on gut reactions, which means they are either overly rejected or fervently embraced. Overall, as they state, this "emotional pull of futuristic scenarios" (p. 1) distracts from the more immediate ethical challenges of the technology [8].

3.3 Enhancement and Transhumanism

During the past decades, there has been discussion about the possibility of using biomedical technologies for enhancement purposes, which means using biomedicine to improve human form or function instead of treating or preventing disease. The term enhancement has primarily been used to characterize biomedical procedures as outside the realm of medicine, as procedures not medically justified but applied to otherwise healthy persons. More broadly, enhancements are interventions that seek to increase an individual's physical or mental capabilities. Attempts to augment brain functions have been characterized as neuroenhancement, neuroaugmentation, brain enhancement, or brain augmentation. Several approaches have been used or suggested, including pharmacological substances and neurotechnologies [18, 30–33].

BCI and BBI applications in non-clinical contexts certainly can be seen in this enhancement paradigm (see [18]). They provide the additional capability of direct brain-to-brain communication. Thus, non-invasive BBI applications in non-clinical contexts could be considered temporary brain enhancements or brain augmentations.

The enhancement idea is embraced by transhumanist authors who argue in favor of making enhancement technologies widely available to improve human nature. Accordingly, enhancement technologies may ultimately enable humans to become transhumans, i.e., beings with properties beyond those humans typically have. These may include characteristics and capabilities like longevity, improved cognitive capabilities, or bodily strength. Several transhumanist authors have discussed this idea of enhancing the human condition within the context of evolution. Proponents argue that enhancement technologies allow us to accelerate human evolution [34]. Accordingly, by integrating technology into the human body, merging with technology, or becoming a cyborg, human evolution can be sped up, as technology develops much faster than biological evolution. For example, Woodrow Barfield argues that it is necessary to move beyond our normal human capabilities in order to keep up with artificial intelligence technology that may soon surpass human intelligence and human capabilities [22, 34].

It seems questionable whether neurotechnologies can effectively influence human evolution, since these modifications would not be passed on to the next generation like genetic modifications would. While a considerable part of philosophical reflections by transhumanists can be characterized as science fiction style speculation, it is certainly true that transhumanist thinking has made neurotechnology in general and BBI technology in particular appear less far off and more plausible.

3.4 Social Network Analogy

In some of the publications on BBIs, especially on BBI networks involving more than two brains, the authors use specific wording, analogies, or metaphors to characterize the BBI technology. For example, Pais-Viera et al. [15] published a study in which four rats collaborated through BBIs and designated the resulting network as Brainet. The Brainet, they write, could provide the core of a new type of computing device, an "organic computer" ([15], p. 1). The authors summarize their research as follows ([15], p. 10): "These results provide a proof of concept for the possibility of creating computational engines composed of multiple interconnected animal brains." In this characterization, the authors use technology metaphors such as "computational engine" or "organic computer" when talking about brains. Also, the terms Brainet and BrainNet used to characterize networks of multiple interconnected brains show similarities with or allude to the term "internet" [4, 15].

In addition, the term "social brain networks" uses an analogy to conventional social networks and social network technology, and the concept of social networks plays a role. The social network analogy is most evident in the publication by Jiang et al. [4], which introduces BrainNet, the first multi-person BBI for direct communication and collective problem solving. This network of three presents proof of principle of direct brain-to-brain communication in brain networks. On various occasions, the authors allude directly to communication in social networks. They write ([4], p. 1): "Our results raise the possibility of future brain-to-brain interfaces that enable cooperative problem solving by humans using a 'social network' of connected brains."

Besides the proof of principle of direct brain-to-brain communication in a brain network in humans, part of the research published by Jiang and colleagues is on whether and how the receiver could find out about the reliability of the sender. They write ([4], p. 2):

> An important feature of communication in social networks is deciding which source of information to prioritize. To investigate whether BrainNet allows such a capability, we

additionally explored whether the Receiver can learn the reliability of each Sender over the course of their brain-to-brain interactions. We varied the reliability of the signal from one Sender compared to the other by injecting noise into the signals from one randomly chosen Sender. Our results show that like conventional social networks, BrainNet allows a Receiver to learn to trust the Sender who is more reliable, i.e., whose signal quality is not affected by our manipulation.

From a practical standpoint, in a brain-to-brain-network, it will be crucial for the receiver to be able to find out about the reliability of the sender, and the analogy to social networks certainly is not far-fetched. The authors stress that the receiver being able to find out about the reliability of the senders brings BrainNet a step closer to conventional social networks in which differential weighting for different sources of information is used.

As Jens Clausen discusses [35], the idea of a social network, in this case, a global emotional network, also plays a central role in Michael Chorost's book "The World Wide Mind: The coming integration of humanity, machines, and the free internet." Published in 2011, this book by Chorost depicts a futuristic scenario in which information is exchanged directly through social networks of brains, making existing social networks like Facebook obsolete.

Overall, the connection made between conventional social networks and direct brain-to-brain networks connects the BBI technology to an existing user experience, something most of us know very well and use in everyday life. In this context, it sounds somehow plausible to take social networks to the next level, so that instead of having to rely on smartphones for communication people can directly connect with their brains.

4 Shaping the Development

BBI technology clearly is at its early experimental stage; it remains to be seen how its future will look. Several potential future applications have been suggested for the complex BBI technology, e.g., in stroke rehabilitation, but the technology comes with a number of practical, technological, and ethical issues.

Conceptually speaking, a considerable part of the ethical issues involved in BBI technology are similar to those in BCI technology [33, 36, 37]. However, the direct information transfer through direct brain-to-brain communication raises particularly complex questions relating to autonomy, shared control, agency, accountability, identity, and self-concept. In larger BBI networks, in which individual participants will probably have the role of being both a sender and a receiver, the complex information flows will result in even more pressing questions. I have discussed these and related ethical issues of BBIs and multi-person BBI networks elsewhere [38, 39]. If at all, BBI technology seems to be affordable for highly technologized, affluent societies at best, which raises questions of social justice and technological divide.

Prospects for the future of BBI technology will not only depend on scientific and technological progress but also on the ethical issues involved and on the extent to which ethical considerations and policy can shape its potential future development. It is crucial to reflect on ethical implications and develop paths for policies and regulations during the early phase of a technology. This allows framing the process and mitigating and avoiding negative effects before a technology is used broadly.

During the process of technological development, it is crucial for the research community to follow research ethics standards including providing adequate informed consent procedures for study participants, being transparent about the state of the art, and reporting not only on the successes but also on the downsides of the technology.

Beginning with the early stages of experimentation involving human subjects, there is a need to devise informed consent procedures that carefully and thoroughly consider issues related to brain-to-brain communication, including privacy, autonomy, independence, and identity. If brain-to-brain communication in the studies becomes more versatile and the number of individuals in multi-person networks increases, the complexity of these questions will increase.

It will be essential to get the public involved. As McGee and Maguire [29] write, technological advances that can impact society require public scrutiny. Researchers need to openly and transparently present and discuss their results in a way that the public can understand. Scientific honesty includes not raising too much hope for immediate technology uses. It involves realistic communication with the public that does not exaggerate potential future uses or successes. In communicating with the public, it will be important for researchers, journalists, and entrepreneurs to clearly distinguish between what is real and what is science fiction in neurotechnology. The Responsible Research and Innovation (RRI) approach provides a framework for public involvement [40].

For complex neurotechnologies like BBIs to have a future, regulation will be required that protects users from potential harm and shapes the development process. A couple of ways this could be achieved are by devising new brain-related rights or neurorights or by reconceptualizing existing rights to better cover brain- and neurotechnology-related contexts. Several authors have suggested brain-related rights including a right to mental privacy, cognitive liberty, psychological continuity, and mental integrity [19, 41, 42].

While these brain-related rights delineate red lines in that they protect privacy, individual autonomy, personal identity, and mental integrity, there clearly is a need for a broader debate on the ethical and legal issues involved. The ethical and policy reflection will have to continuously accompany the scientific and technological development.

5 Conclusion

BBI technology is a highly complex technology for which several potential future applications have been suggested. However, it remains to be seen whether there will ever be a path to future applications of BBI technology. Quite a number of practical, technological, and ethical issues would have to be resolved. Overall, it must be seen and stressed that while the drivers of neurotechnology in general and BBIs in particular depicted above do exist, BBI technology is absolutely experimental at this stage. There is no expectation that this will change any time soon. Even though it is unclear how realistic it is to expect any future BBI applications or a broad use of BBIs or BBI networks, there is a clear role for ethics and policy in shaping the research process and potential future development.

References

1. Grau C, Ginhoux R, Riera A, Nguyen TL, Chauvat H, Berg M, et al. Conscious brain-to-brain communication in humans using non-invasive technologies. PLoS One. 2014;9:e105225. https://doi.org/10.1371/journal.pone.0105225.
2. Rao RPN, Stocco A, Bryan M, Sarma D, Youngquist TM, Wu J, et al. A direct brain-to-brain interface in humans. PLoS One. 2014;9:e111332. https://doi.org/10.1371/journal.pone.0111332.
3. Lee W, Kim S, Kim B, Lee C, Chung YA, Kim L, et al. Non-invasive transmission of sensorimotor information in humans using an EEG/focused ultrasound brain-to-brain interface. PLoS One. 2017;12:e0178476. https://doi.org/10.1371/journal.pone.0178476.
4. Jiang L, Stocco A, Losey DM, Abernethy JA, Prat CS, Rao RPN. BrainNet: a multi-person brain-to-brain interface for direct collaboration between brains. Sci Rep. 2019;9:Article number: 6115. https://doi.org/10.1038/s41598-019-41895-7.
5. Saha S, Mamun KA, Ahmed K, Mostafa R, Naik GR, Darvishi S, Khandoker AH, Baumert M. Progress in brain computer interface: challenges and opportunities. Front Syst Neurosci. 2021;15:578875. https://doi.org/10.3389/fnsys.2021.578875.
6. Shih JJ, Krusienski DJ, Wolpaw JR. Brain-computer interfaces in medicine. Mayo Clin Proc. 2012;87(3):268–79.
7. Vlek RJ, Steines D, Szibbo D, Kübler A, Schneider M-J, Haselager P, Nijboer F. Ethical issues in brain-computer Interface research, development, and dissemination. J Neurol Phys Ther. 2012;36:94–9.
8. Attiah MA, Farah MJ. Minds, motherboards, and money: futurism and realism in the neuroethics of BCI technologies. Front Syst Neurosci. 2014;8:86. https://doi.org/10.3389/fnsys.2014.00086.
9. Lebedev MA, Nicolelis MAL. Brain-machine interfaces: from basic science to neuroprostheses and neurorehabilitation. Physiol Rev. 2017;97:767–837.
10. Demetriades AK, Demetriades CK, Watts C, Ashkan K. Brain-machine interface: the challenge of neuroethics. Surgeon. 2010;8:267–9.
11. Clausen J. Conceptual and ethical issues with brain-hardware interfaces. Curr Opin Psychiatry. 2011;24:495–501.
12. Nijboer F, Clausen J, Allison BZ, Haselager P. The Asilomar survey: stakeholders' opinions on ethical issues related to brain-computer interfacing. Neuroethics. 2013;6:541–78.
13. Stocco A, Prat CS, Losey DM, Cronin JA, Wu J, Abernethy JA, Rao RPN. Playing 20 questions with the mind: collaborative problem solving by humans using a brain-to-brain interface. PLoS One. 2015;10:e0137303. https://doi.org/10.1371/journal.pone.0137303.
14. Nam CS, Traylor Z, Chen M, Jiang X, Feng W, Chhatbar PY. Direct communication between brains: a systematic PRISMA review of brain-to-brain interface. Front Neurorobot. 2021;15:656943. https://doi.org/10.3389/fnbot.2021.656943.
15. Pais-Vieira M, Chiuffa G, Lebedev M, Yadav A, Nicolelis MAL. Building an organic computing device with multiple interconnected brains. Sci Rep. 2015;5:1–15. https://doi.org/10.1038/srep11869.
16. Mashat MEM, Li G, Zhang D. Human-to-human closed-loop control based on brain-to-brain interface and muscle-to-muscle interface. Sci Rep. 2017;7:1–11. https://doi.org/10.1038/s41598-017-10957-z.

17. Maksimenko VA, Hramov AE, Frolov NS, Lüttjohann A, Nedaivozov VO, Grubov VV, Runnova AE, Makarov VV, Kurths J, Pisarchik AN. Increasing human performance by sharing cognitive load using brain-to-brain Interface. Front Neurosci. 2018;12:949. https://doi.org/10.3389/fnins.2018.00949.
18. Trimper JB, Wolpe PR, Rommelfanger KS. When "I" becomes "we": ethical implications of emerging brain-to-brain interfacing technologies. Front Neuroeng. 2014;7:4. https://doi.org/10.3389/fneng.2014.00004.
19. Yuste R, Goering S, et al. Four ethical priorities for neurotechnologies and AI. Nature. 2017;551:159–63.
20. Coates McCall I, Wexler A. Peering into the mind? The ethics of consumer neuromonitoring devices. In: Bard I, Hildt E, editors. Ethical dimensions of commercial and DIY neurotechnologies. Cambridge, MA: Elsevier/Academic Press; 2020. p. 1–22.
21. Kotchetkov IS, Hwang BY, Appelboom G, Kellner CP, Connolly ES. Brain-computer interfaces: military, neurosurgical, and ethical perspective. Neurosurg Focus. 2010;28(5):E25.
22. Barfield W. The process of evolution, human enhancement technology, and cyborgs. Philosophies. 2019;4:10. https://doi.org/10.3390/philosophies4010010.
23. Olds J, et al. Ideas lab for imagining artificial intelligence and augmented cognition in the USAF of 2030. Arlington, VA: Air Force Research Laboratory; 2019.
24. Jain N. Brain-machine Interface: the future is now. Natl Med J India. 2010;23(6):321–3.
25. Haraway D. A manifesto for cyborgs: science, technology, and socialist feminism in the 1980s. In: Nicholson LJ, editor. Feminism/postmodernism. New York/London: Routledge; 1990.
26. Cochrane P. Carbon-silicon convergence. Forbes Magazine, August 1999.
27. Boyce N. Enter the cyborgs: promise and peril in a marriage of brains and silicon. US News World Rep. 2002;132(16):56–8.
28. Warwick K. Cyborg morals, cyborg values, cyborg ethics. Ethics Inf Technol. 2003;5:131–7.
29. McGee EM, Maguire GQ. Becoming Borg to become immortal: regulating brain implant technologies. Camb Q Healthc Ethics. 2007;16:291–302.
30. Juengst ET. What does enhancement mean? In: Parens E, editor. Enhancing human traits. Ethical and social implications. Washington, DC: Georgetown University Press; 1998. p. 29–47.
31. Zehr EP. Future think: cautiously optimistic about brain augmentation using tissue engineering and machine interface. Front Syst Neurosci. 2015;9:72. https://doi.org/10.3389/fnsys.2015.00072.
32. Lebedev MA, Opris I, Casanova MF. Editorial: augmentation of brain function: facts, fiction and controversy. Front Syst Neurosci. 2018;12:45. https://doi.org/10.3389/fnsys.2018.00045.
33. Cinel C, Valeriani D, Poli R. Neurotechnologies for human cognitive augmentation: current state of the art and future prospects. Front Hum Neurosci. 2019;13:13. https://doi.org/10.3389/fnhum.2019.00013.
34. Bostrom N. In defense of posthuman dignity. Bioethics. 2005;19(3):202–14.
35. Clausen J. Social networks for neurons. New Scientist. 2011;209(2800):50.
36. Burwell S, Sample M, Racine E. Ethical aspects of brain computer interfaces: a scoping review. BMC Med Ethics. 2017;18:60. https://doi.org/10.1186/s12910-017-0220-y.
37. Coin A, Mulder M, Dubljevic V. Ethical aspects of BCI technology: what is the state of the art? Philosophies. 2020;5:31. https://doi.org/10.3390/philosophies5040031.
38. Hildt E. What will this do to me and my brain? Ethical issues in brain-to-brain interfacing. Front Syst Neurosci. 2015;9:17.
39. Hildt E. Multi-person brain-to-brain interfaces: ethical issues. Front Neurosci. 2019;13:1177. https://doi.org/10.3389/fnins.2019.01177.
40. von Schomberg R. A vision of responsible innovation. In: Owen R, Heintz M, Bessant J, editors. Responsible innovation. London: John Wiley; 2013.
41. Ienca M, Andorno R. Towards new human rights in the age of neuroscience and neurotechnology. Life Sci Soc Policy. 2017;13:5. https://doi.org/10.1186/s40504-017-0050-1.
42. Lavazza A. Freedom of thought and mental integrity: the moral requirements for any neural prosthesis. Front Neurosci. 2018;12:82. https://doi.org/10.3389/fnins.2018.00082.

Part II

Ethical and Philosophical Issues

A Scoping Review of the Academic Literature on BCI Ethics

Abigail Lang, Allen Coin, and Veljko Dubljević

1 Introduction

Although Brain–Computer Interfaces (BCI) have been used for decades, recent advances in the technology and increased private investment into BCI research have led to rapidly broadening and novel applications that have caught the public's attention. In recent years, BCIs have made headlines and inspired viral coverage of press releases, illustrating the potential applications of the technology and the ambitions of the private and public researchers advancing it [1]. Media coverage has included a monkey playing a video game using its mind alone [2], prototypes of a mass-produced consumer BCI device implanted into a person's skull via an automated surgical robot with aims to introduce economies of scale into the process of implanting an invasive BCI, thus aiding widespread adoption [3], and devices for sale at electronics stores that claim to improve a user's mood or performance in video games by modulating their brain waves [4, 5]. While many people may already personally know someone with a cochlear implant, an example of a BCI device that helps restore one's sense of hearing, the technology has already extended to read the mind of a person with paraplegia to operate a wheelchair, allow direct brain-to-brain

A. Lang
Lake Erie College of Osteopathic Medicine at Seton Hill, Greensburg, PA, USA
e-mail: alang67393@med.lecom.edu

A. Coin
North Carolina State University, Raleigh, NC, USA
e-mail: allen_coin@ncsu.edu

V. Dubljević (✉)
Philosophy, and Science Technology and Society (STS), North Carolina State University, Raleigh, NC, USA
e-mail: veljko_dubljevic@ncsu.edu

V. Dubljević, A. Coin (eds.), *Policy, Identity, and Neurotechnology*, Advances in Neuroethics, https://doi.org/10.1007/978-3-031-26801-4_7

communication, or even enable motor control of an insect with an implanted BCI [6]. These are all examples of applications of BCI technology, which is both a well-established and a quickly growing field of research with the potential for therapeutic medical use as well as a consumer technology.

At its core, BCI is any technology that can read, interpret, and translate brain activity into a format digestible by a computer. The device can then interpret those brain signals as input to create some sort of output in the form of an interaction with the outside world, or pass information about the outside world back to the user as feedback that the user can then act upon. BCI devices may be generally categorized as active, reactive, or passive. Active BCIs are, as the name suggests, action based. They detect and decode mental commands initiated by the user, and many can even translate these signals into motor outputs. Meanwhile, reactive BCIs are sensory based, modulating user-brain activity based on external stimuli. In either of these BCI devices, the user is directing BCI output based on purposeful commands or attentiveness. Passive BCIs, in contrast, work solely to monitor user-brain activity and provide relevant feedback. In these instances, the BCI device is not modulating or reacting to brain activity, aside from reporting arbitrary measurements. As noted above, current examples include the cochlear implant, which detects and transforms sound into electrical signals that stimulate the cochlear nerve, transmitting auditory information to the brain and allowing the user to hear (reactive BCI); or any device that interprets a user's brain activity as a means to control an external prosthesis, such as a robotic arm or wheelchair (active BCI) [7]. BCI devices can be noninvasive, utilizing, for example, an electroencephalogram (EEG) skull cap that can read the brain's electrical activity from outside the skull, or invasive devices that require implantation within the skull and direct contact with the brain. Due to the interference from the intervening tissue, noninvasive BCIs have a poor signal-to-noise ratio, and as a result the more advanced BCI applications usually involve an invasive device, which increases signal quality but also poses greater risks to the user [8]. Proliferation of BCIs has emerged as a trend, as the initial therapeutic BCI technologies are adapted for general public use as "cool" gadgets or in military applications [9], gaming [10], communication [11], and even performance enhancement [12].

2 Ethical Concerns with BCIs

BCI technology is associated with several ethical and societal implications, including issues of safety, stigma and discrimination, autonomy, and privacy, to name but a few. The assessment of the balance of risks and benefits associated with widespread use of this technology is a complex endeavor and must account for concerns about possibly frequent events (e.g., hacking of BCIs and malevolent use of extracted information) as well as relatively rare but catastrophic events (such as a BCI prosthetics failure leading to a fatal traffic accident). Additionally, the use of BCI may contribute to the stigmatization of disability and may even jeopardize autonomy in specific groups.

One specific form of BCI development, Brain-to-Brain Interface (BBI), may lead to particularly novel social and ethical concerns. BBI technology combines BCI with Computer-to-Brain Interfaces (CBI) and, in newer work, multi-brain-to-brain interfaces—such as Jing et al.'s [12] study—real-time transfer of information between two subjects to each other has been demonstrated.

There have been a number of advances in BCI technology in recent years, including commercial ventures that seek to utilize BCI in novel ways. One such example is the company Neuralink, led by entrepreneur Elon Musk, which aims to achieve "a merger with artificial intelligence" [13]. There has been ample skepticism about Neuralink's goals and claims, with some referring to the company's public announcements and demonstrations as "neuroscience theater" [14]. Regardless of whether Neuralink's stated goals are feasible in the near-term future, the existence of commercial ventures like Neuralink in the BCI field certainly signals new areas of active development and may shed some light on where the technology could be heading.

3 Prior Research into BCI Ethics

Prior research on this topic, apart from our own work [15], includes an earlier scoping review of the pertinent academic discussion [16] as well as analysis of print media reports on ethics of BCI [17]. The scoping review, conducted by Burwell and colleagues in 2016 and reported in a 2017 paper, included the selection of 42 academic articles about BCI that were published before 2016. They found that the majority of articles discussed more than one type of ethical issue associated with BCI use, which demonstrated that there is some cause for concern. Among the most common ethical concerns surrounding the use of BCI were user safety, justice, privacy and security, and balance of risks and benefits. Other, less commonly mentioned concerns include military applications, as well as enhancement and uses of BCI that promote controversial ideologies (e.g., transhumanism).

In their first-of-its-kind analysis of the academic literature on BCI ethics, Burwell and colleagues noted the frequency of concerns does not measure the moral or regulatory significance of the issue; ethical concerns that were mentioned once or rarely may be just as pressing as concerns mentioned with high frequency, or even more so. They also found that the articles that mentioned a vast range of issues failed to provide depth of discussion and may be less suitable for use by ethicists and policy makers as guidance to address specific social problems. While numerous concerns are identified in the literature, the authors found the debate up until the year 2016 to be relatively underdeveloped, and few of the analyzed articles made concrete proposals to address social and ethical issues. All in all, Burwell and colleagues conclude that, based on their results, more high-quality work, including empirical studies, should be conducted on this topic.

Kögel and colleagues [18] then conducted a scoping review of empirical BCI studies in fields of medicine, psychology, and the social sciences. They sought to understand empirical methods employed in BCI studies and how the ethical and social issues discussed are associated, while also identifying relevant ethical and social concerns *not* being discussed. With a sample of 73 studies, Kögel and colleagues found that problems of usability and feasibility, such as user opinion and expectations, technical issues, etc., were being frequently addressed. However, potential problems of changes in self-image, user experience, and caregiver perspectives were relatively lacking. Overall, Kögel and colleagues [19] recognize a lack of BCI-centered ethical engagement and exploration among these studies.

A study by Gilbert and colleagues [1] explored how BCI is depicted in the English-speaking media, with emphasis on news outlets. The researchers analyzed 3873 articles by topic and tone. Five major topics were discussed: focus on the future, mention of ethics, sense of urgency, medical applications, and enhancement. As for the tone of print media articles, the researchers contrasted articles that provided positive depictions from those that had negative depictions and reservations about the technology. The authors found that 76.91% ($n = 2979$) of the 3873 total articles portrayed BCI positively including 979 of these articles (25.27%) that had overly positive and enthusiastic narratives. In contrast, 1.6% of articles had a negative tone, with 0.5% of the total articles having an overly negative narrative. Only 2.7% of articles mention issues specific to ethical concerns. In terms of article content, 70.64% of total articles discuss BCI with respect to its future potential, 61.15% of the articles discuss the medical applications of BCIs, and 26.64% of the articles contain claims about BCI enabling enhancement.

Gilbert and colleagues' analysis of the large sample of mass media articles reveals a disproportional bias in favor of a positive outlook on BCI technology. A positive representation of BCI can be a good thing if the purpose is to highlight the potential to help patients and bring attention to the struggles faced by the people who may benefit from the technology. However, there are adverse effects of positive bias on BCI in the media. The technology is far from perfect, and the disproportionate representation of positive articles could overshadow the risks of BCI, therefore not fairly representing the current capabilities and future potential of this technology within the media, intended for mass public consumption. Gilbert and colleagues suggested that the positive bias in the media misrepresents the state of the technology by disregarding ethical issues, risks, and shortcomings. They conclude that the media seems to lack objective information regarding risks and adverse effects of BCI and disregard the potential impacts of the technology on key topics such as agency, autonomy, responsibility, privacy, and justice.

4 Recent Trends in BCI Ethics

In order to elucidate the ethical issues inherent in the development of BCI technology, we revisit our previously published work [15] that began an in-depth scoping review of the ethics literature concerning BCI. This work has updated the mapping of the BCI ethics literature published since 2016, when the last review of this nature was conducted [16], and updated the coding strategy accordingly. Revisiting the academic literature around BCI ethics just 4 years after the publication of Burwell and colleagues' first-of-its-kind review was necessary given the rapid growth of the technology in recent years; when reproducing Burwell et al.'s search methodology in 2020, we found that almost as many relevant academic papers discussing BCI ethics had been published in the years *since* 2016 (n = 34) as Burwell and colleagues had identified in all years *prior* to 2016 (n = 42). This indicates the body of academic literature on BCI ethics is rapidly growing, and an updated analysis is warranted. Previously, we reviewed a randomly selected statistically significant sample (n = 7, 20.6%) of the 34 academic papers addressing the ethical and social issues of BCI technology published between 2016 and 2020 following a systematic search with inclusion and exclusion criteria. In that paper, we established the continued utility of the coding schema developed by Burwell and colleagues that can continue to be used, with some modifications, to understand the landscape of the academic literature around the ethical, legal, and social issues (ELSI) inherent to BCI technology.

In this chapter, we outline the methodology and findings for the next phase of this work, which systematically categorizes the entire sample. The aim of this work is to collect and synthesize all of the pertinent academic scholarship into the ELSI of BCI technology in order to provide a foundation for future scholars, ethicists, and policy makers to understand the landscape of the relevant ELSI concepts and pave the way for assessing the need for regulatory action.

In this endeavor, we are guided by Blank's [20] taxonomy of regulatory responses (See Table 1), which mirrors the familiar distinctions in moral philosophy between things that are (a) morally required (and thus should be made mandatory), (b) morally desirable and permissible (and thus should be encouraged), (c) morally neutral and permissible (and thus should be left to the unfettered operation of the market), (d) bad but nevertheless still morally permissible (and thus should be discouraged), and (e) morally impermissible (and thus should be prohibited). Blank's work provides a specialized outline regarding Neuropolicy, particularly factors guiding regulatory responses of the government. We use this framework to contextualize our discussion of ELSI of BCI technology and how future regulations may be considered. In this work, we do not seek to make specific recommendations about the regulatory response that may be appropriate for the different ethical and social issues that arise from BCI technology, but instead hope to gather and synthesize the relevant salient facts and normative positions in order to propel the debate to a more mature state where policy action is more informed and feasible.

It should be noted that there are different levels of background regulation of BCI technology. For instance, research and development in BCI (both invasive and

Table 1 Blank's [20] taxonomy of regulatory responses

That which is…	Should be…
Morally required	Mandatory
Morally desirable and permissible	Encouraged
Morally neutral and permissible	Left to the unfettered operation of the market
Bad, but still morally permissible	Discouraged
Morally impermissible	Prohibited

noninvasive) are currently encouraged via government incentives (e.g., DARPA-supported Brain Initiative and BrainGate in the USA) while certain forms of invasive BCI use are discouraged via the gate-keeper medical model and noninvasive forms are left to market forces. The issue of whether policy change is necessary should reflect an open public discussion where ethical and policy concerns are not only thoroughly mapped, but also ranked for importance (see, e.g., the study by Voarino and colleagues [21]).

5 Materials and Methods

In this study, we have completed an expansion upon our previous research [15] into the academic discussion of BCI technology, which in turn followed from Burwell et al.'s 2016 study. While Burwell et al.'s 2016 study was a first-of-its-kind literature review of the ethics scholarship around BCI, our 2020 study adopted a similar methodology, but looked at the rapidly growing body of literature published in the time since Burwell et al.'s study was conducted. We [15] analyzed a randomly selected pilot sample (20.6%, $n = 7$) from the original pool of articles ($n = 34$) published since 2016 in order to identify recent trends in research and ethical debates regarding BCI technology and to assess the continued utility of the coding structure previously established by Burwell, et al. We now evaluate the entire sample. Thus, the same search information and criteria were used. The search was conducted in April of 2020 using PubMed and PhilPapers. Search queries included:

PubMed: ((“brain computer interface” OR “BCI” OR “brain machine interface” OR “Brain-computer Interfaces”[Mesh]) AND ((“personhood” OR “Personhood”[Mesh]) OR “cyborg” OR “identity” OR (“autonomy” OR “Personal autonomy”[Mesh]) OR (“liability” OR “Liability, Legal”[Mesh]) OR “responsibility” OR (“stigma” OR “Social stigma”[Mesh]) OR (“consent” OR “Informed Consent”[Mesh]) OR (“privacy” OR “Privacy”[Mesh]) OR (“justice” OR “Social Justice”[Mesh]))).

PhilPapers: ((brain-computer-interface|bci|brain-machine-interface)&(person hood|cyborg| identity|autonomy|legal|liability|responsibility|stigma|consent|pri vacy|justice)).

Following our prior work [15], we seek to elaborate on and identify changes in academic literature on BCI ethics since 2016 regarding new ethical discussions identified in the pilot sample. Further, we hope to better understand and quantify the preliminary trends observed within the literature using Burwell and colleagues' ethical framework. As in our prior work [15], the slightly modified search yielded 34 articles since 2016, as compared with Burwell et al.'s original 42 articles. At the full text screening phase, one article was excluded as tangential, leaving a sample of 33 texts. The search was conducted using a similar, slightly modified methodology and exclusion/inclusion criteria as Burwell et al. expanded to include applications involving animals and other subjects, such as brain organoids.

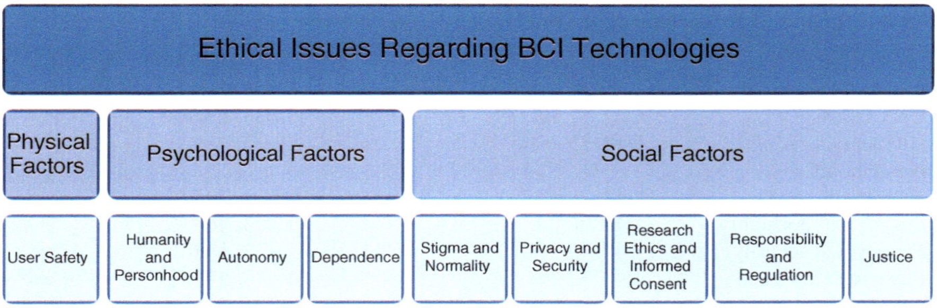

Fig. 1 Overarching themes in BCI ethics. **Note**: Adapted from our prior work [15]

As with the pilot sample, the abductive inference approach to qualitative research [22] was applied such that Burwell et al.'s framework was used to identify and map the overarching themes of ethical issues posed by BCIs (see Fig. 1). The map identifies eight specific ethical concerns that define the conceptual space of the ethics of BCI as a field of research. Only one of the ethical concerns refers to physical factors specifically: User Safety. Two are explicitly about psychological factors: Humanity/Personhood and Autonomy, while the remaining five focus on social factors: Stigma and Normality; Responsibility and Regulation; Research Ethics and Informed Consent; Privacy and Security; and Justice. While coding the texts with an eye toward any additional discussions of BCI-related ethical issues not identified in Burwell and colleagues' framework, we found a recurring theme of Dependence on Technology among our sample. Thus, we decided to add this as an additional ethical concern under the overarching theme "psychological factors."

6 Results

Similar to our prior published work [15], analysis of the fully updated sample includes discussion of all eight original ethical categories identified by Burwell and colleagues [16], with a notable addition of Dependence on Technology as a growing ethical theme not seen in the BCI literature prior to 2016. Table 2 summarizes our findings in 2021 compared to Burwell et al.'s findings reported in their 2017 paper.

The most frequently discussed categories in our 2021 analysis were *Autonomy* ($n = 26$, 78.8%) and *Responsibility and Regulation* ($n = 26$, 78.8%), which appeared at equal frequencies in the sample. Discussions of autonomy were primarily concerned with the level of control those with BCI devices have over their actions and decisions. This idea is especially relevant to BCI design as the BCI developers are the ones who decide whether users should be able to override or ignore BCI-mediated behavioral or physiological responses. As one study [21] notes, "Even while performing an action, the users themselves might be uncertain about being the (only) agent of an action, with systems that make autonomous decisions additionally decreasing the users' own autonomy."

Table 2 Our 2021 distribution of 33 selected papers vs. Burwell et al.'s 2017 distribution

Ethical issue discussed	Burwell et al.'s distribution out of 42 selected papers (2017)	Our distribution out of 33 selected papers (2021)
User safety	24/42, 57.1%	19/33, 57.6%
Humanity and personhood	15/42, 35.7%	15/33, 45.5%
Autonomy	12/42, 28.6%	26/33, 78.8%
Stigma and normality	11/42, 26.2%	12/33, 36.4%
Privacy and security	19/42, 45.2%	21/33, 63.6%
Research ethics and informed consent	14/42, 33.3%	19/33, 57.6%
Responsibility and regulation	13/42, 31.0%	26/33, 78.8%
Justice	20/42, 47.6%	13/33, 39.4%
Enhancement	NA	7/33, 21.2%
Military applications	NA	3/33 9.1%
Dependence on technology	NA	6/33, 18.2%
Other	NA	3/33, 9.1%

While most articles discussed benefits in terms of the increases in autonomy and independence gained from using a BCI [10, 17, 23–25], the potential for autonomy to be compromised was also discussed. For example, Hildt [10] mentions the possibility of taking the information gained from BCI—or in this case, Brain-to-Brain Interface (BBI)—from the individual and using it without their consent or knowledge:

> Participants in BBI networks depend heavily on other network members and the input they provide. The role of recipients is to rely on the inputs received, to find out who are the most reliable senders, and to make decisions based on the inputs and past experiences. In this, a lot of uncertainty and guessing will be involved, especially as it will often be unclear where the input or information originally came from. For recipients in brain networks, individual or autonomous decision-making seems very difficult if not almost impossible [10].

These ideas were often tied into discussions of *Responsibility and Regulation*, which was largely concerned with who should be held responsible in the cases of adverse consequences of BCI-mediated actions. The issue at the heart of *Responsibility and Regulation* can be understood with the hypothetical question: if a negative action was to be carried out by someone using a BCI, would liability fall upon the user of the technology, the technology itself, or perhaps the developers of the technology? For instance, if someone were to use a BCI-controlled prosthetic arm to pull the trigger on a gun and kill another person in the process, is there an argument to be made that the manufacturer of the BCI-prosthetic bears some responsibility for the action? What if the user of the BCI claims that they did not intend to fire the gun, and it was a malfunction of the BCI device? Many researchers claim that our legal system is not yet equipped to deal with such a situation. In the sample of articles, there was contention as to not only the moral and legal challenges associated with determining accountability in these instances, but also as to how to differentiate between responsibility on the part of the user, the machine, or the BCI developer. Rainey and colleagues, for example, state that "on the one hand, having limited control of devices seems to suggest device users ought to be considered less responsible for their actions mediated via BCIs. On the other hand, it is predictable that devices will be only partially controllable" [19]. This then relates to issues of regulation, not only in the development, distribution, and use of BCI devices, but also in how to enforce legal accountability in situations such as these.

Privacy and Security (n = 21, 63.6%) was another commonly discussed issue. The nature of a BCI sending brain signals directly to a computer raises the possibility of hacking, and many sources acknowledged the potential of brain hacking, in which control of a BCI device or access to its data (including the user's brain activity signals and the BCI's interpretation of those signals) might be seized by an unauthorized party. This could lead to a host of potential harms and privacy concerns for the user, especially if personal information—including their mental state or truthfulness at a given time—may be accessed in this way. On a related note, there are notable concerns that EEG data might be used to identify users and gain access to sensitive information. This then introduces concerns as to how neural data should be "gathered, collected, and stored," [26] in addition to "data ownership and privacy concerns" [26, p8]. Some articles [10, 24, 27] talked about the risks of extracting private information from people's brains and using it without their knowledge or consent, which is a significant concern for BCI technologies. Müller & Rotter connected this issue to User Safety, arguing that the increased fidelity of BCI data yields inherently more sensitive data, and that the "impact of an unintended manipulation of such brain data, or of the control policy applied to them, could be potentially harmful to the patient or his/her environment" [24].

The theme of *User Safety* (n = 19, 57.6%) tied into this discussion, as concerns for the psychological harms that might arise from brain hacking and privacy breaches were discussed on both an individual and societal level. As Müller and Rotter explain, "The impact of an unintended manipulation of such brain data, or of the control policy applied to them, could be potentially harmful to the patient or his/her environment" [24]. Physical harms were also mentioned as a point of ethical contention under this category, as detrimental consequences of BCI malfunctions and risks associated with implantation were taken into account. There was also discussion of the harms that might befall others aside from the BCI user, as in cases of adverse behavioral outcomes resulting from BCI malfunction or user mistakes. In these scenarios, as one source claims, "BCI-mediated action that deviates from standard norms or that leads to some kind of harm ought to be accommodated" [19]. Thus, both psychological and physical harm were explained as serious possibilities that need to be considered [10, 19, 24, 25]. One article discussed the impacts of harm on the results of a BCI study, stressing the importance of stopping a clinical trial if the risks to the individual participants begin to outweigh the potential benefits to science [23].

Research Ethics and Informed Consent (n = 19, 57.6%) were addressed at the same frequency as *User Safety*. Discussions surrounding this topic were primarily about whether subjects had an in-depth and comprehensive understanding of all associated risks, including the potential for psychological, social, and physical adverse effects involved. Indeed, as Yuste and colleagues point out, current consent practices may become problematic due to their "focus only on the physical risks of surgery, rather than the possible effects of a device on mood, personality or sense of self" [28]. Another point of concern was whether clinicians and researchers were providing an accurate representation of the limitations of BCI, taking care to avoid overhyping its potential among vulnerable or desperate populations. Few mentioned the particular challenges in obtaining informed consent from those in locked-in states. The main consensus among the ethicists that discussed this theme was that it is very important to obtain informed consent and make sure that the subjects are aware of all possible implications of BCI technology before consenting to its use. Additionally, some ethicists warned against the possibility of exploiting potentially vulnerable BCI research subjects. As Klein and Higger note: "the inability to communicate a desire to participate or decline participation in a research trial—when the capacity to form and maintain that desire is otherwise intact—undermines the practice of informed consent. Individuals cannot give an informed consent for research if their autonomous choices cannot be understood by others" [23].

Humanity and Personhood ($n = 15$, 45.5%) was the next most commonly discussed category. The largest consideration within this topic was the potential for changes to user identity and "sense of self" resulting from BCI use, contributing to a "pressing need to explore and address the potential effects of BCIs as they may impinge on concepts of self, control and identity" [17]. Many sources describe how users grapple with changes to their self-image following therapeutic use of BCI technology, both in terms of their disorder and associated limitations, and the extent to which the BCI technology is a part of them. Some sources also cited changes to personality as a risk associated with BCI technology, a concern arising from the finding that "some people receiving deep-brain stimulation...have reported feeling an altered sense of agency and identity" [28].

This is an important concern since BCIs could impact one's sense of self. In one specific study of BCI technology used in patients with epilepsy, there were a variety of resulting perspectives on sense of self, with some individuals saying that it made them feel more confident and independent, while others felt like they were not themselves anymore. One patient expressed that the BCI was an "…extension of herself and fused with part of her body…" [17]. Other articles more generally discussed the possibility of the sense of self changing and the ways BCI technology could contribute to this. Sample and colleagues categorized three ways in which one's sense of self and identity could change: altering the users' interpersonal and communicative life; altering their connection to legal capacity; and by way of language associated with societal expectations of disability [25]. Meanwhile, Müller and Rotter argue that BCI technology constitutes a fusion of human and machine, stating that "the direct implantation of silicon into the brain constitutes an entirely new form of mechanization of the self… [T]he new union of man and machine is bound to confront us with entirely new challenges as well" [24].

Justice ($n = 13$, 39.4%) was less frequently discussed among the sample, with the central concern being the potential for the technology to exacerbate existing inequalities, both in inherent design flaws and in distribution processes and barriers to access. There was also some discussion as to the potential for the technology to restore basic human rights among populations experiencing debilitating diseases and disorders, prompting the need for fair and ethical advancement of BCI technology. Two texts specifically discussed healthcare coverage of BCI access, noting how "unequal access to BCIs because of personal variations in BCI proficiency might raise questions of healthcare justice" [29]. One final concern was that "through algorithmic discrimination, existing inequalities might be reinforced" [29], disproportionately affecting disadvantaged populations. An additional concern related to inequality and injustice arose within the BCI research itself. These discussions often related back to the aforementioned questions of when the trials would end and if the participants were permitted to subsequently keep the BCI technologies [23].

Stigma and Normality ($n = 12$, 36.4%) was discussed to nearly the same degree as Justice. These conversations were largely centered around concerns that visible BCIs might further target their users for discrimination, leading to biased interactions. This could contribute to social isolation and exclusion among these populations. One source, for example, theorizes the technology that "confers disability group identity on the user might validate or otherwise reinforce harmful stigmas that often accompany that disability group identity and isolate, dominate, devalue, and generally oppress disabled people" [30]. Thus, stigma was mainly discussed from the perspective of the device itself having a negative stigma around it, and the device itself being what is stigmatizing about the individual [23]. However, it was also mentioned that perhaps universalizing the technology instead of only targeting it toward a group that is considered "disabled" could reduce or eliminate stigma [25].

Surprisingly, *Enhancement* ($n = 7$, 21.2%) was discussed at a greater frequency than *Military Applications*, diverging from Burwell and colleagues' sample [16]. Sources mentioned a potential "extended mind" and augmentation capabilities, one going so far as to suggest a future in which "powerful computational systems linked directly to people's brains aid their interactions with the world such that their mental and physical abilities are greatly enhanced" [28].

Dependence on Technology ($n = 6$, 18.2%) was a category unique to our sample that seems to have emerged in the ethical discussion of BCI technology since 2016. These discussions were dominated by concerns that BCI users might become dependent on their devices and fail to recognize potential errors in the machine's decision-making capabilities. Gilbert and colleagues explain that "the ethical problem with over-reliance is that the device ends up supplanting agency rather than supplementing it" [31], which relates back to challenges associated with autonomy. Alternatively, some might become dependent on BCI technology that they are unable to continue using beyond study participation.

Consistent with the findings of Burwell and colleagues [16], *Military Applications* ($n = 3$, 9.1%) was a relatively infrequent consideration within our sample, with sources briefly touching on the idea of the military as a relevant target population. Additionally, in our sample, *Other Ethical Issues* ($n = 3$, 9.1%) were similarly infrequent, briefly citing the potential for "therapeutic misconception and unrealistic expectations" [32], issues with advance directives among BCI users, and the ethical implications of slowing technology advancement.

7 Discussion

While there have been notable advancements in BCI and BBI technology and the body of literature on the ethical aspects of BCI technology has grown substantially since the original publication of Burwell and colleagues' research, these findings suggest that the original taxonomy developed by Burwell and colleagues remains a useful framework for understanding the body of literature, specifically on the social factors of the ethics of BCI. Ethicists can use this taxonomy—with some slight modifications, which we outline below—to understand how the body of literature on the ethics of BCI is grappling with ethical issues arising from the applications of this rapidly advancing technology. Articles published since 2016 still mostly conform to the taxonomy and can be categorized using it in future iterations of the scoping review methodology (Fig. 2).

There are, however, some areas within the growing body of literature on BCI ethics that have arisen since the original research was published that need to be incorporated into the taxonomy. We recommend the following modifications to the conceptual mapping outlined in Fig. 1. First, expanding the discussion of the physical (e.g., harms to test animals) and psychological (e.g., radical psychological distress) effects of BCI technology. The publicly available information on commercial BCI endeavors frequently mentions experiments with increasingly

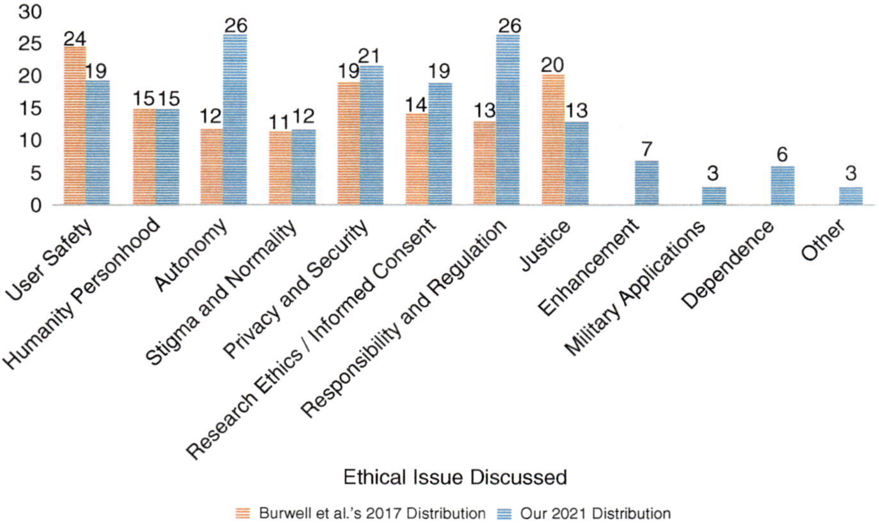

Fig. 2 Changes in discussion of the relevant ethical issues of BCI in recent years

complex and even sentient animals, such as Neuralink's demonstration of their technology on live pigs [14]. The lack of ethical scrutiny of these studies is an essential cause for concern [33]. Thus, ethical discussions should be expanded to include public awareness of private industry research into BCI using animals. Secondly, while the risks of physical harm from BCI are fairly well understood and covered in the literature, further research is needed to understand emerging psychological factors in BCI ethics, examining how human–AI intelligence symbiosis, brain-to-brain networking, and other novel applications of the technology may affect psychological well-being in humans. For instance, in the interview study by Gilbert and colleagues, one patient mentioned that "she was unable to manage the information load returned by the device," which led to radical psychological distress [17].

Going forward, it is imperative to expand on the connection between ethics and policy in discussions of BCI technology and conduct more empirical studies that will help separate non-urgent policy concerns, which are based on theoretical effects of BCI, from the more urgent concerns based on the current state of science in regard to BCI technology. In this, we echo Voarino and colleagues [21], in stating that we must advance the discussion from merely mapping ethical issues into an informed debate that explains which ethical concerns are high priority, which issues are moderately important, and establishing what constitutes a low priority discussion of possible future developments.

That said, it is important to make sure that the ethics literature keeps pace with engineering advances and that policy does not lag behind. In that vein, following Dubljević [34], we propose that the key ethical question for future work on BCI ethics is:

What would be the most legitimate public policies for regulating the development and use of various BCI neurotechnologies in a reasonably just, though not perfect, democratic society?

Additionally, ethicists need to distinguish between ethical questions regarding BCI technology that engineers and social scientists can answer for policy makers, versus those that cannot be resolved even with extensive research funding [35]. Therefore, following Dubljević and colleagues [36], we posit that these four additional questions need to be answered to ensure that discussions of BCI technology are realistic:

1. What are the criteria for assessing the relevance of BCI cases to be discussed?
2. What are the relevant policy options for targeted regulation (e.g., of research, manufacture, use)?
3. What are the relevant external considerations for policy options (e.g., international treaties)?
4. What are the foreseeable future challenges that public policy might have to contend with?

By providing answers to such questions (and alternate or additional guiding questions proposed by others), ethicists can systematically analyze and rank issues in BCI technology based on an as-yet to be determined measure of importance to society. While we have not completed such analyses yet, we do provide a blueprint above, based on conceptual mapping and newly emerging evidence, of how this can be done.

8 Conclusion

This chapter builds on, and updates, previous research conducted by Burwell and colleagues [16] to review relevant literature published since 2016 on the ethics of BCI. Although their article is now somewhat outdated in terms of specific references to and details from the relevant literature, the thematic framework, and the map we created—with the eight specific categories that it provides—and the nuanced discussion of overarching social factors have withstood the test of time and remain a valuable tool to scope BCI ethics as an area of research. A growing body of literature focuses on each of the eight categories, contributing to further clarification of existing problems concerning BCI technology. BCI ethics is still in its early stages, and more work needs to be done to provide solutions for how these social and ethical issues should be addressed.

Despite seeing evidence that these eight categories continue to be significant in more recent research, it is worth noting that we have found that the distribution of the eight categories was different in recent years, compared with the distribution previously identified by Burwell and colleagues [16] in the literature published before 2016. For instance, in the full sample of articles, we found that the two categories discussed most frequently were Autonomy (n = 26, 78.8%) and Responsibility and Regulation (n = 26, 78.8%), with Privacy and Security being discussed in 63.6% (n = 21) of articles, and User Safety, and Research Ethics and Informed Consent each discussed in 19 out of the 33 articles analyzed [57.6%]. However, despite Responsibility and Regulation being mentioned in 26 of the 33 papers [78.8%], it was not frequently discussed at length. Three of the four most frequently discussed categories identified in this distribution were not among Burwell and colleagues' top four most frequently mentioned (see Table 2). It seems that while the eight issues mapped are still ethically significant with regard to BCI research, the emphasis among them may be shifting toward concerns of psychological impact.

On that note, psychological effects (e.g., radical psychological distress) need to be carefully scrutinized in future research on BCI ethics. Additionally, one aspect that was not explicitly captured in the original thematic framework or the map we reconstructed from it is physical harm to animals used in BCI experimentation [33]. Finally, more detailed proposals for BCI policy have not yet become a frequent point of discussion in the relevant literature on BCI ethics, and this should be addressed in future work. We have provided guiding questions that will help ethicists and policy makers grapple with the most important issues first.DisclosuresNo funding (industry or otherwise) was received for this work.

References

1. Gilbert F, Pham C, Viana J, Gillam V. Increasing brain-computer interface media depictions: pressing ethical concerns. Brain Computer Interfaces. 2019;6(3):49–70.
2. Wakefield J. Elon Musk's Neuralink 'shows monkey playing pong with mind'. BBC News. 2021, April 9. https://www.bbc.com/news/technology-56688812.
3. Etherington D. Take a closer look at Elon Musk's Neuralink Surgical Robot. TechCrunch. 2020, August 28. https://techcrunch.com/2020/08/28/take-a-closer-look-at-elon-musks-neuralink-surgical-robot/.
4. Childers N. The video game Helmet that can hack your brain. Vice. 2013, June 6. https://www.vice.com/en/article/d77bmx/the-video-game-helmet-that-can-hack-your-brain.
5. Grush L. Those 'mind-reading' EEG headsets definitely can't read your thoughts. The Verge. 2016, January 12. https://www.theverge.com/2016/1/12/10754436/commercial-eeg-headsets-video-games-mind-control-technology.
6. Li G, Zhang D. Brain-computer interface controlled cyborg: establishing a functional information transfer pathway from human brain to cockroach brain. PLoS One. 2016;11(3):e0150667.
7. Aricò P, Borghini G, Di Flumeri G, Sciaraffa N, Babiloni F. Passive BCI beyond the lab: current trends and future directions. Physiol Meas. 2018;39(8):08TR02.
8. Shih J, et al. Brain-computer interfaces in medicine. Mayo Clin Proc. 2012;87(3):268–79.
9. Trimper JB, et al. When 'I' Becomes 'we': ethical implications of emerging brain-to-brain interfacing technologies. Front Neuroeng. 2014;7:4. www.frontiersin.org/articles/10.3389/fneng.2014.00004/full
10. Hildt E. Multi-person brain-to-brain interfaces: ethical issues. Front Neurosci. 2019;13:1177. www.frontiersin.org/articles/10.3389/fnins.2019.01177/full
11. Schmitz S. The communicative phenomenon of brain-computer interfaces. In: Mattering feminism, science, and materialism. New York/London: New York University Press; 2016. p. 140–58.
12. Coin A, Dubljević V. The authenticity of machine-augmented human intelligence: therapy, enhancement, and the extended mind. Neuroethics. 2020;14(2):283–90. https://doi.org/10.1007/s12152-020-09453-5.
13. Marsh S. Neurotechnology, Elon Musk and the goal of human enhancement. The Guardian. 2018, January 1. https://www.theguardian.com/technology/2018/jan/01/elon-musk-neurotechnology-human-enhancement-brain-computer-interfaces.
14. Regalado A. Elon Musk's Neuralink is neuroscience theater. In MIT Technology Review. https://www.technologyreview.com/2020/08/30/1007786/elon-musks-neuralink-demo-update-neuroscience-theater/. Accessed 30 Aug 2020.
15. Coin A, Mulder M, Dubljević V. Ethical aspects of BCI technology: what is the state of the art? Philosophies. 2020;5(4):31. https://doi.org/10.3390/philosophies5040031.
16. Burwell S, et al. Ethical aspects of brain computer interfaces: a scoping review. BMC Med Ethics. 2017;18:60.
17. Gilbert F, Cook M, O'Brien T, Illes J. Embodiment and estrangement: results from a first-in-human "intelligent BCI" trial. Sci Eng Ethics. 2019;25(1):83–96.
18. Kögel J, Schmid JR, Jox RJ, et al. Using brain-computer interfaces: a scoping review of studies employing social research methods. BMC Med Ethics. 2019;20:18. https://doi.org/10.1186/s12910-019-0354-1.
19. Rainey S, Maslen H, Savulescu J. When thinking is doing: responsibility for BCI-mediated action. AJOB Neurosci. 2020;11(1):46–58.
20. Blank R. Globalization: pluralist concerns and contexts. In: Giordano J, Gordijn B, editors. Scientific and philosophical perspectives in neuroethics. Cambridge: Cambridge University Press; 2010. p. 321–42.
21. Voarino N, Dubljević V, Racine E. tDCS for memory enhancement: a critical analysis of the speculative aspects of ethical issues. Front Hum Neurosci. 2017;10:678. https://doi.org/10.3389/fnhum.2016.00678.

22. Timmermans S, Tavory I. Theory construction in qualitative research: from grounded theory to abductive analysis. Sociol Theory. 2012;30:167–86.
23. Klein E, Higger M. Ethical considerations in ending exploratory brain–computer interface research studies in locked-in syndrome. Camb Q Healthc Ethics. 2018;27(4):660–74.
24. Müller O, Rotter S. Neurotechnology: current developments and ethical issues. Front Syst Neurosci. 2017;11:93. https://doi.org/10.3389/fnsys.2017.00093.
25. Sample M, Aunos M, Blain-Moraes S, Bublitz C, Chandler JA, Falk TH, et al. Brain–computer interfaces and personhood: interdisciplinary deliberations on neural technology. J Neural Eng. 2019;16(6):063001.
26. Naufel S, Klein E. Brain-computer interface (BCI) researcher perspectives on neural data ownership and privacy. J Neural Eng. 2020;17(1):016039.
27. Agarwal A, Dowsley R, McKinney ND, Wu D, Lin CT, De Cock M, Nascimento AC. Protecting privacy of users in brain-computer interface applications. IEEE Trans Neural Syst Rehabil Eng. 2019;27(8):1546–55.
28. Yuste R, Goering S, Arcas BA, Bi G, Carmena JM, Carter A, Fins JJ, Friesen P, Gallant J, Huggins JE, Illes J, Kellmeyer P, Klein E, Marblestone A, Mitchell C, Parens E, Pham M, Rubel A, Sadato N, et al. Four ethical priorities for neurotechnologies and AI. Nature. 2017;551:159–63.
29. Wolkenstein A, Jox RJ, Friedrich O. Brain-computer interfaces: lessons to be learned from the ethics of algorithms. Camb Q Healthc Ethics. 2018;27(4):635–46.
30. Stramondo JA. The distinction between curative and assistive technology. Sci Eng Ethics. 2019;25(4):1125–45.
31. Gilbert F, O'Brien T, Cook M. The effects of closed-loop brain implants on autonomy and deliberation: what are the risks of being kept in the loop? Camb Q Healthc Ethics. 2018;27(2):316–25.
32. Klein E. Informed consent in implantable BCI research: identifying risks and exploring meaning. Sci Eng Ethics. 2016;22(5):1299–317.
33. Johnson SL, Fenton A, Shriver A, editors. Neuroethics and nonhuman animals. Berlin/Heidelberg: Springer; 2020.
34. Dubljević V. Neuroethics, justice and autonomy: public reason in the cognitive enhancement debate, vol. 19. Springer; 2019.
35. Parens E, Johnston J. Does it make sense to speak of neuroethics? Three problems with keying ethics to hot new science and technology. EMBO Rep. 2007;8(S1):S61–4.
36. Dubljević V, Trettenbach K, Ranisch R. The socio-political roles of neuroethics and the case of Klotho. AJOB Neurosci. 2021;13(1):10–22.

Having the Ability to Have a Good Life: What Might Be the Impact of BCIs?

Brielle Lillywhite and Gregor Wolbring

1 Introduction

Neuro-advancements including neuro/cognitive enhancements raise many ethical, legal, and social issues [1–14]. Brain computer interfaces (BCIs) are one category of neuro-products that are recognized to pose many ethical, legal, and social issues [15–33] including those related to cognitive enhancement enabled by BCI [34–36], brain to brain interfacing [37, 38], and BCI enabled by artificial intelligence and machine learning [38–40]. The health and well-being of people and society are a main part of the ability to have a good life [41–44], and many social determinants for health and well-being are identified [45–50]. Marginalized individuals and groups are known to encounter problems in relation to many social determinants of health and well-being and with that the ability to experience a good life [41, 51–54]. Disabled people are one marginalized group that faces many problems in their ability to obtain a good life [55, 56], and disabled people are at the same time covered extensively as beneficiaries of BCIs [57]. Being aware of the social implications of scientific and technological products and with that the impact of these products on the ability to have a good life is a vital part of STEM education [58, 59]. Furthermore, STEM education has a social impact [60]. Therefore, in this exploratory study, we ascertained the views of STEM students on the impact of BCIs on the ability have a good life using the indicators of four composite measures of health and well-being (Social Determinants of Health [45, 46]; Canadian Index of Wellbeing [47],

B. Lillywhite
Department of Chemical and Petroleum Engineering, Schulich School of Engineering, University of Calgary, Calgary, AB, Canada
e-mail: brielle.lillywhite@ucalgary.ca

G. Wolbring (✉)
Community Rehabilitation and Disability Studies, Department of Community Health Sciences, Cumming School of Medicine, University of Calgary, Calgary, AB, Canada
e-mail: gwolbrin@ucalgary.ca

V. Dubljević, A. Coin (eds.), *Policy, Identity, and Neurotechnology*, Advances in Neuroethics, https://doi.org/10.1007/978-3-031-26801-4_8

Organization for Economic Cooperation and Development (OECD) Better Life Index [48], and World Health Organization (WHO) Community Based Rehabilitation Matrix [49]). We also asked participants how they think BCIs will impact the ability of various social groups including disabled people to have a good life both now and in the future.

2 The Ability to Have a Good Life

The health and well-being of people and society are a main part of the ability to have a good life [41–44]. As it is stated: "A person's well-being is what is 'good for' them"[43]. The ability to have a good life has many social determinants [50]. In a recent study [50], we found that of the following tools that exist to measure well-being (Social Determinants of Health, OECD Better Life Index, Canadian Index of Wellbeing, Community Based Rehabilitation Matrix, WHO Quality of Life measure (WHOQoL), The Quality of Being Scale, Assessment of Quality of Life (Aqol), Calvert-Henderson Quality of Life, Satisfaction With Life Scale, Perceived Life Satisfaction Scale, Flourishing Scale, Scale of Positive and Negative Experience, Comprehensive Inventory of Thriving, Brief Inventory of Thriving, The Disability and Wellbeing Monitoring Framework and Indicators), only the Social Determinant of Health tool was mentioned to a significant extent in conjunction with neuro and other technologies [50]. In the same study, we found that the academic literature that focuses on 50 neurotechnologies, which included BCIs, engages very unevenly with the individual indicators of four of these composite measures (Social Determinants of Health; Canadian Index of Wellbeing, OECD Better Life Index and World Health Organization Community Based Rehabilitation Matrix).

3 BCIs and a Good Life

The ability to have a good life is part of the social implications [61, 62] category and the language of the social good which is linked to the ability to have a good life is a language that has the potential to resonate with students [63]. However, when we searched for the co-occurrence of the phrase "good life" and "brain computer," or "brain machine," or "human computer," using the same databases as in [50] we found no hits. Using the term "better life," we found seven hits. Within these seven hits, only two were relevant; one author argued that BCIs were "developed to help the disabled to lead a better life" [64]; a sentiment also put forward in [65]. The term "social good" only generated five hits, but all with the phrase human–computer and none with brain computer or brain machine [66–69]. The term "social implication" only generated three hits with "brain computer," one hit with "brain machine" and 17 hits with "human computer." In regard to the BMI/BCI hits, the phrase "social implications" is mentioned in relation to the company "Neuralink." Neuralink's use of BMI goes beyond a medical target group instigating evaluation of the social implications of the "Neuralink" product [70]. Secondly, a case study was used to explore the social implications of using BCIs for workers [71]. Thirdly, a study argued that BCIs raise issues of "identity, normality, authority, responsibility, privacy, and justice," but that study does not label them as social implications. Rather, it simply states that they are part of a project that investigates the ethical and social implications [72]. Finally, a fourth study looked at BCIs in relation to paralyzed patients stating "ethical, psychological and social implications of BCIs concern mainly hasten death decisions and euthanasia" [73].

Content in tables 4–7 in [50] suggests that many terms that could be seen as indicators of the ability to have a good life were rarely or not covered in the academic literature engaging with 50 neurotechnologies including BCIs. As [50] covers BCI as part of the list of 50 neurotechnologies, we did some searches for indicators from the four measures used in [50] that we consider directly linked to the ability to have a good life in relation to BCIs. When we searched "brain computer" or "brain machine" in the abstracts of Scopus and EBSCO-Host as done in [50] and obtained 18,324 hits. However, adding some of the indicators that we see as fitting the category of social implications, the numbers dropped drastically as to be expected given the data from [50] for 50 neurotechnologies. "Discrimination" was mentioned in 384 abstracts (most were false positive so they were not linked to social discrimination), stress 91, literacy 13, "Social Relationships" five, "Social Exclusion," "Empowerment," "Social engagement" and "Social relationship" two, "Living standard," "Livelihood" once, "Self-Employment," "Financial services," "Wage employment," "Social protection," "Social Situation," "Social mobilization," "Political participation," "Self-help groups," "Disabled people's organizations," "Social Support," "Community safety," "Social norms," "Attitudes toward others," "Democratic Engagement," "Personal wellbeing," "Social Safety Network," "Advocacy," "Social engagement," "Social status," "Civic engagement," and "Life satisfaction" not at all.

4 The Case of Disabled People

Disabled people are a marginalized group and their already precarious ability to have a good life is made evident in, for example, the *United Nations Convention on the Rights of Persons with Disabilities* and the *United Nations 2018 flagship report on disability and development: realization of the Sustainable Development Goals by, for and with persons with disabilities* [55, 56]. Furthermore, many of the indicators of the composite measures (Social Determinants of Health; Canadian Index of Wellbeing, OECD Better Life Index, and World Health Organization Community Based Rehabilitation Matrix), can be seen as lacking a positive reality for disabled people if one looks at [55, 56].

The ability of disabled people to have a good life can be impacted by BCIs in various ways:

1. By potential non-therapeutic use of a product (consumer angle).
2. By potential therapeutic use of a product (patient angle).

3. By changing societal parameters caused by humans using a technology (e.g. changes in ability expectations).
4. By changing societal parameters demanded and caused by BCI governance and activism.
5. By being a potential argument used in BCI governance and activism.

5 STEM Education

The action of looking for and being aware of the social implications of technology is part of STEM education [59], and, overall, STEM education has a social impact [60]. At the same time, it is noted that there are problems in engaging with the topic of social implications in certain subareas of engineering education such as civil engineering [58] and STEM education in general [74]. Literature covers many competencies students are to obtain from their engineering degree [75] and these competencies include the twenty-first century skills listed by the Organization for Economic Co-operation and Development (OECD) and categorized in three groups: knowledge (research and problem solving skills; identifying, searching, evaluating, selecting, organizing, analyzing, and interpreting information), communication, and ethical and social impact [76]. Furthermore, it is noted that to have STEM students "graduate with these skills has become one of the most important questions awaiting for an answer all over the world" [76]. Language of the social good and the goal of STEM contributing to the social good resonates with students [63] as does "the potential to have a positive social impact" [77]. Social Awareness Curriculum has an impact on the Engineering Identity Formation of High School Girls [78]. However, at the same time it is noted that techno-determinism and techno-optimism are recognized as biased forms of reporting within the STEM education literature [79–82].

6 Method

Given the impact of BCIs on many indicators of a good life, the possible impact that BCIs may have for disabled people on experiencing a good life, and that STEM students are to be aware of social implications of science and technology, we used indicators of the four composite measures (Social Determinants of Health; Canadian Index of Wellbeing; OECD Better Life Index; WHO Community Based Rehabilitation Matrix) we used before in [50] to investigate two research questions: a) how do STEM students perceive BCIs to impact the indicators of these four measures and b) how do STEM students think BCIs will impact the ability for various groups, including disabled people, to have a good life both now and in the future.

Data was collected through a survey delivered online using the Survey Monkey platform. The survey received ethics approval from the University of Calgary's

Health Research Ethics board (REB17-0785) on 23 February 2018. The link to the online survey was sent to the students through personal contacts after ethics approval was received. The survey data was collected between March and April 2021.

An online survey was chosen to reach as many student participants as possible [83] and to give students the flexibility to participate in this study at their convenience. The survey was distributed to four cohorts of individuals from four different University STEM related groups engaged in STEM and engineering extracurricular activities. The online survey was set in such a way that we could not identify the participants or their IP addresses. The consent form alerted participants that the US government could access data as survey monkey falls under U.S jurisdiction. Participants could stop the survey at any time and be free to choose which questions they want to answer or not.

The survey included 23 questions, with simple yes or no and Likert questions with the opportunity for participants to add comments, along with open-ended questions. In the here presented study, we provide the results of questions 1–6 and 17–23 which covered: a) demographics; b) the abilities participants view as needed for a good life; c) participants' familiarity with BCIs; d) how participants perceive the impact of BCIs on various social groups; and e) how participants perceive the impact of BCIs on the indicators of four measures (Social Determinants of Health, Better Life Index, Canadian Index of Wellbeing, and Community Based Rehabilitation Matrix).

Frequency counts and percentage measures of the descriptive quantitative data were extracted and analyzed using Survey Monkey's intrinsic frequency distribution analysis capability.

7 Limitations

Our study has various limitations. Given that we used an online delivered survey instrument, we could not ask for clarifications of answers. Also, there might be a selection bias in the sense that only students that were already interested in the topic might have chosen to answer the survey. Further, the distribution of the participants' year of study was not equally distributed between years one to four. This may lead to participants not having enough education or knowledge around BCIs to validate the findings. However, participants were asked if they know what BCIs are to understand the data collected. Lastly, there may be selection bias in that about 92% of the participants were females, while the field of engineering specifically is male dominated. From our understanding, the gender composition of each cohort the survey was distributed to varied. The large proportion of female respondents was unexpected, so given that the number of responses from each cohort the survey was sent to is unknown, there was selection bias in that mostly females chose to answer the survey.

8 Results

8.1 Demographic Results

The response rate from the students we accessed reflects 13.14% (51 from 388) of the students contacted. Q1–6 received 48 responses, while Q17/18 had 28, Q19/22/23 had 27, and Q21 had 25 responses. 91.67% were females and 8.33% were males. 97.92% were 18 to 30 years of age and 2.08% were 30 to 65 years of age. 97.92% of the participants were undergraduate students, while 2.08% were PhD students. As for abilities, 95.8% identified as able bodied, and 4.2% as disabled.

More specifically, 27.08% were first-year undergraduate students, 33.33% second-year undergraduate students, 29.17% third-year undergraduate students, and 8.33% fourth-year undergraduate students (Fig. 1).

The population consisted of a majority STEM students, specifically 60.42% engineering students (6.25% biomedical engineering, 6.25% chemical engineering, 10.42% civil engineering, 6.25% electrical engineering, 18.75% mechanical engineering, 6.25% software engineering, and 6.25% common first year engineering), 2.08% computer science students, 2.08% mathematics and statistics students, 18.75% biological sciences, 4.17% health sciences, 4.17% neurological sciences, 2.08% physiology, 2.08% kinesiology, 2.08% business, and 2.08% other (dual degree in mechanical engineering and business) (Fig. 2).

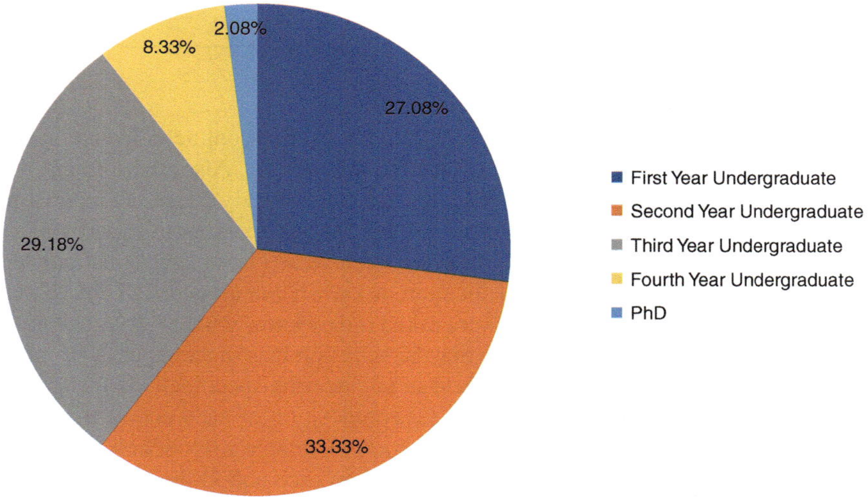

Fig. 1 Year of study of participants

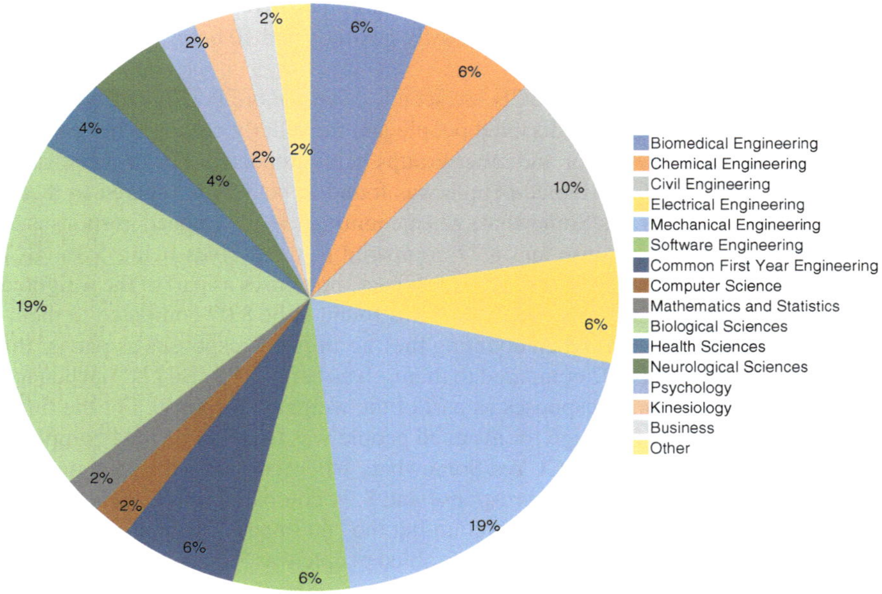

Fig. 2 Degree

9 Familiarity with BCI

Survey Q17 asked: "Are you familiar with brain computer interfaces? If not, one description is: A BCI allows brain signals to control an external activity linked to the BCI. For example, a BCI could be used to control a prosthetic limb or games using brain signals."

We found that 10.71% said yes, 46.43% said somewhat, and 42.86% said no.

Looking at the answer to Q17 by years of study (years 1–4) we found no trend in how Q17 was answered from year 1–4 of the undergraduate degree.

As for the responses to the remaining questions, we divided the responses based on whether they identified themselves as having or not having knowledge on the topic of BCI (tables presenting the exact results for Q18–23 can be found in the appendix).

10 Participants' Views on the Impact of BCI on Various Social Groups

In questions 18 and 19, we asked participants how they perceive BCIs to impact the ability to have a good life at the moment (Q18) and in the future (Q19) for the following groups: participants themselves, disabled people, women, the elderly, youth, men, non-binary people, people of ethnic background (minority groups), people of low income, people of high income, countries of the north, countries of the south, animals, and nature. Participants were to indicate the impact on a scale: "0 being not impacted, 1 being purely negative, 2-4 being more negative than positive, 5 being equal positive and negative, 6-9 being mostly positive, and 10 being purely positive impact."

The responses revealed that BCIs are seen to have a more positive effect on the ability to have a good life for disabled people and the elderly compared to how the participants see themselves or and other groups being impacted. For the timeframe of today, the average for disabled people was found to be 8.27 compared to themselves being 3.93 (BCI Yes/Somewhat) when counting the "no impact" responses as part of the weighted response and 8.27 compared to themselves being 5.36 (BCI Yes/Somewhat) when not counting the "no impact" responses as part of the weighted response. The average for disabled people was found to be 8.07 compared to themselves being 4.0 (BCI NO) when counting the "no impact" responses as part of the weighted response and 8.07 compared to themselves being 6.88 (BCI NO) when not counting the "no impact" responses as part of the weighted response. For the timeframe of the future, the average for disabled people was found to be 8.94 compared to themselves being 5.14 (BCI Yes/Somewhat) when counting the "no impact" responses as part of the weighted response and 8.94 compared to themselves being 6.31. (BCI Yes/Somewhat) when not counting the "no impact" responses as part of the weighted response. The average for disabled people was 8.64 compared to themselves being 5.27. (BCI No) when counting the "no impact" responses as part of the weighted response and 8.64 compared to themselves being 7.25 (BCI NO) when not counting the "no impact" responses as part of the weighted response. Furthermore, for Q18/19 there were also differences in groups seen as not being impacted by BCIs. Although the "no impact" option was answered by only a few for all of the groups including themselves, disabled people and the elderly were the only groups where no one saw "no impact" (independent of Yes/Somewhat or No as BCI knowledge and timeframe of today and future). As for other groups, the BCI No group showed more "no impact" responses for today and the future than the Yes/Somewhat group whereby the "no impact" went down for the Q18 (today) compared to Q19 (the future).

The responses varied with participants' year of study. For the timeframe of today, the average for disabled people was found to increase with increasing year of study, found to be 8 for first and second years and 8.42 for third and fourth years (without counting the "no impact" responses). For the timeframe of the future, the average for disabled people increased, while also increasing with year of study for years 2–4, found to be 8.75 for second years and 9.5 for third and fourth years (without counting the "no impact" responses). This trend also follows for the elderly comparing the timeframe of now and the future. For the timeframe of today, the averages were relatively similar between the years of study found to be 7.71 for first years, 7.25 for second years, and 7.46 for third and fourth years (without counting the "no impact" responses). For the timeframe of the future, the average for elderly people increased, while also increasing with year of study for years 2–4, found to be 8.13 for second years and 8.17 for third and fourth years (without counting the "no impact" responses). This can be compared to the averages for themselves, which were found to be much lower. For the timeframe of today, the average for themselves was found to be 5.29 for first years, 2.63 for second years, and 4.08 for third and fourth years (without counting the "no impact" responses). For the timeframe of the future, the average for themselves increased for all years of study, found to be 6.43 for first years, 3.25 for second years, and 5.75 for third and fourth years (without counting the "no impact" responses). There were no trends found in relation to the knowledge of BCI or not in relation to the year of study.

11 Participants Views on the Impact of BCIs on the Indicators from the Social Determinants of Health, OECD Better Life Index, Canadian Index of Wellbeing, and Community Based Rehabilitation Matrix

Survey Q20-Q23 asked: "There are various measures that try to ascertain people's situation in life, how well they are. How will the below indicators be impacted by brain computer interfaces?" The differences being, Q20 focused on the indicators of the community based rehabilitation matrix, Q21 on the indicators of the Canadian Index of Wellbeing, Q22 on the indicators of the Social Determinants of Health, and Q23 on the indicators of the Better Life Index.

Table 2 summarizes the number of indicators that are within the % agreement in each category of impact. For many of the indicators in Q20–23, the impacted "only positive" response was significant or sometimes the majority, while there were fewer or no responses for impacted "purely negative." The BCI Yes/Somewhat participants were found to be more techno-optimistic than the BCI No participants for some indicators and vice versa for other indicators. As for the "no impact" option, there was no indicator where the majority clicked "no impact." For both food security and housing, the number was exactly 50%. Most of the answers were between 0 and 16.66% for the "no impact" option. For some, the "no impact" responses decreased from the BCI No group to the Yes/Somewhat group, such as "discrimination" from 30% to 18.25%, as well as stress, social status, civic engagement to list just a few.

The responses varied with participants' year of study. For health-related indicators such as health, healthcare, and health prevention, along with indicators such as assistive technology and rehabilitation within the Community Based Rehabilitation Matrix, the first-year students perceived there to be a greater "only positive" impact than the second to fourth-year students. However, the third- and fourth-year students indicated that disabled people's organizations will be impacted "only positive" at 61.54%, which is greater than the first and second years which showed 57.14% and 25.00%, respectively, for "only positive." Overall, for the indicators of measure included in the Canadian Index of Wellbeing, the third- and fourth-year students indicated a greater "only positive" impact for almost all indicators compared to participants in first and second year. Responses for the indicators within the Social Determinants of Health and the Better Life Index did not show much variation between the years of study other than the health services indicator, which third and fourth years indicated 72.73% as "only positive," compared to first and second years indicating 42.86% and 25.00%, respectively, for "only positive." A similar trend was found for the safety and health indicator within the Better Life Index.

12 Discussion

Our study has two main take-home messages. Regarding questions that asked how BCIs impact the ability of different social groups to have a good life, participants indicated that disabled people and the elderly are impacted much more positively compared to how the participants saw themselves or other groups being impacted. That remained true for the timeframe of today and in the future and whether the participants indicated that they are familiar with BCIs or not. As for the indicators of the composite measures, the "only positive" option was selected substantially, while very few indicators were considered to be impacted by BCIs in a "purely negative" way. As for the "no impact" category, there was not one indicator for which the majority selected the "no impact" option. In general, our results reveal that participants had a techno-optimistic view of the impact of BCIs on the ability to have a good life.

Our techno-optimistic findings in relation to disabled people might be a consequence of disabled people being mostly mentioned within the BCI academic literature in their role of patient and the perception of having to be fixed [36]. The bias is also present in the coverage of other scientific and technological advancements such as social robotics [84] and artificial intelligence/machine learning [85]. It might also be a consequence of recognized techno-optimism biased forms of reporting of scientific advancement in relation to disabled people and within the STEM education literature [79–82], and media [86, 87], and the lack of coverage of many of the well-being, the ability to have a good life, indicators in the academic literature focusing on neurotechnologies [50] and BCI (see introduction section).

The techno-optimistic tone is problematic. A techno-optimistic tone does not lend itself to cover the possibility of advancement of science and technology including BCIs impacting the social, the quality of life, the good life of disabled people and others in a negative way, which is a possibility we know exists [36, 84, 85, 88–90]. Not only that, but it also influences what social implications of science and technology are seen to be an issue for disabled people including what topics are researched in relation to disabled people [91]. Furthermore, although not thinking about science and technology but still applicable, it was noted by disabled participants that took part in online fora focusing on the 2030 sustainable development goals that the medical imagery is one reason why disabled people are excluded from many policy discourses that are seen to focus on non-medical problems such as sustainable development [92–96].

13 Adding Indicators

The findings related to the indicators of measure in our prior study [50] and in our short research with just using "brain computer" and "brain machine" (see introduction section) fit with other studies noted that social factors such as justice, stigma, normality, and the phrase "social implication" were rarely mentioned in the literature focusing on BCIs [38]. In our study, we used only indicators that are linked to the four composite measures. But one could add indicators to the list such as stigma, normality, stereotype, justice, independence (with the meaning of doing it yourself), independence (with the meaning of being in control), autonomy, self-determination, and interdependence mentioned by others as important in relation to the impact of BCIs [14, 32, 38]. Associating BCIs with disabled people is seen to hinder others from buying non-medical consumer type BCI products [32, 36, 97] and is seen to further the stigmatization of disabled wearers [89]. This is also noted in [38]: "stigma was mainly discussed from the perspective of the device itself having a negative stigma around it, and the device itself being what is stigmatizing about the individual." It is noted that disabled persons might wish to use a BCI but reject the medical/deficiency label linked to it [31]. The benefits of BCIs are frequently discussed in conjunction with an increase in the ability of autonomy and independence by using a BCI [38]. However, the very word "independence" as a selling point is questioned as it is used around disabled people and BCIs [36], but also other technologies such as social robots [84]. Independence is often not used to indicate that one is in control of their decisions but it comes with the ability expectation of having to do things by oneself and not needing others [84], which is supported by the supercrip imagery [98–104] including its use in BCIs [105] and supports the stigmatization of disabled people as soon as they cannot do it alone. Indeed independence as a term that has been questioned for some time by the disability rights movement [106]. Independent living was originally "the ability of disabled people to participate actively in society: to work, have a home, raise a family if they wish, in sum to decide their own futures according to the cultural context within which they live" [107]. As such, in many places, the term self-determination is used instead [108] and interdependence is seen as a goal versus independence as in, do-it-yourself [109]. As such, adding such terms to the list of indicators we used will give important insight into the social implication of BCIs and other technologies on the ability to have a good life. Finally one can merge our list of indicators used with other lists of existing well-being related indicators [50, 110].

14 Future Research

Our study suggests various research possibilities. Using interviews, one could ask participants to answer the same questions we did but with the opportunity for follow-up questions. One could, for example, ask participants in relation to question 18–19 (Tables 1, 3, 4, 5, and 6) why they chose or did not choose "no impact" for a given group. One could also use the answers specifically from tables 4–7 in [50] and

Table 1 Views on the impact of BCI on the ability to have a good life for various groups

	Weighted averages							
Timeframe	In the moment				In the future			
Knowledge of BCI	Yes/somewhat		No		Yes/somewhat		No	
Group	With 0	Without 0	With 0	Without 0	With 0	Without 0	With 0	Without 0
Myself	3.93	5.36	4.00	6.86	5.14	6.31	5.27	7.25
Disabled people	8.27	8.27	8.08	8.08	8.94	8.94	8.64	8.64
Women	4.88	5.57	4.27	6.71	5.44	5.80	5.91	7.22
The elderly	7.44	7.44	7.50	7.50	7.81	7.81	8.00	8.00
Youth	5.56	5.56	4.92	6.56	6.13	6.13	6.64	7.30
Men	4.88	5.57	4.09	6.43	5.56	5.93	5.82	7.11
Non-binary	4.81	5.50	4.50	6.75	5.69	6.07	5.73	7.00
People of ethnic background (minority groups)	4.80	5.54	4.25	6.38	5.88	6.27	5.55	6.78
People of low income	4.40	5.08	2.92	4.38	5.19	5.53	5.91	6.50
People of high income	6.44	6.87	5.33	6.40	7.19	7.19	6.73	7.40
Countries of the North	6.31	6.73	4.33	6.50	6.19	6.19	5.73	7.00
Countries of the South	4.94	5.27	3.83	5.75	5.75	5.75	5.27	6.44
Animals	4.40	6.00	4.08	6.13	4.69	6.25	5.18	7.13
Nature	3.93	5.00	2.58	4.43	4.33	5.42	3.00	4.13

Tables 2 and 7, 8, 9, and 10 in this study to focus on specific indicators. Indeed, given our answers to the "no impact" option, it might be useful to give the survey to participants, read the answers as an investigator, and then ask for more clarifications as to the reasoning. One could also use all the indicators but use different groups of participants or again STEM students and see whether the key trajectories are the same. One could also use different social groups for the measures and look at whether the impact perception of participants in relation to the indicators differs for groups we did not list.

Related to disabled people, one could differentiate based on why they are labeled as a disabled person, as one can expect BCI visions to impact disabled people with different characteristics in different ways. It would be interesting how participants judge the impact for different groups of disabled people. It is recognized that data

Table 2 Summary of the sentiment toward indicators

	Number of indicators							
Knowledge of BCI	Yes/somewhat				No			
Sentiment toward indicators % agreeing	Only positive	Only negative	Both	No impact	Only positive	Only negative	Both	No impact
Community based rehabilitation matrix (34 indicators)								
0%	0	12	0	9	7	16	1	14
1–25%	12	22	4	25	8	13	18	18
26–50%	18	0	23	0	11	5	15	2
51–75%	3	0	7	0	7	0	0	0
76–100%	1	0	0	0	1	0	0	0
Canadian index of wellbeing (30 indicators)								
0%	1	14	0	8	0	30	0	0
1–25%	4	16	1	19	9	0	8	18
26–50%	20	0	18	3	21	0	21	12
51–75%	5	0	10	0	0	0	1	0
76–100%	0	0	1	0	0	0	0	0
Social determinants of health (30 indicators)								
0%	2	0	0	0	0	18	0	5
1–25%	15	30	12	7	15	12	0	16
26–50%	12	0	18	23	12	0	21	9
51–75%	1	0	0	0	3	0	9	0
76–100%	0	0	0	0	0	0	0	0
Better life index (12 indicators)								
0%	0	9	0	3	0	11	0	0
1–25%	5	3	0	5	9	1	0	4
26–50%	6	0	6	4	3	0	8	8
51–75%	1	0	6	0	0	0	4	0
75–100%	0	0	0	0	0	0	0	0

segregated for different categories of disabled persons are missing such as in equity, diversity, and inclusion (EDI) discourses [111]. In many cases, such as employment numbers [112], the data is generic, although disabled people are not a homogeneous group, and given the broad characteristics that nowadays are labeled as disabled people, the numbers are very likely different for different bodily realities. In relation to disabled people, a study highlighted that there was "little to no intersectionality between disabled people and other groups such as gender, race (using the term ethnic due to false positive with the term race not covering ethnicity), and various terms used to depict indigenous people ("Aboriginal" OR 'first nations' OR 'Metis' OR 'indigenous people' OR 'Inuit')." The term "women with disabilities" and "disabled women" were not mentioned once" [50]. Indeed, in the same way that the characteristic an individual has that leads one to be defined as a disabled person are impacted differently by BCI advancements. This occurs if the disabled person belongs to more than one marginalized group. As such, one could make the questions related to the impact on disabled people more granular by asking about disabled people linked to other marginalized groups such as disabled indigenous people.

One study suggests some differences between disabled and non-disabled participants in relation to agent-related concerns and consequence related concerns [31]. One could categorize the measures we used for the analysis as agent related or consequence related, and one could be more granular in the disabled person category.

One could use the questions we asked for different BCI applications, for example, medical versus non-medical use of BCIs. This, for example, would allow one to evaluate whether our suggestion that the more techno-optimistic sentiment toward disabled people and the elderly is because the participants envisioned them in a medical framework and other groups is true or not. One could give futuristic examples for BCIs use like what Mark Zuckerberg and others have in mind when they cover the metaverse [113–118]. The metaverse-BCI linkage and BCI Virtual Reality linkage [119], especially of non-invasive versions of BCI, are especially interesting to explore given the shift to a virtual setting in education and work during Covid-19 and the dominant narrative that argues that face-to-face in the same physical space delivery of education is desirable although others see the virtual setting as an opportunity [120].

Our approach might be useful to connect the groups and individuals engaged with the measures, the science and technology focused groups and individuals in general, and the governance of science and technology focused groups and individuals, given that specific groups and individuals are linked to the four composite well-being measures, but are not often if at all, linked to the governance of neurotechnologies such as BCI discussions. Our study's approach might also be useful for people that employ the framework validator questions [32] as many of the answers could be investigated through the lens of the questions we asked in our study. Our results and follow-up studies could also inform the answer to these posted five questions. What would be the most legitimate public policies for regulating the development and use of various BCI neurotechnology's by healthy adults in a reasonably just, though not perfect, democratic society? What are the criteria for assessing the relevance of BCI cases to be discussed? What are the relevant policy options for targeted regulation (e.g., of research, manufacture, use)? What are the relevant external considerations for policy options (e.g., international treaties)? What are the foreseeable future challenges that public policy might have to contend with? [38]. Filling the gap also fits with the goals of many universities to advance the EDI agenda at universities [111]. One cannot have EDI if the social situation of an EDI targeted group, like disabled people, is not part of the academic discourse. Indeed, we suggest our findings are a reflection of a problematic EDI discourse in relation to disabled people in universities [111, 121].

15 Conclusion

Our exploratory study reveals a techno-optimistic sentiment in how participants answered both research questions. We suggest that the surveys provide numerous insights in how participants view the impact of BCIs. Also, our results could be used to generate many follow-up studies using BCIs but also other technologies with a multitude of opportunities for inter, intra, and transdisciplinary and intersectional research collaborations. Finally, one can use the surveys we used and create modified versions in classrooms as pedagogical tools by discussing the results of how the students from that given class filled out the surveys.

Acknowledgments This study was funded by Government of Canada, Canadian Institutes of Health Research, Institute of Neurosciences, Mental Health and Addiction ERN 155204 in cooperation with ERA-NET NEURON JTC 2017.

The authors have no relevant financial or non-financial interests to disclose.

The authors have no competing interests to declare that are relevant to the content of this book chapter.

All authors certify that they have no affiliations with or involvement in any organization or entity with any financial interest or non-financial interest in the subject matter or materials discussed in this manuscript.

The authors have no financial or proprietary interests in any material discussed in this article.

Appendix

Table 3 Q18 How do you see brain computer interface technology impacting ... ability to have a good life in the moment? 0 = no impact, 1 purely negative impact, 2–4 more negative impact, 5 equal negative and positive impact, 6–9 more positive impact, and 10 = only positive impact. **BCI Yes/somewhat**

Group	0	1	2	3	4	5	6	7	8	9	10	Total	Weighted average with 0	Weighted average without 0
Myself	26.67% 4	0.00% 0	6.67% 1	0.00% 0	0.00% 0	33.33% 5	20.00% 3	13.33% 2	0.00% 0	0.00% 0	0.00% 0	15	3.93	5.36
Disabled people	0.00% 0	0.00% 0	0.00% 0	0.00% 0	0.00% 0	6.67% 1	13.33% 2	13.33% 2	13.33% 2	20.00% 3	33.33% 5	15	8.27	8.27
Women	12.50% 2	0.00% 0	0.00% 0	6.25% 1	0.00% 0	31.25% 5	37.50% 6	12.50% 2	0.00% 0	0.00% 0	0.00% 0	16	4.88	5.57
The elderly	0.00% 0	0.00% 0	0.00% 0	0.00% 0	0.00% 0	6.25% 1	12.50% 2	31.25% 5	37.50% 6	6.25% 1	6.25% 1	16	7.44	7.44
Youth	0.00% 0	0.00% 0	6.25% 1	0.00% 0	0.00% 0	37.50% 6	37.50% 6	18.75% 3	0.00% 0	0.00% 0	0.00% 0	16	5.56	5.56
Men	12.50% 2	0.00% 0	6.25% 1	0.00% 0	0.00% 0	25.00% 4	43.75% 7	12.50% 2	0.00% 0	0.00% 0	0.00% 0	16	4.88	5.57
Non-binary	12.50% 2	0.00% 0	6.25% 1	0.00% 0	0.00% 0	31.25% 5	37.50% 6	12.50% 2	0.00% 0	0.00% 0	0.00% 0	16	4.81	5.50
People of ethnic background (minority groups)	13.33% 2	0.00% 0	0.00% 0	6.67% 1	0.00% 0	33.33% 5	33.33% 5	13.33% 2	0.00% 0	0.00% 0	0.00% 0	15	4.80	5.54
People of low income	13.33% 2	6.67% 1	0.00% 0	6.67% 1	0.00% 0	40.00% 6	20.00% 3	13.33% 2	0.00% 0	0.00% 0	0.00% 0	15	4.40	5.08
People of high income	6.25% 1	0.00% 0	0.00% 0	6.25% 1	0.00% 0	12.50% 2	18.75% 3	18.75% 3	18.75% 3	18.75% 3	0.00% 0	16	6.44	6.87
Countries of the North	6.25% 1	0.00% 0	0.00% 0	0.00% 0	0.00% 0	12.50% 2	37.50% 6	25.00% 4	6.25% 1	6.25% 1	6.25% 1	16	6.31	6.73
Countries of the South	6.25% 1	0.00% 0	6.25% 1	6.25% 1	0.00% 0	37.50% 6	31.25% 5	12.50% 2	0.00% 0	0.00% 0	0.00% 0	16	4.94	5.27
Animals	26.67% 4	0.00% 0	0.00% 0	6.67% 1	6.67% 1	13.33% 2	20.00% 3	20.00% 3	0.00% 0	0.00% 0	6.67% 1	15	4.40	6.00
Nature	21.43% 3	0.00% 0	0.00% 0	7.14% 1	14.29% 2	35.71% 5	14.29% 2	7.14% 1	0.00% 0	0.00% 0	0.00% 0	14	3.93	5.00

Table 4 Q18 BCI No

Group	0	1	2	3	4	5	6	7	8	9	10	Total	Weighted average with 0	Weighted average without 0
Myself	41.67% 5	0.00% 0	0.00% 0	0.00% 0	0.00% 0	25.00% 3	8.33% 1	8.33% 1	0.00% 0	0.00% 0	16.67% 2	12	4.00	6.86
Disabled people	0.00% 0	0.00% 0	0.00% 0	0.00% 0	0.00% 0	25.00% 3	0.00% 0	8.33% 1	16.67% 2	8.33% 1	41.67% 5	12	8.08	8.08
Women	36.36% 4	0.00% 0	0.00% 0	0.00% 0	0.00% 0	27.27% 3	9.09% 1	9.09% 1	0.00% 0	9.09% 1	9.09% 1	11	4.27	6.71
The elderly	0.00% 0	0.00% 0	0.00% 0	0.00% 0	0.00% 0	33.33% 4	0.00% 0	8.33% 1	16.67% 2	25.00% 3	16.67% 2	12	7.50	7.50
Youth	25.00% 3	0.00% 0	0.00% 0	0.00% 0	0.00% 0	25.00% 3	25.00% 3	8.33% 1	0.00% 0	8.33% 1	8.33% 1	12	4.92	6.56
Men	36.36% 4	0.00% 0	0.00% 0	0.00% 0	0.00% 0	36.36% 4	9.09% 1	0.00% 0	0.00% 0	9.09% 1	9.09% 1	11	4.09	6.43
Non-binary	33.33% 4	0.00% 0	0.00% 0	0.00% 0	0.00% 0	25.00% 3	16.67% 2	8.33% 1	0.00% 0	0.00% 0	16.67% 2	12	4.50	6.75
People of ethnic background (minority groups)	33.33% 4	0.00% 0	0.00% 0	0.00% 0	0.00% 0	33.33% 4	8.33% 1	8.33% 1	8.33% 1	0.00% 0	8.33% 1	12	4.25	6.38
People of low income	33.33% 4	8.33% 1	8.33% 1	0.00% 0	0.00% 0	33.33% 4	16.67% 2	0.00% 0	0.00% 0	0.00% 0	0.00% 0	12	2.92	4.38
People of high income	16.67% 2	0.00% 0	0.00% 0	0.00% 0	8.33% 1	25.00% 3	16.67% 2	16.67% 2	0.00% 0	8.33% 1	8.33% 1	12	5.33	6.40
Countries of the North	33.33% 4	0.00% 0	0.00% 0	0.00% 0	0.00% 0	25.00% 3	16.67% 2	8.33% 1	8.33% 1	0.00% 0	8.33% 1	12	4.33	6.50
Countries of the South	33.33% 4	0.00% 0	8.33% 1	0.00% 0	0.00% 0	25.00% 3	16.67% 2	8.33% 1	0.00% 0	0.00% 0	8.33% 1	12	3.83	5.75
Animals	33.33% 4	0.00% 0	0.00% 0	0.00% 0	8.33% 1	16.67% 2	16.67% 2	8.33% 1	16.67% 2	0.00% 0	0.00% 0	12	4.08	6.13
Nature	41.67% 5	0.00% 0	16.67% 2	0.00% 0	8.33% 1	25.00% 3	0.00% 0	0.00% 0	8.33% 1	0.00% 0	0.00% 0	12	2.58	4.43

Table 5 Q19 How do you see brain computer interface technology impacting ... ability to have a good life in the future? 0 = no impact, 1 purely negative impact, 2–4 more negative impact, 5 equal negative and positive impact, 6–9 more positive impact, and 10 = only positive impact. **BCI Yes/somewhat**

Group	0	1	2	3	4	5	6	7	8	9	10	Total	Weighted average with 0	Weighted average without 0
Myself	18.75% 3	0.00% 0	0.00% 0	0.00% 0	0.00% 0	25.00% 4	25.00% 4	12.50% 2	18.75% 3	0.00% 0	0.00% 0	16	5.14	6.31
Disabled people	0.00% 0	0.00% 0	0.00% 0	0.00% 0	0.00% 0	6.25% 1	6.25% 1	6.25% 1	6.25% 1	18.75% 3	56.25% 9	16	8.94	8.94
Women	6.25% 1	0.00% 0	6.25% 1	0.00% 0	6.25% 1	31.25% 5	25.00% 4	6.25% 1	12.50% 2	6.25% 1	0.00% 0	16	5.44	5.80
The elderly	0.00% 0	0.00% 0	0.00% 0	0.00% 0	0.00% 0	12.50% 2	0.00% 0	25.00% 4	25.00% 4	31.25% 5	6.25% 1	16	7.81	7.81
Youth	0.00% 0	0.00% 0	6.25% 1	0.00% 0	0.00% 0	25.00% 4	31.25% 5	18.75% 3	12.50% 2	6.25% 1	0.00% 0	16	6.13	6.13
Men	6.25% 1	0.00% 0	6.25% 1	0.00% 0	0.00% 0	31.25% 5	31.25% 5	6.25% 1	12.50% 2	6.25% 1	0.00% 0	16	5.56	5.93
Non-binary	6.25% 1	0.00% 0	0.00% 0	0.00% 0	6.25% 1	31.25% 5	31.25% 5	6.25% 1	12.50% 2	6.25% 1	0.00% 0	16	5.69	6.07
People of ethnic background (minority groups)	6.25% 1	0.00% 0	0.00% 0	0.00% 0	0.00% 0	31.25% 5	31.25% 5	12.50% 2	12.50% 2	6.25% 1	0.00% 0	16	5.88	6.27
People of low income	6.25% 1	0.00% 0	6.25% 1	6.25% 1	0.00% 0	31.25% 5	25.00% 4	18.75% 3	6.25% 1	0.00% 0	0.00% 0	16	5.19	5.53
People of high income	0.00% 0	0.00% 0	0.00% 0	0.00% 0	0.00% 0	31.25% 5	6.25% 1	18.75% 3	12.50% 2	18.75% 3	12.50% 2	16	7.19	7.19
Countries of the North	0.00% 0	0.00% 0	0.00% 0	0.00% 0	6.25% 1	43.75% 7	12.50% 2	12.50% 2	12.50% 2	12.50% 2	0.00% 0	16	6.19	6.19
Countries of the South	0.00% 0	0.00% 0	0.00% 0	0.00% 0	12.50% 2	37.50% 6	25.00% 4	12.50% 2	12.50% 2	0.00% 0	0.00% 0	16	5.75	5.75
Animals	25.00% 4	0.00% 0	0.00% 0	0.00% 0	6.25% 1	25.00% 4	12.50% 2	18.75% 3	6.25% 1	0.00% 0	6.25% 1	16	4.69	6.25
Nature	20.00% 3	0.00% 0	0.00% 0	6.67% 1	13.33% 2	33.33% 5	0.00% 0	20.00% 3	6.67% 1	0.00% 0	0.00% 0	15	4.33	5.42

Table 6 Q19 **BCI No**

Group	0	1	2	3	4	5	6	7	8	9	10	Total	Weighted average with 0	Weighted average without 0
Myself	27.27% 3	0.00% 0	9.09% 1	0.00% 0	0.00% 0	9.09% 1	9.09% 1	9.09% 1	9.09% 1	0.00% 0	27.27% 3	11	5.27	7.25
Disabled people	0.00% 0	0.00% 0	0.00% 0	0.00% 0	0.00% 0	18.18% 2	0.00% 0	0.00% 0	18.18% 2	9.09% 1	54.55% 6	11	8.64	8.64
Women	18.18% 2	0.00% 0	9.09% 1	0.00% 0	0.00% 0	9.09% 1	9.09% 1	9.09% 1	18.18% 2	9.09% 1	18.18% 2	11	5.91	7.22
The elderly	0.00% 0	0.00% 0	0.00% 0	0.00% 0	0.00% 0	18.18% 2	9.09% 1	0.00% 0	36.36% 4	0.00% 0	36.36% 4	11	8.00	8.00
Youth	9.09% 1	0.00% 0	9.09% 1	0.00% 0	0.00% 0	9.09% 1	18.18% 2	9.09% 1	9.09% 1	9.09% 1	27.27% 3	11	6.64	7.30
Men	18.18% 2	0.00% 0	9.09% 1	0.00% 0	0.00% 0	9.09% 1	9.09% 1	9.09% 1	27.27% 3	0.00% 0	18.18% 2	11	5.82	7.11
Non-binary	18.18% 2	0.00% 0	9.09% 1	0.00% 0	0.00% 0	9.09% 1	9.09% 1	18.18% 2	18.18% 2	0.00% 0	18.18% 2	11	5.73	7.00
People of ethnic background (minority groups)	18.18% 2	0.00% 0	9.09% 1	0.00% 0	0.00% 0	9.09% 1	9.09% 1	18.18% 2	27.27% 3	0.00% 0	9.09% 1	11	5.55	6.78
People of low income	9.09% 1	0.00% 0	9.09% 1	0.00% 0	0.00% 0	9.09% 1	27.27% 3	18.18% 2	18.18% 2	0.00% 0	9.09% 1	11	5.91	6.50
People of high income	9.09% 1	0.00% 0	9.09% 1	0.00% 0	0.00% 0	9.09% 1	9.09% 1	9.09% 1	18.18% 2	18.18% 2	18.18% 2	11	6.73	7.40
Countries of the North	18.18% 2	0.00% 0	9.09% 1	0.00% 0	0.00% 0	9.09% 1	9.09% 1	9.09% 1	27.27% 3	9.09% 1	9.09% 1	11	5.73	7.00
Countries of the South	18.18% 2	0.00% 0	9.09% 1	0.00% 0	0.00% 0	18.18% 2	9.09% 1	18.18% 2	18.18% 2	0.00% 0	9.09% 1	11	5.27	6.44
Animals	27.27% 3	0.00% 0	0.00% 0	0.00% 0	9.09% 1	9.09% 1	18.18% 2	0.00% 0	9.09% 1	18.18% 2	9.09% 1	11	5.18	7.13
Nature	27.27% 3	9.09% 1	27.27% 3	0.00% 0	0.00% 0	18.18% 2	9.09% 1	0.00% 0	0.00% 0	0.00% 0	9.09% 1	11	3.00	4.13

Table 7 Q20 There are various measures that try to ascertain people's situation in life, how well they are. How will the below indicators be impacted by brain computer interfaces? These are the indicators of the community based rehabilitation matrix

	BCI No				BCI Yes/somewhat			
	BCI: impacted only positive	BCI: impacted only negative	BCI: impacted positive and negative	BCI: not impacted	BCI: impacted only positive	BCI: impacted only negative	BCI: impacted positive and negative	BCI: not impacted
Health	66.67%	0.00%	25.00%	0.00%	68.75%	0.00%	31.25%	0.00%
	8	0	3	0	11	0	5	0
Healthcare	58.33%	8.33%	25.00%	0.00%	75.00%	0.00%	25.00%	0.00%
	7	1	3	0	12	0	4	0
Assistive	66.67%	0.00%	16.67%	0.00%	87.50%	0.00%	6.25%	0.00%
Technology/assistive	8	0	2	0	14	0	1	0
Technologies/ assistive								
Device								
Health promotion	58.33%	0.00%	33.33%	0.00%	50.00%	0.00%	31.25%	12.50%
	7	0	4	0	8	0	5	2
Health prevention	50.00%	0.00%	16.67%	16.67%	43.75%	0.00%	25.00%	25.00%
	6	0	2	2	7	0	4	4
Rehabilitation	83.33%	0.00%	16.67%	0.00%	75.00%	0.00%	18.75%	0.00%
	10	0	2	0	12	0	3	0
Education	25.00%	8.33%	41.67%	25.00%	50.00%	0.00%	43.75%	0.00%
	3	1	5	3	8	0	7	0
Childhood education	33.33%	0.00%	50.00%	8.33%	37.50%	12.50%	37.50%	12.50%
	4	0	6	1	6	2	6	2
Primary education	33.33%	0.00%	50.00%	8.33%	37.50%	12.50%	37.50%	12.50%
	4	0	6	1	6	2	6	2
Secondary education	33.33%	0.00%	50.00%	16.67%	43.75%	6.25%	43.75%	0.00%
	4	0	6	2	7	1	7	0
Non-formal education	25.00%	0.00%	50.00%	16.67%	37.50%	6.25%	31.25%	18.75%
	3	0	6	2	6	1	5	3
Life-long learning	33.33%	0.00%	50.00%	8.33%	43.75%	0.00%	31.25%	6.25%

	4	0	6	1	7	0	5	1
Livelihood	66.67%	0.00%	33.33%	0.00%	37.50%	0.00%	56.25%	0.00%
	8	0	4	0	6	0	9	0
Skills development	58.33%	0.00%	41.67%	0.00%	50.00%	0.00%	50.00%	0.00%
	7	0	5	0	8	0	8	0
Self-employment	50.00%	25.00%	16.67%	8.33%	37.50%	6.25%	31.25%	25.00%
	6	3	2	1	6	1	5	4
Financial services	41.67%	25.00%	16.67%	16.67%	13.33%	13.33%	46.67%	20.00%
	5	3	2	2	2	2	7	3
Wage employment	25.00%	16.67%	41.67%	25.00%	18.75%	6.25%	43.75%	12.50%
	3	2	5	3	3	1	7	2
Social protection	18.18%	9.09%	45.45%	18.18%	12.50%	6.25%	43.75%	18.75%
	2	1	5	2	2	1	7	3
Social situation	25.00%	0.00%	41.67%	25.00%	25.00%	12.50%	43.75%	6.25%
	3	0	5	3	4	2	7	1
Social relationship	25.00%	0.00%	41.67%	25.00%	18.75%	6.25%	56.25%	6.25%
	3	0	5	3	3	1	9	1
Family	41.67%	0.00%	25.00%	25.00%	18.75%	0.00%	62.50%	12.50%
	5	0	3	3	3	0	10	2
Personal assistance	58.33%	0.00%	25.00%	8.33%	46.67%	13.33%	40.00%	0.00%
	7	0	3	1	7	2	6	0
Culture	25.00%	8.33%	50.00%	16.67%	6.25%	18.75%	68.75%	6.25%
	3	1	6	2	1	3	11	1
Arts	33.33%	8.33%	25.00%	33.33%	13.33%	20.00%	60.00%	6.67%
	4	1	3	4	2	3	9	1
Recreation	33.33%	8.33%	25.00%	33.33%	37.50%	12.50%	37.50%	12.50%
	4	1	3	4	6	2	6	2

(continued)

Table 7 (continued)

	BCI No				BCI Yes/somewhat			
	BCI: impacted only positive	BCI: impacted only negative	BCI: impacted positive and negative	BCI: not impacted	BCI: impacted only positive	BCI: impacted only negative	BCI: impacted positive and negative	BCI: not impacted
Leisure	33.33%	8.33%	33.33%	16.67%	31.25%	12.50%	37.50%	18.75%
	4	1	4	2	5	2	6	3
Sport	0.00%	8.33%	16.67%	0.00%	31.25%	12.50%	37.50%	18.75%
	0	1	2	0	5	2	6	3
Access to justice	0.00%	41.67%	25.00%	0.00%	20.00%	6.67%	53.33%	20.00%
	0	5	3	0	3	1	8	3
Empowerment	0.00%	33.33%	16.67%	0.00%	33.33%	6.67%	46.67%	13.33%
	0	4	2	0	5	1	7	2
Communication	0.00%	33.33%	0.00%	0.00%	46.67%	6.67%	33.33%	6.67%
	0	4	0	0	7	1	5	1
Social mobilization	0.00%	16.67%	16.67%	0.00%	18.75%	6.25%	50.00%	25.00%
	0	2	2	0	3	1	8	4
Political participation	8.33%	33.33%	25.00%	8.33%	18.75%	6.25%	56.25%	18.75%
	1	4	3	1	3	1	9	3
Self-help groups	0.00%	33.33%	8.33%	0.00%	25.00%	6.25%	37.50%	25.00%
	0	4	1	0	4	1	6	4
Disabled people's organizations	0.00%	16.67%	8.33%	0.00%	50.00%	0.00%	37.50%	6.25%
	0	2	1	0	8	0	6	1

Table 8 Q21 There are various measures that try to ascertain people's situation in life, how well they are. How will the below indicators be impacted by brain computer interfaces? These are the indicators of the Canadian Index of Wellbeing

	BCI No				BCI Yes somewhat			
	BCI: impacted only positive	BCI: impacted only negative	BCI: impacted positive and negative	BCI: not impacted	BCI: impacted only positive	BCI: impacted only negative	BCI: impacted positive and negative	BCI: not impacted
Social	20.00%	0.00%	40.00%	30.00%	33.33%	6.67%	46.67%	13.33%
Relationships	2	0	4	3	5	1	7	2
Social	20.00%	0.00%	50.00%	30.00%	33.33%	6.67%	53.33%	6.67%
Engagement	2	0	5	3	5	1	8	1
Social support	30.00%	0.00%	40.00%	20.00%	33.33%	0.00%	60.00%	6.67%
	3	0	4	2	5	0	9	1
Community	20.00%	0.00%	40.00%	30.00%	26.67%	6.67%	46.67%	20.00%
Safety	2	0	4	3	4	1	7	3
Social norms	10.00%	0.00%	70.00%	10.00%	26.67%	0.00%	66.67%	6.67%
	1	0	7	1	4	0	10	1
Attitudes	30.00%	0.00%	40.00%	20.00%	20.00%	6.67%	60.00%	13.33%
Toward others	3	0	4	2	3	1	9	2
Democratic	30.00%	0.00%	20.00%	40.00%	26.67%	13.33%	33.33%	26.67%
Engagement	3	0	2	4	4	2	5	4
Participation	30.00%	0.00%	30.00%	30.00%	26.67%	0.00%	46.67%	20.00%
	3	0	3	3	4	0	7	3
Communication	50.00%	0.00%	20.00%	20.00%	53.33%	0.00%	40.00%	6.67%
	5	0	2	2	8	0	6	1
Leadership	30.00%	0.00%	20.00%	40.00%	42.86%	14.29%	35.71%	7.14%
	3	0	2	4	6	2	5	1
Education	30.00%	0.00%	40.00%	20.00%	53.33%	6.67%	33.33%	0.00%
	3	0	4	2	8	1	5	0
Competencies	50.00%	0.00%	30.00%	10.00%	60.00%	13.33%	20.00%	0.00%
	5	0	3	1	9	2	3	0
Knowledge	50.00%	0.00%	20.00%	20.00%	53.33%	6.67%	33.33%	6.67%
	5	0	2	2	8	1	5	1

(continued)

Table 8 (continued)

	BCI No				BCI Yes somewhat			
	BCI: impacted only positive	BCI: impacted only negative	BCI: impacted positive and negative	BCI: not impacted	BCI: impacted only positive	BCI: impacted only negative	BCI: impacted positive and negative	BCI: not impacted
Skill	50.00%	0.00%	20.00%	20.00%	57.14%	0.00%	42.86%	0.00%
	5	0	2	2	8	0	6	0
Environment	20.00%	0.00%	30.00%	40.00%	0.00%	20.00%	40.00%	33.33%
	2	0	3	4	0	3	6	5
Healthy	40.00%	0.00%	30.00%	20.00%	40.00%	6.67%	40.00%	6.67%
Population	4	0	3	2	6	1	6	1
Personal	40.00%	0.00%	40.00%	10.00%	33.33%	0.00%	53.33%	0.00%
Wellbeing	4	0	4	1	5	0	8	0
Physical health	50.00%	0.00%	30.00%	10.00%	40.00%	0.00%	53.33%	0.00%
	5	0	3	1	6	0	8	0
Life	40.00%	0.00%	30.00%	20.00%	46.67%	0.00%	46.67%	6.67%
Expectancy	4	0	3	2	7	0	7	1
Mental health	30.00%	0.00%	50.00%	10.00%	20.00%	0.00%	80.00%	0.00%
	3	0	5	1	3	0	12	0
Functional	50.00%	0.00%	30.00%	10.00%	35.71%	0.00%	57.14%	0.00%
Health	5	0	3	1	5	0	8	0
Lifestyle	40.00%	0.00%	30.00%	20.00%	26.67%	6.67%	60.00%	6.67%
	4	0	3	2	4	1	9	1
Public health	40.00%	0.00%	40.00%	10.00%	42.86%	0.00%	50.00%	7.14%
	4	0	4	1	6	0	7	1
Healthcare	50.00%	0.00%	30.00%	10.00%	40.00%	6.67%	53.33%	0.00%
	5	0	3	1	6	1	8	0
Culture	20.00%	0.00%	40.00%	30.00%	7.14%	7.14%	71.43%	14.29%
	2	0	4	3	1	1	10	2
Leisure	40.00%	0.00%	20.00%	30.00%	40.00%	0.00%	40.00%	13.33%
	4	0	2	3	6	0	6	2

Living standard	50.00%	0.00%	20.00%	20.00%	40.00%	0.00%	46.67%	6.67%
	5	0	2	2	6	0	7	1
Income	20.00%	0.00%	30.00%	40.00%	40.00%	6.67%	33.33%	13.33%
	2	0	3	4	6	1	5	2
Economic security	20.00%	0.00%	30.00%	40.00%	33.33%	0.00%	46.67%	13.33%
	2	0	3	4	5	0	7	2
Time	30.00%	0.00%	20.00%	40.00%	21.43%	7.14%	28.57%	35.71%
	3	0	2	4	3	1	4	5

Table 9 Q22 There are various measures that try to ascertain people's situation in life, how well they are. How will the below indicators be impacted by brain computer interfaces? These are indicators of the social determinants of health

	BCI No				BCI Yes somewhat			
	BCI: impacted only positive	BCI: impacted only negative	BCI: impacted positive and negative	BCI: not impacted	BCI: impacted only positive	BCI: impacted only negative	BCI: impacted positive and negative	BCI: not impacted
Income	30.00%	20.00%	20.00%	30.00%	43.75%	0.00%	37.50%	18.75%
	3	2	2	3	7	0	6	3
Education	20.00%	10.00%	40.00%	20.00%	56.25%	0.00%	43.75%	0.00%
	2	1	4	2	9	0	7	0
Unemployment	30.00%	10.00%	10.00%	40.00%	31.25%	18.75%	37.50%	6.25%
	3	1	1	4	5	3	6	1
Job security	20.00%	10.00%	30.00%	30.00%	31.25%	6.25%	62.50%	0.00%
	2	1	3	3	5	1	10	0
Employment	30.00%	10.00%	30.00%	20.00%	31.25%	6.25%	50.00%	12.50%
	3	1	3	2	5	1	8	2
Early Childhood Development	10.00%	10.00%	50.00%	20.00%	31.25%	12.50%	50.00%	6.25%
	1	1	5	2	5	2	8	1
Food Insecurity	20.00%	10.00%	10.00%	50.00%	25.00%	0.00%	37.50%	31.25%
	2	1	1	5	4	0	6	5
Housing	20.00%	10.00%	10.00%	50.00%	13.33%	6.67%	40.00%	40.00%
	2	1	1	5	2	1	6	6
Social Exclusion	10.00%	20.00%	20.00%	40.00%	37.50%	6.25%	56.25%	0.00%
	1	2	2	4	6	1	9	0
Social safety Network	10.00%	10.00%	30.00%	40.00%	26.67%	0.00%	60.00%	13.33%
	1	1	3	4	4	0	9	2
Health Services	40.00%	10.00%	30.00%	10.00%	62.50%	0.00%	37.50%	0.00%
	4	1	3	1	10	0	6	0
"Aboriginal" OR "first Nations" OR "Metis" OR "Indigenous People" OR "Inuit"	10.00%	10.00%	30.00%	40.00%	12.50%	12.50%	56.25%	18.75%
	1	1	3	4	2	2	9	3
Gender	10.00%	10.00%	30.00%	40.00%	12.50%	0.00%	37.50%	50.00%
	1	1	3	4	2	0	6	8
Disabled People	40.00%	10.00%	30.00%	10.00%	50.00%	0.00%	50.00%	0.00%
	4	1	3	1	8	0	8	0
Ethnic people	30.00%	10.00%	30.00%	20.00%	12.50%	0.00%	56.25%	31.25%
	3	1	3	2	2	0	9	5
Immigration	30.00%	10.00%	20.00%	30.00%	12.50%	0.00%	37.50%	50.00%
	3	1	2	3	2	0	6	8
Globalization	10.00%	10.00%	40.00%	30.00%	37.50%	0.00%	31.25%	31.25%
	1	1	4	3	6	0	5	5
Coping	20.00%	10.00%	20.00%	40.00%	25.00%	0.00%	56.25%	18.75%
	2	1	2	4	4	0	9	3
Discrimination	0.00%	10.00%	50.00%	30.00%	25.00%	18.75%	43.75%	18.75%
	0	1	5	3	4	3	7	3

Table 9 (continued)

	BCI No				BCI Yes somewhat			
	BCI: impacted only positive	BCI: impacted only negative	BCI: impacted positive and negative	BCI: not impacted	BCI: impacted only positive	BCI: impacted only negative	BCI: impacted positive and negative	BCI: not impacted
Genetic	20.00% 2	10.00% 1	30.00% 3	30.00% 3	25.00% 4	6.25% 1	50.00% 8	12.50% 2
Stress	30.00% 3	10.00% 1	20.00% 2	30.00% 3	12.50% 2	6.25% 1	62.50% 10	12.50% 2
Transportation	30.00% 3	10.00% 1	10.00% 1	40.00% 4	25.00% 4	0.00% 0	43.75% 7	31.25% 5
Vocational Training	30.00% 3	10.00% 1	20.00% 2	30.00% 3	12.50% 2	0.00% 0	37.50% 6	50.00% 8
Social Integration	18.18% 2	9.09% 1	27.27% 3	36.36% 4	31.25% 5	0.00% 0	43.75% 7	18.75% 3
Advocacy	18.18% 2	9.09% 1	27.27% 3	36.36% 4	25.00% 4	0.00% 0	43.75% 7	25.00% 4
Literacy	36.36% 4	9.09% 1	18.18% 2	27.27% 3	37.50% 6	0.00% 0	31.25% 5	25.00% 4
Walkability	54.55% 6	9.09% 1	27.27% 3	9.09% 1	62.50% 10	0.00% 0	31.25% 5	6.25% 1
Physical environment	18.18% 2	9.09% 1	18.18% 2	45.45% 5	18.75% 3	6.25% 1	31.25% 5	43.75% 7
Social engagement	27.27% 3	9.09% 1	27.27% 3	27.27% 3	26.67% 4	0.00% 0	53.33% 8	20.00% 3
Social status	0.00% 0	9.09% 1	36.36% 4	45.45% 5	12.50% 2	6.25% 1	62.50% 10	18.75% 3

Table 10 Q23 There are various measures that try to ascertain people's situation in life, how well they are. How will the below indicators be impacted by brain computer interfaces? These are the indicators of the Better Life Index

	BCI No				BCI Yes somewhat			
	BCI: impacted only positive	BCI: impacted only negative	BCI: impacted positive and negative	BCI: not impacted	BCI: impacted only positive	BCI: impacted only negative	BCI: impacted positive and negative	BCI: not impacted
Housing	9.09%	9.09%	27.27%	45.45%	12.50%	0.00%	43.75%	43.75%
	1	1	3	5	2	0	7	7
Income	18.18%	0.00%	36.36%	36.36%	31.25%	0.00%	50.00%	18.75%
	2	0	4	4	5	0	8	3
Jobs	18.18%	0.00%	36.36%	27.27%	31.25%	0.00%	56.25%	12.50%
	2	0	4	3	5	0	9	2
Community	9.09%	0.00%	45.45%	36.36%	31.25%	6.25%	62.50%	0.00%
	1	0	5	4	5	1	10	0
Education	18.18%	0.00%	54.55%	18.18%	43.75%	0.00%	56.25%	0.00%
	2	0	6	2	7	0	9	0
Environment	9.09%	0.00%	54.55%	27.27%	6.25%	6.25%	50.00%	37.50%
	1	0	6	3	1	1	8	6
Physical	9.09%	0.00%	45.45%	36.36%	12.50%	6.25%	43.75%	37.50%
Environment	1	0	5	4	2	1	7	6
Civic	27.27%	0.00%	27.27%	36.36%	12.50%	0.00%	68.75%	18.75%
Engagement	3	0	3	4	2	0	11	3
Health	36.36%	0.00%	45.45%	9.09%	62.50%	0.00%	37.50%	0.00%
	4	0	5	1	10	0	6	0
Life	18.18%	0.00%	63.64%	9.09%	43.75%	0.00%	50.00%	6.25%
Satisfaction	2	0	7	1	7	0	8	1
Safety	27.27%	0.00%	45.45%	18.18%	37.50%	0.00%	56.25%	6.25%
	3	0	5	2	6	0	9	1
Work life	9.09%	0.00%	54.55%	27.27%	6.25%	0.00%	62.50%	31.25%
Balance	1	0	6	3	1	0	10	5

References

1. Roskies A. Neuroethics for the new millenium. Neuron. 2002;35(1):21–3.
2. Farah MJ. Neuroethics: the practical and the philosophical. Trends Cogn Sci. 2005;9(1):34–40.
3. Levy N. Introducing neuroethics. Neuroethics. 2008;1(1):1–8.
4. Wallach W. From robots to techno sapiens: ethics, law and public policy in the development of robotics and neurotechnologies. Law Innov Technol. 2011;3(2):185–207.
5. Fins JJ. Neuroethics, neuroimaging, and disorders of consciousness: promise or peril? Trans Am Clin Climatol Assoc. 2011;122:336–46. PMC3116331
6. Stahnisch FW. History of neuroscience and neuroethics: introduction. In: Handbook of neuroethics. Dordrecht: Springer; 2015. p. 461–6.
7. Wolbring G. Hearing beyond the normal enabled by therapeutic devices: the role of the recipient and the hearing profession. Neuroethics. 2013;6(3):607–16. https://doi.org/10.1007/s12152-011-9120-x.
8. Ball N, Wolbring G. Cognitive enhancement: perceptions among parents of children with disabilities. Neuroethics. 2014;7(3):345–64. https://doi.org/10.1007/s12152-014-9201-8.
9. Wolbring G, Martin A, Tynedal J, Ball N, Yumakulov S. Exploring discourse surrounding therapeutic enhancement of veterans and soldiers with injuries. Work. 2015;50(1):149–60. https://doi.org/10.3233/WOR-141936.

10. Yuste R, Goering S, Bi G, Carmena JM, Carter A, Fins JJ, et al. Four ethical priorities for neurotechnologies and AI. Nat News. 2017;551(7679):159.
11. Aicardi C, Fothergill BT, Rainey S, Stahl BC, Harris E. Accompanying technology development in the human brain project: from foresight to ethics management. Futures. 2018;102:114–24.
12. Aicardi C, Reinsborough M, Rose N. The integrated ethics and society programme of the human brain project: reflecting on an ongoing experience. J Respons Innovat. 2018;5(1):13–37.
13. Ienca M. Democratizing cognitive technology: a proactive approach. Ethics Info Tech. 2019;21(4):267–80. https://doi.org/10.1007/s10676-018-9453-9.
14. Burwell S, Sample M, Racine E. Ethical aspects of brain computer interfaces: a scoping review. BMC Med Ethics. 2017;18:60. https://doi.org/10.1186/s12910-017-0220-y.
15. Nijboer F, Clausen J, Allison B, Haselager P. The Asilomar survey: stakeholders' opinions on ethical issues related to brain-computer interfacing. Neuroethics. 2013;6(3):541–78. https://doi.org/10.1007/s12152-011-9132-6.
16. Clausen J. Conceptual and ethical issues with brain–hardware interfaces. Curr Opin Psychiatry. 2011;24(6):495–501.
17. Pham M, Goering S, Sample M, Huggins JE, Klein E. Asilomar survey: researcher perspectives on ethical principles and guidelines for BCI research. Brain Comput Interfac. 2018;5(4):97–111.
18. Goering S, Klein E. Neurotechnologies and justice by, with, and for disabled people. In: Cureton A, Wasserman DT, editors. The Oxford handbook of philosophy and disability. Oxford: Oxford Press; 2019.
19. Goering S, Klein E. Fostering neuroethics integration with neuroscience in the BRAIN initiative: comments on the NIH neuroethics roadmap. AJOB Neurosci. 2020;11(3):184–8.
20. Kögel J, Schmid JR, Jox RJ, Friedrich O. Using brain-computer interfaces: a scoping review of studies employing social research methods. BMC Med Ethics. 2019;20:18. https://doi.org/10.1186/s12910-019-0354-1.
21. Sample M, Aunos M, Blain-Moraes S, Bublitz C, Chandler JA, Falk TH, et al. Brain-computer interfaces and personhood: interdisciplinary deliberations on neural technology. J Neural Eng. 2019;16(6):063001. https://doi.org/10.1088/1741-2552/ab39cd.
22. Kögel J. Performing a disembodied mind: neurotechnology between empowerment and normalization. International Conference on Human-Computer Interaction: Springer; 2021. p. 239–251.
23. Kögel J. Brain-computer Interface use as materialized crisis management. Clinical neurotechnology meets artificial intelligence: philosophical, ethical, legal and social implications. 2021:101.
24. Soekadar SR, Birbaumer N. Brain–machine interfaces for communication in complete paralysis: ethical implications and challenges. In: Handbook of neuroethics. Oxford: Oxford University Press; 2015. p. 705–24.
25. Jebari K. Brain machine Interface and human enhancement–an ethical review. Neuroethics. 2013;6(3):617–25.
26. Schermer M. The mind and the machine. On the conceptual and moral implications of brain-machine interaction. NanoEthics. 2009;3(3):217–30.
27. Clausen J. Bonding brains to machines: ethical implications of electroceuticals for the human brain. Neuroethics. 2013;6(3):429–34.
28. Tamburrini G. Philosophical Reflections on Brain–Computer Interfaces. Brain-Computer-Interfaces in their ethical, social and cultural contexts. Dordrecht: Springer; 2014. p. 147–62.
29. Wolbring G, Diep L, Yumakulov S, Ball N, Yergens D. Social robots, brain machine interfaces and neuro/cognitive enhancers: three emerging science and technology products through the lens of technology acceptance theories, models and frameworks. Technologies. 2013;1(1):3–25.
30. Wolbring G, Diep L, Yumakulov S, Ball N, Leopatra V, Yergens D. Emerging therapeutic enhancement enabling health technologies and their discourses: what is discussed within the health domain? Healthcare. 2013;1(1):20–52.

31. Sample M, Sattler S, Blain-Moraes S, Rodríguez-Arias D, Racine E. Do publics share experts' concerns about brain–computer interfaces? A trinational survey on the ethics of neural technology. Sci Technol Hum Values. 2020;45(6):1242–70.
32. Hosseini N, Kumar P. Gaps in Neuroethics in relation to brain computer interfaces: systematic literature review. International Conference on Human-Computer Interaction. Springer; 2020. p. 448–474.
33. Grübler G, Al-Khodairy A, Leeb R, Pisotta I, Riccio A, Rohm M, et al. Psychosocial and ethical aspects in non-invasive EEG-based BCI research - A survey among BCI users and BCI professionals. Neuroethics. 2014;7(1):29–41. https://doi.org/10.1007/s12152-013-9179-7.
34. Devlin M. Cultivating better brains: transhumanism and its critics on the ethics of cognitive enhancement via brain-computer interfacing (Thesis format: Monograph). The University ICS of Western Ontario; 2014.
35. Limerick H, Coyle D, Moore JW. The experience of agency in human-computer interactions: a review. Front Hum Neurosci. 2014;8:643.
36. Wolbring G, Diep L. Cognitive/neuroenhancement through an ability studies lens. In: Jotterand F, Dubljevic V, editors. Cognitive enhancement. Oxford: Oxford University Press; 2016. p. 57–75.
37. Trimper JB, Wolpe PR, Rommelfanger KS. When "I" becomes "we": ethical implications of emerging brain-to-brain interfacing technologies. Front Neuroeng. 2014;7:4.
38. Coin A, Mulder M, Dubljević V. Ethical aspects of BCI technology: what is the state of the art? Philosophies. 2020;5(4):1–31. https://doi.org/10.3390/philosophies5040031.
39. Miller A. The intrinsically linked future for human and artificial intelligence interaction. J Big Data. 2019;6(1):1–9.
40. Zhang X, Ma Z, Zheng H, Li T, Chen K, Wang X, et al. The combination of brain-computer interfaces and artificial intelligence: applications and challenges. Ann Transl Med. 2020;8(11).
41. Braveman P, Gruskin S. Defining equity in health. J Epidemiol Community Health. 2003;57(4):254–8. https://doi.org/10.1136/jech.57.4.254.
42. Backholer K, Baum F, Finlay S. Australia in 2030 What is our path to health for all? 2021. https://www.mja.com.au/system/files/2021-05/MJA%20supplement_214_8_3%20May.pdf. Accessed 3 Oct 2022.
43. Crisp R. Well-Being. 2021. https://plato.stanford.edu/entries/well-being/. Accessed 3 Oct 2022.
44. Centers for Disease Control and Prevention: Well-Being Concepts. 2021. https://www.cdc.gov/hrqol/wellbeing.htm. Accessed 3 Oct 2022.
45. World Health Organization: Social determinants of health. 2020. https://www.who.int/social_determinants/en/. Accessed 3 Oct 2022.
46. Raphael D, Bryant T, Mikkonen J, Raphael A. Social Determinants of Health: The Canadian Facts. 2020. https://thecanadianfacts.org/. Accessed 3 Oct 2022.
47. Canadian Index of Wellbeing Organization: What is Wellbeing? 2019. https://uwaterloo.ca/canadian-index-wellbeing/what-wellbeing. Accessed 3 Oct 2022.
48. OECD: OECD Better Life Index. 2020. http://www.oecdbetterlifeindex.org/#/11111111111. Accessed 3 Oct 2022.
49. World Health Organization: About the community-based rehabilitation (CBR) matrix. 2011. http://www.who.int/disabilities/cbr/matrix/en/. Accessed 3 Oct 2022.
50. Wolbring G. Auditing the impact of neuro-advancements on health equity. J Neurol Res. 2022;12(2):54–68. https://doi.org/10.14740/jnr695.
51. National Academies of Sciences Engineering Medicine: Communities in action: Pathways to health equity. 2017. https://www.nap.edu/catalog/24624/communities-in-action-pathways-to-health-equity. Accessed 3 Oct 2022.
52. Zimmerman FJ. A robust health equity metric. Public Health. 2019;175:68–78.
53. Braveman P, Arkin E, Orleans T, Proctor D, Alonzo P. What is health equity? And what difference does a definition make? 2017. https://nccdh.ca/images/uploads/comments/RWJ_Foundation_-_What_Is_Health_Equity.pdf. Accessed 3 Oct 2022.

54. Manuel T. How does one live the good life? Assessing the state of intersectionality in public policy. The Palgrave handbook of intersectionality in public policy. Springer; 2019. p. 31–58.
55. United Nations: Convention on the Rights of Persons with Disabilities (CRPD). 2015. https://www.un.org/development/desa/disabilities/convention-on-the-rights-of-persons-with-disabilities.html. Accessed 3 Oct 2022.
56. United Nations: United Nations 2018 flagship report on disability and development: realization of the Sustainable Development Goals by, for and with persons with disabilities. 2018. https://www.un.org/development/desa/disabilities/publication-disability-sdgs.html#:~:text=%E2%80%9CThe%20UN%20Flagship%20Report%20on,can%20create%20a%20more%20inclusive. Accessed 3 Oct 2022.
57. Kögel J, Wolbring G. What it takes to be a Pioneer: ability expectations from brain-computer interface users. NanoEthics. 2020;14(3):227–39. https://doi.org/10.1007/s11569-020-00378-0.
58. Josa I, Aguado A. Social sciences and humanities in the education of civil engineers: current status and proposal of guidelines. J Clean Prod. 2021;311:127489. https://doi.org/10.1016/j.jclepro.2021.127489.
59. Kelley TR, Knowles JG. A conceptual framework for integrated STEM education. Int J STEM Educ. 2016;3(1):1–11.
60. Ramirez Velazquez M. Not Just Teaching How: Supporting a Culture Shift in STEM Education. 2021. https://scholarship.tricolib.brynmawr.edu/handle/10066/23046. Accessed 3 Oct 2022.
61. Vesnic-Alujevic L, Nascimento S, Polvora A. Societal and ethical impacts of artificial intelligence: critical notes on European policy frameworks. Telecommunications Pol 2020;44(6):Article 101961. doi: https://doi.org/10.1016/j.telpol.2020.101961.
62. Salgado-Criado J, Fernández-Aller C. A wide human-rights approach to artificial intelligence regulation in Europe. IEEE Technol Soc Mag. 2021;40(2):55–65.
63. Rodriguez-Nikl T. Technology, uncertainty, and the good life: a stoic perspective. Engineering and Philosophy Springer; 2021. p. 219–233.
64. Rao PS, Soumya A. A study on music based audio and brain signal processing. In: 2019 4th international conference on computational systems and information Technology for Sustainable Solution. CSITSS: IEEE; 2019. p. 1–6.
65. Hosni SM, Shedeed HA, Mabrouk MS, Tolba MF. EEG-EOG based virtual keyboard: toward hybrid brain computer interface. Neuroinformatics. 2019;17(3):323–41.
66. Frid-Jimenez A, Carson J, Scott A, Khantidhara P, Elza D. Designing participedia: a collaborative research platform. Proceedings of the 16th Participatory Design Conference 2020-Participation(s) Otherwise-Volume 2020. p. 21–5.
67. Lee HR, Cheon E, De Graaf M, Alves-Oliveira P, Zaga C, Young J. Robots for Social Good: Exploring Critical Design for HRI. ACM/IEEE International Conference on Human-Robot Interaction. 2019. p. 681–682.
68. Avouris N, Sintoris C, Katsini C. Studying human-computer interaction for social good: the case of digital government evaluation and re-design project. ACM International Conference Proceeding Series. 2018. p. 230–235.
69. Pal J. CHI4Good or Good4CHI. Conference on Human Factors in Computing Systems—Proceedings; 2017. p. 709–721.
70. Armstrong W, Michael K. The implications of Neuralink and Brain Machine Interface Technologies. 2020 IEEE International Symposium on Technology and Society (ISTAS): IEEE; 2020. p. 201–3.
71. Wong RY, Merrill N, Chuang J. When BCIs have APIs: Design fictions of everyday brain-computer interface adoption. Proceedings of the 2018 Designing Interactive Systems Conference 2018. p. 1359–71.
72. Klein E, Brown T, Sample M, Truitt AR, Goering S. Engineering the brain: ethical issues and the introduction of neural devices. Hastings Cent Rep. 2015;45(6):26–35. https://doi.org/10.1002/hast.515.

73. Nijboer F, Matuz T, Kübier A, Birbaumer N. Ethical, psychological and social implications of brain-computer interface application in paralyzed patients. AAAI Workshop—Technical Report 2006. p. 48–50.
74. Garibay JC. STEM students' social agency and views on working for social change: are STEM disciplines developing socially and civically responsible students? JRScT. 2015;52(5):610–32.
75. Hunt C, Collins B, Wardrop A, Hutchings M, Heaslip V, Pritchard C. First- and second-generation design and engineering students: experience, attainment and factors influencing them to attend university. High Educ Res Dev. 2018;37(1):30–43. https://doi.org/10.1080/07294360.2017.1342607.
76. Korkmaz Ö, Çakir R, Erdoğmuş FU. Secondary school students' basic STEM skill levels according to their self-perceptions: a scale adaptation. Participat Educ Res. 2021;8(1):423–37.
77. Bennett D, Knight E, Bawa S, Dockery AM. Understanding the career decision making of university students enrolled in STEM disciplines. Aust J Career Dev. 2021;30(2):95–105.
78. Burks G, Clancy KB, Hunter CD, Amos JR. Impact of ethics and social awareness curriculum on the engineering identity formation of high school girls. Educ Sci. 2019;9(4):250.
79. Vigdor L. A techno-passion that is not one: rethinking marginality, exclusion, and difference. Int J Gend Sci Technol. 2011;3(1):4–37.
80. Collett C, Dillon S. AI and Gender: Four Proposals for Future Research. 2019. http://lcfi.ac.uk/media/uploads/files/AI_and_Gender___4_Proposals_for_Future_Research_210619_p8qAu8L.pdf. Accessed 3 Oct 2022.
81. Cormier D, Jandrić P, Childs M, Hall R, White D, Phipps L, et al. Ten years of the Postdigital in the 52group: reflections and developments 2009–2019. Postdig Sci Educ. 2019;1:475–506.
82. Garcia P, Scott K. Traversing a political pipeline: An intersectional and social constructionist approach toward technology education for girls of color. 2016. http://stelar.edc.org/sites/stelar.edc.org/files/Garcia%20%26%20Scott%202016.pdf. Accessed 3 Oct 2022.
83. Schmidt WC. World-wide web survey research: benefits, potential problems, and solutions. Behav Res Methods. 1997;29(2):274–9.
84. Yumakulov S, Yergens D, Wolbring G. Imagery of disabled people within social robotics research. In: Ge S, Khatib O, Cabibihan J-J, Simmons R, Williams M-A, editors. Social robotics. Lecture notes in computer science. Berlin, Heidelberg: Springer; 2012. p. 168–77.
85. Lillywhite A, Wolbring G. Coverage of artificial intelligence and machine learning within academic literature, Canadian newspapers, and twitter tweets: the case of disabled people. Societies. 2020;10(1):1–27. https://doi.org/10.3390/soc10010023.
86. Inclezan D, Pradanos LI. A critical view on smart cities and AI. J Artif Intell Res. 2017;60(November):681–6.
87. Einsiedel EF. Framing science and technology in the Canadian press. PUS. 1992;1:89–102.
88. Nierling L, João-Maia M, Hennen L, Bratan T, Kuuk P, Cas J, et al.. Assistive technologies for people with disabilities Part III: Perspectives on assistive technologies. 2018. http://www.europarl.europa.eu/RegData/etudes/IDAN/2018/603218/EPRS_IDA(2018)603218(ANN3)_EN.pdf. Accessed 3 Oct 2022.
89. Diep L, Wolbring G. Who needs to fit in? Who gets to stand out? Communication technologies including brain-machine interfaces revealed from the perspectives of special education school teachers through an ableism lens. Educ Sci. 2013;3(1):30–49.
90. Diep L, Wolbring G. Perceptions of brain-machine interface technology among mothers of disabled children. Disabil Stud Quart. 2015;35(4)
91. Lillywhite A, Wolbring G. Undergraduate disabled students as knowledge producers including researchers: perspectives of disabled students. Educ Sci. 2022;12(2):77. https://doi.org/10.3390/educsci12020077.
92. Wolbring G, Mackay R, Rybchinski T, Noga J. Disabled people and the post-2015 development goal agenda through a disability studies lens. Sustainability. 2013;5(10):4152–82.
93. Participants of the Global Online Discussion on Science Technology and Innovation for SDGs: Global Online Discussion on Science, Technology and Innovation for SDGs. 2016. https://sustainabledevelopment.un.org/forum/?forum=20. Accessed 3 Oct 2022.

94. Participants of the UN Department of Economic and Social Affairs (UNDESA) and UNICEF organized Online Consultation - 8 March - 5 April Disability inclusive Development Agenda Towards 2015 & Beyond: Disability Inclusive Development Agenda Towards 2015 & Beyond. 2013. http://www.un.org/en/development/desa/news/social/disability-inclusive-development.html. Accessed 3 Oct 2022.
95. Diep L. Anticipatory Governance, Anticipatory Advocacy, Knowledge Brokering, and the State of Disabled People's Rights Advocacy in Canada: Perspectives of Two Canadian Cross-Disability Rights Organizations. 2017. https://prism.ucalgary.ca/bitstream/handle/11023/4051/ucalgary_2017_diep_lucy.pdf?sequence=3&isAllowed=y. Accessed 3 Oct 2022.
96. World Bank, World Health Organization: World Report on Disability. 2011. https://www.who.int/disabilities/world_report/2011/report.pdf. Accessed 3 Oct 2022.
97. Aas S, Wasserman D. Brain–computer interfaces and disability: extending embodiment, reducing stigma? J Med Ethics. 2015;42(1):37–42.
98. Harnett A. Escaping the evil avenger and the supercrip: images of disability in popular television. Irish Communic Rev. 2000;8(1):21–9.
99. Kama A. Supercrips versus the pitiful handicapped: reception of disabling images by disabled audience members. Communications. 2004;29(4):447–66.
100. Howe PD. Cyborg and supercrip: the Paralympics Technology and the (Dis) empowerment of disabled athletes. Sociology. 2011;45(5):868–82.
101. Wolbring G, Litke B. Superhip to supercrip: the 'trickle-down' effect of the Paralympics. 2012. https://theconversation.com/superhip-to-supercrip-the-trickle-down-effect-of-the-paralympics-9009. Accessed 3 Oct 2022.
102. Brylla C. Bypassing the supercrip trope in documentary representations of blind visual artists. Disabil Stud Quart. 2018;38(3):11. https://doi.org/10.18061/dsq.v38i3.6485.
103. Lourens H. Supercripping the academy: the difference narrative of a disabled academic. Disabil Soc. 2020.;latest articles:1–16.; https://doi.org/10.1080/09687599.2020.1794798.
104. Fahn CW. Marketing the prosthesis: supercrip and superhuman narratives in contemporary cultural representations. Philosophies. 2020;5(3):11.
105. Wolbring G. Media coverage of Cybathlon 2016: implication for ParaSport. In: Brittain I, Beacom A, editors. The Palgrave handbook of paralympic studies. Basingstoke: Palgrave Macmillan; 2018. p. 439–59.
106. Chappell J. A movement towards independence: One perspective on the disability rights movement. 1991. http://citeseerx.ist.psu.edu/viewdoc/download?doi=10.1.1.833.7414&rep=rep1&type=pdf#page=23. Accessed 3 Oct 2022.
107. Scotch RK. Politics and policy in the history of the disability rights movement. Milbank Q. 1989;67(Suppl 2 Pt 2):380–400.
108. Köbsell S. Towards self-determination and equalization: A short history of the German Disability Rights Movement. 2006. https://dsq-sds.org/article/view/692/869. Accessed 3 Oct 2022.
109. Friedman C, Van Puymbrouck L. Ageism and ableism: unrecognized biases in occupational therapy students. Phys Occupat Therap Geriatr. 2021;39(4):354–69. https://doi.org/10.1080/02703181.2021.1880531.
110. Henley L. The quantification and visualisation of human flourishing. 2015. https://ir.canterbury.ac.nz/handle/10092/10441. Accessed 3 Oct 2022.
111. Wolbring G, Lillywhite A. Equity/equality, diversity, and inclusion (EDI) in universities: the case of disabled people. Societies. 2021;11(2):49. https://doi.org/10.3390/soc11020049.
112. Bureau of Labor Statistics United States Department of Labor (USA): The employment situation — February 2020. 2020. https://www.bls.gov/news.release/pdf/empsit.pdf. Accessed 3 Oct 2022.
113. McKinnon A. From brain-computer interfaces to digital humans: how these technologies are bringing us closer to the metaverse. 2020. Accessed 3 Oct 2022.

114. Plotkin J. The Metaverse, the Mind and our Cybernetic Future. 2020. https://plotkinjack15.medium.com/the-metaverse-the-mind-and-our-cybernetic-future-438bafbd1b71. Accessed 3 Oct 2022.
115. 0 B: The tech that's going to link us to the metaverse. 2021. https://www.next-mind.com/bci-tech-link-metaverse/. Accessed 3 Oct 2022.
116. Spencer MS. The Precursor to the Metaverse Occurs in 2022. 2021. https://www.linkedin.com/pulse/precursor-metaverse-occurs-2022-michael-spencer-/. Accessed 3 Oct 2022.
117. Weldon T: Why Facebook Ditched Its Mind-Reading Neural Interface. 2021. https://www.fool.com/investing/2021/07/31/why-facebook-ditched-its-mind-reading-neural-inter/. Accessed 3 Oct 2022.
118. Facebook. Introducing Horizon Workrooms: Remote Collaboration Reimagined. 2021. https://about.fb.com/news/2021/08/introducing-horizon-workrooms-remote-collaboration-reimagined/. Accessed 3 Oct 2022.
119. Dugdale M. Cognixion One, where brain-computer interfacing meets augmented reality, is coming this year. 2021. https://vrworldtech.com/2021/02/25/cognixion-one-where-brain-computer-interfacing-meets-augmented-reality-is-coming-this-year/. Accessed 3 Oct 2022.
120. Duan H, Li J, Fan S, Lin Z, Wu X, Cai W. Metaverse for social good: a University Campus Prototype. arXiv preprint arXiv:210808985. 2021.
121. Lillywhite A, Wolbring G. Undergraduate disabled students as knowledge producers including researchers: a missed topic in academic literature. Educ Sci. 2019;9(4):259.

Cyborg Virtues: Using Brain Stimulation for Moral Enhancement

James Hughes

1 Introduction

Drawing links between brain structure and moral behavior has been a focus of research since at least the unfortunate case of Phineas Gage [1]. In 1848, Gage was a foreman overseeing the laying of tracks, when a freak accident drove an iron bar through his skull, destroying his left frontal lobe. After his improbable survival, Gage's personality was alleged to have changed for the worse, although he apparently recovered all his social and emotional capacities in later years. The Gage case was subsequently enlisted by both proponents and critics of the theory that specific mental capacities were localized in specific brain regions. The debate over the degree of brain localization of cognitive functions continues to this day, and for good reason since it was implicated in the rise of pseudoscience and unethical neurosurgeries. The pseudoscience of phrenology, for instance, used the idea of brain localization to attempt to identify correlations between moral traits and the shape of the skull. The concept of brain localization led to psychosurgeries and frontal lobotomies as treatments for behavioral disorders [2].

Given the fraught history of pseudoscience and horrifying medical practices associated with brain localization, this chapter's proposal that we may be able to enhance moral behavior by stimulating specific brain regions is rightfully approached with a good deal of caution. Nonetheless, decades of research on brain lesions, brain imaging experiments, and brain stimulation studies have shown that, while any cognitive function enlists multiple brain areas in complex ways, functions are more or less localized to specific areas.

J. Hughes (✉)
University of Massachusetts Boston, Boston, MA, USA

Institute for Ethics and Emerging Technologies, Willington, CT, USA
e-mail: jamesj.hughes@umb.edu

V. Dubljević, A. Coin (eds.), *Policy, Identity, and Neurotechnology*, Advances in Neuroethics, https://doi.org/10.1007/978-3-031-26801-4_9

The debate over whether cognitive functions are localized is confined to neuroscience, but there is an enormous widespread interest in the prospect of brain stimulation and brain–computer interfaces. Since researchers first demonstrated in the 1950s that electrodes in the brain's pleasure center could be used to control behavior, even to the detriment of self-preservation [3, 4], the cyborg with brain–computer interfaces has elicited both horror and enthusiasm. The term cyborg was coined in 1960 in an essay proposing that astronauts' physical and mental state should be monitored remotely so that ground control could administer psychiatric drugs if necessary [5]. Despite a thousand science-fictional images of humans used as batteries by the Matrix or assimilated by the Borg, there are now hundreds of thousands of people with brain stimulation devices implanted to treat depression and epilepsy [6], and many more experimenting with magnetic and electrical stimulation of the brain, as I review in the next section. The entrepreneur Elon Musk has tapped into this popular enthusiasm with his Neuralink project.

This chapter is also situated in the debate over "moral enhancement" in the neuroethics literature [7–9]. As moral neuroscience ballooned in the last two decades, bioethicists have proposed multiple ways that moral sentiments, cognition, and behavior, such as empathy, could be improved through pharmacological and genetic interventions. However, few have addressed using direct brain stimulation for moral enhancement. The techniques are relatively new and taking pills is more practical than non-invasive brain stimulation, not to mention brain surgery. On the other hand, drugs impact the entire brain and body, while targeted brain stimulation can have a much more precise impact. In the next section, I review some of the existing brain stimulation modalities and make the case that emerging neurotechnologies will soon allow for more precise control of the moral brain.

2 Brain Stimulation and Brain–Computer Interfaces

The impact of brain stimulation on cognition and behavior is partly a function of the kind of stimulation being used. Transcranial electrical and magnetic stimulation, from outside the skull, can be focused but not very precisely and not very deeply. Implanted electrodes just impact the neurons they are directly in contact with although those neurons can trigger activity in many parts of the brain. Stimulation methods can both inhibit and induce emotion and cognition and have transient or longer-lasting effects. Brain stimulation can change the expression of neurochemicals and genes, induce the growth of new neurons [10], or destroy tissue permanently.

Noninvasive methods include

- **Transcranial Direct Current Stimulation (tDCS)** involves passing a weak electrical current directly through the scalp into the brain between two or more electrodes. The electrical current can excite or inhibit neuronal signaling in the targeted area depending on whether the current generated has a positive or negative charge. In 2020, the US Food and Drug Administration (FDA) approved the experimental use of tDCS for depression, and research is ongoing on its use for attention deficit disorder, brain injuries and stroke, language and movement disorders, pain, and addiction. A recent meta-analysis of tDCS studies found that it is "definitely effective" in treating depression, and "probably effective" for pain, fibromyalgia, migraine, Parkinson's, stroke rehabilitation, epilepsy, schizophrenia, and alcohol addiction [11]. Patients can apply tDCS at home under a doctor's direction, and tDCS devices are commercially available for the adventurous.

- **Transcranial Magnetic Stimulation (TMS)** involves passing an electrical current through wire coils on the scalp, inducing magnetic fields that excite or inhibit neurons' electrical activity. The FDA approved the use of TMS as a treatment for depression in 2013, for obsessive-compulsive disorder in 2018, and for nicotine addiction in 2020. Unlike tDCS, there is not yet a commercially available portable TMS apparatus although they are being developed [12]. TMS can be focused to a roughly 2 mm diameter, exciting about 130 neurons [13], down to a depth of 3 cm [14]. Side effects are rare and mild, such as a transient headache. Treatments can be one-off or repeated, and research is ongoing on TMS' effects on depression, pain, dystonia, and epilepsy.
- **Transcranial Focused Ultrasound Stimulation (tFUS)** has been developed more recently, with intended applications for depression [15] and the kinds of applications being investigated for TMS and tDCS. Focused sound waves inhibit or excite neurons via a mechanical effect on ion channel gating, rather than electrical modulation of neural signaling, and thereby tFUS produces less heat and potential cell damage than tDCS and TMS. The FDA approved tFUS as a therapy for tremors in 2016. The sound waves can be tuned to target either excitatory or inhibitory neurons [16]. An advantage of tFUS over tDCS and TMS is that it can be focused down to 1 mm [16] and can reach deeper parts of the brain.

Invasive methods of stimulating targeted areas of the brain include

- **Vagus Nerve Stimulation (VNS)** involves placing electrodes in the vagus nerve, which runs from the abdomen, through the neck, and into the brainstem. The FDA approved VNS as a treatment for intractable epilepsy and depression in 2005. Research is ongoing in using vagus nerve stimulators to control obesity, manage pain, and reduce systemic inflammation [17], even for reviving people in vegetative states [18]. While the electrodes are in the neck, VNS impacts many cortical functions [19], for instance, by changing activity in the anterior cingulate cortex [20].
- **Deep Brain Stimulation (DBS)** uses electrodes placed directly in the brain. The FDA approved DBS for tremor and Parkinson's disease in 1997, dystonia in 2003, obsessive-compulsive disorder (OCD) in 2009, epilepsy in 2018, and Parkinson's disease in 2020 [21]. DBS has been explored as a therapy for pain, Tourette's syndrome, depression, and obesity. As with vagus nerve stimulators, the DBS electrode has a wire connected to a pulse generator and battery, usually implanted in the clavicle or abdomen.

The next stage of brain stimulation involves more "closed-loop" integration of stimulation with sensors, more computing power, and more miniaturization.

Closed Loop Integration. An example of closed-loop feedback is using EEG to detect when a driver is falling asleep, triggering a tDCS helmet to wake them up [22]. Sensors implanted alongside or as part of a DBS electrode can detect the characteristic cascades of neural firing that indicate an imminent epileptic seizure [23], tremors [24], or the onset of depression, triggering the DBS electrode to stop them [25]. A system developed at NYU detects pain signals in the anterior cingulate cortex, triggering stimulation of the prefrontal cortex that provides pain relief [26].

Onboard Computing. Researchers have been implanting computer chips with electrodes and communication capabilities into the brain and other neural tissue since the 1990s. Since the interpretation of neural signals requires complex software, the goal is specialized, microscopic chips [27] the most publicized of which is the Neuralink technology being developed by Elon Musk. The current Neuralink unit reports neural firings through 1000 electrodes, each about 4–6 microns wide, to a specialized coin-sized computer chip that is 23 mm wide and 8 mm deep [28], sitting on the surface of the cortex under the skull. Although current models connect these chips to computers outside the skull using wires, they will eventually connect wirelessly with another implanted unit that will then connect wirelessly with devices outside the skull. The devices are powered by a daily, wireless inductive charge from outside the skull. Neuralink's electrode "threads" are not only much smaller than previous electrodes but also more flexible and thus less likely to damage tissue. By comparison, DBS uses 4–8 electrodes, each about 800 times bigger than Neuralink's threads. The Neuralink system is currently being evaluated by the FDA, with expectations for approval in 2022.

Miniaturization. "Neural dust" is an example of the advancing miniaturization of brain–computer interfaces. Proposed in 2011 and now being developed for medical use [29], neural dust combines sensors and communication links into a device the size of a grain of sand, powered by piezoelectric crystals that turn ultrasound energy into electricity. Compared to electrodes, neural dust can be introduced with minor invasive surgery and a much larger brain area. Researchers are already creating neural models from networks of dozens of these units in rodent brains, with plans to scale up to hundreds [30]. The new NeuroSWARM system does not require any power source. It uses devices only 63 nanometers wide—smaller than the average virus—to convert neural electrical signals into near-infrared optical signals that can be detected outside the skull [31, 32].

External and internal brain stimulation is already capable of changing moral sentiment, cognition, and behavior, as I review below. Given the rapid progress in brain stimulation and brain–computer interfaces, however, it seems likely that these systems will eventually also be able to recognize emotions and behaviors, and selectively enhance or suppress them. For instance, just as a seizure has a discernible neural cascade signature, so might the brain have discernible signatures for depression, addictive relapse, or explosive anger, which an implant could then suppress. Before reviewing which parts of the brain might be targeted for moral enhancement, however, we require a short review of the relationship of theories of virtue to moral neuroscience.

3 A Neurologically Grounded Model of the Virtues

In previous work [33–35], I have proposed a model of six moral virtues that bears a rough correspondence to cross-cultural taxonomies of virtue, personality psychology, and the emerging neuroscience of morality. The six virtues I have proposed are self-control, caring, fairness, intelligence, positivity, and transcendence. Each of these virtues correlates with the "five-factor" or OCEAN personality traits, as well as with specific neurotransmitters and neuroanatomical areas. For instance, self-control is correlated with the personality trait of conscientiousness, variations in dopaminergic genes, and the size and activity of the prefrontal cortex. Positive mood is correlated with the personality trait of neuroticism, variation in serotonergic genes, and the function, and connectivity of multiple brain regions [36]. The personality trait of open-mindedness, which is a component of the "intellectual virtues," is correlated with fairness, intelligence, and transcendence [37–40].

Not all six virtues are recognized by every religious or secular virtue model, and the model has only an indirect relationship to virtues like faith, filial piety, or loyalty. I introduce the model here only as a valuable heuristic for the project of moral enhancement. One advantage of such a model is that it suggests the importance of the prudential balance of multiple virtues in a mature moral character. Much of the debate has charged that one or the other form of moral enhancement will be inadequate or have perverse effects without acknowledging that the project of character building proposed by theologians and philosophers has always involved the maturation of multiple virtues that balance one another. Virtue can become a vice if practiced without self-control and intelligence, or "prudence." Unchecked positivity can lead to recklessness, and intelligence can be sterile without empathy and social intelligence.

In "Virtue Theory for Moral Enhancement," for instance, Fabiano [41] agrees with the importance of a balanced approach to moral enhancement, noting that a multi-virtue model also reduces the likelihood of someone becoming so different after enhancement that they have committed identity suicide. "An increase in a currently desirable moral trait would constantly be evaluated against a wider background of other traits and contexts to be considered a true moral enhancement." But then Fabiano proposes the Social Value Orientations (SVO) model as a framework for moral enhancement, a model of four moral types, individualistic, competitive, cooperative, and altruistic. Johnson [42] points out that SVO really only addresses one dimension of virtue, self-centeredness vs. other orientation, and thus fails the test of articulating the critical balance of virtues.

A second virtue of defining moral enhancement through the lens of virtue theory is that it suggests that moral enhancement can be beneficial to both the individual and society. Moral enhancement advocates coming from a more hedonic utilitarian framework, like Perrson and Savulescu [43], identify ethics with altruism and argue that moral enhancement requires self-sacrifice for collective well-being. Virtue models tend to argue for a eudaemonic understanding of happiness over a hedonic one; the rewards of a mature moral character are superior to hedonic gratification. In general, people with more self-control, empathy, or intelligence are both better citizens and have more fulfilled lives.

3.1 Self-Control and Addiction

Since the origins of Greek and Indian philosophy, the capacity for self-control has been considered a fundamental moral virtue. Enacting every other virtue requires self-control, and many of the classical vices—lust, greed, anger, sloth—are an absence of self-control in the face of overwhelming urges. The treatments for some psychiatric disorders, such as attention deficit disorder, have the direct or indirect goal of enhancing self-control. The prefrontal cortex (PFC) is the newest part of the mammal brain. It is the locus for self-awareness, executive functioning, planning and supervising action, such as moral decisions, and the regulation of emotions [44]. The PFC occupies about 10% of the volume of the cerebral cortex and has many substructures with their own localized functions, such as self-monitoring, suppressing impulses, and switching attention from task to task.

When the PFC is impaired or weakened in relation to the other brain parts, it can lead to risk-taking, impulsiveness, criminality, and aggression. It is hard to escape the conclusion that the PFC is the seat of reason, constantly attempting to rein in the animalistic impulses and emotional responses from the other parts of the brain. While this is a useful model for much of moral neuroscience, which often involves a balance between the fast, hot impulses from the limbic or other systems and the slow, cool work of the PFC [45], we need to remind ourselves again that the PFC is the agent of the passions, long term or short term, and not a rational actor struggling to free itself from the cortical mob [46].

Within the PFC, the dorsolateral region (dlPFC) lies behind the right and left sides of the forehead and is the part most often implicated in executive functions, such as planning, abstract reasoning, impulse inhibition, working memory, and the ability to switch tasks. The dlPFC is central to inhibiting selfish impulses to act following prosocial norms [47]. The dlPFC is also a central structure in the "dual process" model of moral neuroscience proposed by Greene et al. [48], which is similar to the slow and fast thinking model advanced by Kahneman [45]. The dual process model focuses on the dlPFC's "slow," deliberative role in making moral judgments, balanced against the "fast" impulses "driven by automatic, intuitive emotional heuristics that are relatively insensitive to the consequences of an action" [49]. Using tDCS to excite the dlPFC helps insulate reasoning from emotion [50, 51]. From this perspective, a large part of character development involves strengthening the dlPFC's deliberative role until the fast impulses from the amygdala, limbic system, and temporoparietal junction (TPJ) are in better accord with deliberative judgments, the turning of conscious moral effort into automatic moral habits [52].

While many reject widespread neuromodulation as a threat to autonomy and self-control, self-applied neuromodulation would enhance our autonomy by allowing us to align our short-term preferences with our long-term ones [53]. For instance, stimulation of the dlPFC contributes to self-control as a treatment for addiction. Bolloni et al. [54] and Antonelli et al. [55] reviewed more than two dozen "encouraging" studies on treating addictions to food, cocaine, nicotine, alcohol, heroin, and amphetamines with TMS excitation of the dlPFC. Lapenta et al. [56] likewise reviewed dozens of studies of tDCS applied to the dlPFC and concluded that it generally helped treat addiction, although the methods employed vary widely [56]. A meta-analysis of a dozen addiction treatments using either tDCS or TMS, applied to the dlPFC, found "a large positive main effect" on reducing addictive cravings [57]. Using TMS, tDCS, tFUS, and DBS to stimulate the anterior cingulate cortex (ACC), a structure below the PFC that evaluates how rewarding something will be, or the nucleus accumbens (nAC), which pumps out dopamine in response to addictions, are also proving to be effective targets for treating addiction [58–61].

While too much of most virtues becomes a vice, there is less risk from too much self-control, and thus there is a little less concern about the side-effects of brain stimulation for self-control. Some researchers attribute problems like obsessive-compulsive disorder or eating disorders to excessive self-control, but the evidence suggests that these problems are just another example of *lack* of self-control, in this case over one's own controlling behaviors. In a 2011 review, Grant and Schwartz argued that there is little evidence that there is any cost to high levels of self-control although excessive delaying of gratification might be a candidate. "Individuals with extreme self-control may never consume and thus never experience pleasure" [62].

3.2 Intelligence, Memory, and Learning

Much of the widespread enthusiasm for brain stimulation and implants like Neuralink stem from hopes that they will allow the enhancement of cognitive speed, learning, and memory in the healthy. Recent meta-analyses and systematic reviews of tDCS' impacts on cognition show that it can enhance processing speed, working memory, and executive functions in patients with psychiatric disorders [63, 64], and improve working and episodic memory, and reaction time and accuracy, in the healthy [65–69]. A 2020 literature review found that tDCS was effective in many studies in improving the cognitive deficits of ADHD, including response inhibition, working memory, attention, and cognitive flexibility [70]. The dlPFC is the preferred target for cognitive enhancement, followed by the TPJ. In one study, the positive effects of stimulation on memory lasted up to a month [65].

Some people experience transient itching, tingling, headaches, or burning sensations when using tDCS [71] but a 2017 review found no *serious* adverse effects have been reported in tDCS experiments [72]. There have been cases in which deep brain stimulation for Parkinson's caused cognitive decline [73], and there is the possibility of adverse consequences from too much attention, memory, or speed from brain stimulation. Stimulants, for instance, have a U-shaped relationship with cognitive performance, with optimal dosing depending on the person; too much stimulant, or any stimulants at all for some, degrades cognitive performance [74, 75]. As targeting becomes more precise, use more continuous, and especially when the stimulation is directly into the brain through electrodes, there will need to be careful calibration to avoid adverse side effects.

3.3 Empathy and Pro-Social Behavior

There are at least two kinds of empathy, emotive and cognitive. Emotive empathy stems from old mammalian brain structures that generate sympathetic emotions in us when we see others stub their toes or get a hug. On the other hand, cognitive empathy is more of a prefrontal phenomenon, requiring a sophisticated "theory of mind" that gives us insight into what others are feeling even if we do not directly witness their emotions. The prefrontal cortex in general, and the dlPFC in particular, is key to pro-social behavior by recognizing and suppressing impulses such as anger and aggression. Damage to the dlPFC is tied to increased aggression, and reduced empathy and pro-social behavior. Stimulating the dlPFC with tDCS or TMS increases trust and cooperation [76, 77] and decreases anger and aggression [78, 79]. Moreover, stimulating the dlPFC with anodal tDCS excitation increases empathy and pro-social behavior, while *inhibiting* the dlPFC with cathodal tDCS *decreases* empathy and pro-social behavior [80].

Among the other parts of the brain important for controlling anger and aggression, or promoting empathy and pro-social behavior, are the ventromedial prefrontal cortex (vmPFC) and the temporo-parietal junction (TPJ). While the dlPFC and TPJ mediate the reasoning component of cognitive empathy, the vmPFC—through its connections to the amygdala among other bits—mediates whether you can understand and predict other people's emotions, "affective theory of mind" [81, 82]. Damage to the vmPFC impairs the ability to recognize emotions in other people's faces, for instance [83], and stimulating the vmPFC calms the amygdala and reduces fear [84]. A meta-analysis of studies applying tDCS to the vmPFC also found an increase in empathy and a decrease in aggression [85, 86]. As for the TPJ, which is key to altruism and theory of mind, a meta-analysis found that anodal, excitatory tDCS applied to the TPJ improves cognitive empathy in healthy adults [87] while inhibiting the TPJ with TMS reduces attention to other people's beliefs and interests in moral decision-making [88].

Again, regulators, clinicians, and users should pay close attention to any side effects of these therapies, and every virtue needs to be balanced and tempered by the rest. As Aristotle warned, too much compassion can become a vice. In *Against Empathy* [89] Bloom argued that emotional empathy, as opposed to cognitive empathy, often leads moral decision-making astray, prioritizing a baby in a well over a hundred thousand victims of a natural disaster. Excessive visceral empathy can also make us trust the untrustworthy, underestimate bad actors, and lead to burn-out and distress. Boosting oxytocin not only increases trust in members of one's in-group, but also aggression against out-groups [90–93].

3.4 Fairness

There are two aspects of fairness's virtue: internal and external, or metacognitive and distributional preferences. The metacognitive part involves self-awareness of one's biases, and habits of mind like "intellectual humility." The distributional part is related to our willingness to sacrifice for more equal outcomes, and our willingness to judge friends and foes by the same moral yardstick and utilitarian or egalitarian preferences involve many parts of the brain, including the prefrontal cortex, parietal and temporal lobes, ACC, and insula [94–97].

However, the favorite target for the neuromodulation studies of fairness has been the dlPFC. Exciting the dlPFC with tDCS enhances metacognition [98], reduces emotional and implicit biases [67], and (usually) reduces the willingness to accept unfair offers in laboratory game experiments [99]. Applying tDCS on the right dlPFC can enhance (with positive anodal excitation) or depress (with negative cathodal inhibition) the willingness to distribute benefits to the least well-off, or in Rawlsian terms, to put oneself "behind the veil of ignorance" [97, 100–103]. Likewise, *stimulating* the dlPFC tips moral decision-making from emotive empathy for individuals to utilitarian reasoning [84, 88, 89] while *inhibiting* the dlPFC with TMS increases sensitivity to harming individuals even if justice or the utilitarian calculus requires it [103].

The insula, coupled to the amygdala, is key to processing disgust, such as the disgust components of racial bias [104], and empathic reactions to others' pain. Applying anodal and cathodal tDCS to the left insula, respectively, enhances and decreases self-reported feelings of disgust, and the absolutist, deontological moral judgments associated with disgust [105, 106].

The cognitive domains also assess self-interest, complicating the model that stimulating prefrontal control is always good for fairness. Sometimes stimulating the PFC enhances preferences for fair outcomes, but sometimes it doesn't [96, 107]. Sometimes, it just makes us more sensitive to being the victim of unfairness without wanting more fairness for others [108]. Many consider pure utilitarian reasoning without any empathy for harm to individuals (pushing the fat man onto the tracks in the trolley problem, for instance) to be psychopathic. Indeed, psychopaths are more consistent utilitarians [109]. One could imagine that a brain stimulation for pure fairness would be more welcome for judges in the courtroom and generals on the battlefield than when among friends and family. In short, simply being more rational and less emotive does not guarantee fairness without a larger ensemble of moral values that steer us from self-interest to prosociality, informed but not governed by empathy and moral reasoning.

3.5 Positive Mood and Depression

Happiness has a lot of different meanings in philosophy, psychology, and neuroscience. For instance, being in a positive mood can be distinguished from feeling a sense of meaning and purpose. Most religious and philosophical systems see happiness or positive mood as a benefit of living a virtuous life rather than a virtue in itself. Often they will distinguish the contentment that results from virtue as a higher order of happiness, *eudaemonia*. Nonetheless, many philosophies recognize aspects or correlates of positive mood as virtues. "Hope" in "faith, hope, and charity" is a positive future orientation correlated with positive mood [110]. Likewise, one of the core virtues in Buddhism is *viriya or* vigor. While being depressed makes it more challenging to achieve one's own goals or help others, possessing hope, optimism, or vigor, and generally being positive, is not only rewarding in itself but makes it more likely you will be productive and helpful to others [111].

While people living with chronic pain or depression can be subjectively happy in other ways, neuromodulation to treat pain and depression is one powerful way to contribute to subjective well-being. Many studies now show that enhancing prefrontal control with DBS electrodes reduces the effect of pain [112]. Systems are being developed that stimulate the PFC only after detecting the unique signature of pain from sensors in the ACC [113].

As with pain, meta-analyses show that brain stimulation is an effective treatment for severe depression [101–103]. Focusing ultrasound on the right inferior frontal gyrus (rIFG), another key mood and emotional regulation area, enhances mood and emotional regulation [15]. TMS applied to the frontal lobes increased perseverance by shortening the giving-up response [114]. As with pain, progress is also being made in closed-loop neuromodulation using sensors to detect the onset of depression and disrupt it with DBS electrodes in the ventral capsule/ventral striatum [115].

Treating chronic pain and depression raises fewer flags than the eventual use of neuromodulation to enhance mood in the healthy. Initially coined by science fiction author Larry Niven in the 1960s [116], the term "wireheading" has come to refer to people addicted to inducing pleasure with brain electrodes. Michael Chrichton's 1972 novel *The Terminal Man* imagined an epilepsy patient with DBS electrodes who becomes addicted to the euphoria the electrodes induce until he is driven to a murderous rage [117]. Nonetheless, with appropriate technical safeguards in place to control the risk of overuse and adverse side-effects, neuromodulation for moderate enhancement of mood in the healthy, which appears to be safe and effective [118], would likely have many positive effects for individuals and society [111].

3.6 Selflessness and Transcendent Experiences

A final complement to the other virtues, and a capstone to character formation, is the capacity to experience altered states of consciousness that turn off the default mode network, our constant stream of self-referential thoughts [119–124]. Mindfulness meditation and psychedelics, for instance, both disrupt the "default mode network" with lasting positive impacts such as reducing anxiety and addictive cravings. These transcendent states give people distance from their habitual thoughts and behavior and boost equanimity and pro-social behavior [120–123, 125, 126]. Neuroimaging and neuromodulation are identifying which parts of the brain are key to such experiences.

As with the other virtues, multiple brain regions are implicated in experiences of awe or oneness, but the most common foci in studies of the spiritual brain are the parietal cortex, insula, and temporo-parietal junction (TPJ). Imaging shows that the right parietal cortex is less active during spiritual experiences [124, 127–130], and damage in the parietal region can cause spontaneous transcendent experiences and radical changes in religiosity [131, 132]. The insula and TPJ integrate physical sensations into a model of the body in space, anchoring our subconscious sense of self, while damage to or inhibition of the insula or TPJ can create out-of-body or "oneness" experiences [133]. Stimulating the right TPJ with tDCS reduces egocentric perspective-taking [134]. Deep brain stimulation of the dorsal anterior insula can induce ecstatic experiences in epilepsy patients [135]. "Flow" states involve reducing the interference of the default mode network with behavior, and getting into flow states can be facilitated by applying tDCS to the medial prefrontal cortex (mPFC) [136], the dlPFC, and the parietal cortex [137].

It is possible that brain stimulation for transcendent experiences could become habit-forming and disabling although studies of psychedelic use suggest the risk is low for those without mental health problems, and psychedelic use can be beneficial for those with mental health problems [126, 138]. Even long-term subclinical use of psychedelics or "microdosing" appears to be safe [139]. Nonetheless we don't want people accidentally entering a higher plane of being while driving or cooking, or having disorienting "flashbacks," so there will need to be close scrutiny of the side effects of and contextual regulation of transcendent brain stimulation.

4 The Ethics of Neuromodulating for Moral Enhancement

We will soon have technologies that allow the neuromodulation of many parts of the brain, complementing and probably going farther than psychopharmaceuticals. As we continue applying neuromodulation to the treatment of psychiatric disorders we will be obliged to regulate their potential use in criminal rehabilitation and enthusiasts' self-application of these technologies [140, 141]. Models of the multiple virtues to be cultivated in a mature moral character can hopefully address some of downsides of enhancing single virtues, and point to the multiple areas of the brain that will require sensors, chips and electrodes for "virtue engineering" [142].

All neuromodulation therapies require regulation to determine efficacy and side effects, and the more invasive the technology, the higher the safety and efficacy bar they will need to meet. Non-invasive brain stimulation is already widely available for consumers. While the severely disabled may be permitted to consent to brain implants, devices that could be permitted for use inside healthy brains will take some time. However, the most pressing ethical issues with moral neuromodulation are less regulatory and more philosophical and phenomenological [143, 144]. Under what conditions can someone consent to use brain stimulation to change their most fundamental thoughts and emotions? Do people using such devices feel less authentic [145, 146]? Addressing these questions will be increasingly relevant as neuromodulation becomes more common.

References

1. MacMillan M. An odd kind of fame: stories of Phineas Gage. Cambridge: MIT Press; 2022
2. Haas LF. Phineas Gage and the science of brain localisation. J Neurol Neurosurg Psychiatry. 2001;71:761.
3. Milner PM. Brain-stimulation reward: a review. Can J Psychol. 1991;45:1–36.
4. Olds J, Milner P. Positive reinforcement produced by electrical stimulation of septal area and other regions of rat brain. J Comp Physiol Psychol. 1954;47:419–27.
5. Clynes ME, Kline NS. Cyborgs and space. Astronautics. 1960;14:26–31.
6. Graat I, Figee M, Denys D. The application of deep brain stimulation in the treatment of psychiatric disorders. Int Rev Psychiatry. 2017;29:178–90.
7. Earp BD. Psychedelic moral enhancement. In: Hauskeller M, Coyne L, editors. Royal Institute of philosophy supplement. Cambridge: Cambridge University Press; 2018. p. 1–21
8. Persson I, Savulescu J. Moral hard-wiring and moral enhancement. Bioethics. 2017;31:286–95
9. Paulo N, Bublitz JC. How (not) to argue for moral enhancement: reflections on a decade of debate. Topoi. 2019;38:95–109.
10. Cavaleiro C, Martins J, Gonçalves J, Castelo-Branco M, Marsili L. Memory and cognition related neuroplasticity enhancement by transcranial direct current stimulation in rodents: a systematic review. Neural Plast. 2020;2020:1–23.
11. Fregni F, El-Hagrassy MM, Pacheco-Barrios K, et al. Evidence-based guidelines and secondary meta-analysis for the use of transcranial direct current stimulation in neurological and psychiatric disorders. Int J Neuropsychopharmacol. 2021;24:256–313.
12. Badran BW, Caulfield KA, Lopez JW, Cox C, Stomberg-Firestein S, DeVries WH, McTeague LM, George MS, Roberts D. Personalized TMS helmets for quick and reliable TMS administration outside of a laboratory setting. Brain Stimul. 2020;13:551–3.
13. Romero MC, Davare M, Armendariz M, Janssen P. Neural effects of transcranial magnetic stimulation at the single-cell level. Nat Commun. 2019;10:1–11.
14. Rodriguez R. Evaluating transcranial magnetic stimulation technologies: deep vs traditional psychiatry advisor. New York: Psychiatry Advisor; 2021. p. 1–2.
15. Sanguinetti JL, Hameroff S, Smith EE, Sato T, Daft CMW, Tyler WJ, Allen JJB. Transcranial focused ultrasound to the right prefrontal cortex improves mood and alters functional connectivity in humans. Front Hum Neurosci. 2020;14:1–13.
16. Vaccar S. Focused ultrasound enables precise noninvasive therapy. Pittsburgh: Carnegie Mellon News; 2021. p. 1–2.
17. Johnson RL, Wilson CG. A review of vagus nerve stimulation as a therapeutic intervention J Inflamm Res. 2018;11:203.
18. Corazzol M, Lio G, Lefevre A, et al. Restoring consciousness with vagus nerve stimulation Curr Biol. 2017;27:R994–6.
19. Collins L, Boddington L, Steffan PJ, McCormick D. Vagus nerve stimulation induces widespread cortical and behavioral activation. Curr Biol. 2021;31:2088–2098.e3.
20. Cao B, Wang J, Shahed M, Jelfs B, Chan RHM, Li Y. Vagus nerve stimulation alters phase synchrony of the anterior cingulate cortex and facilitates decision making in rats. Sci Rep 2016;6(1):35135.
21. FDA. FDA approves new DBS device that measures brain activity. Parkinson's disease Silver Spring: FDA; 2020.
22. Li G, Chung WY. Combined EEG-gyroscope-TDCS brain machine interface system for early management of driver drowsiness. IEEE Trans Hum Mach Syst. 2018;48:50–62.
23. Kim T, Nguyen P, Pham N, Bui N, Truong H, Ha S, Vu T. Epileptic seizure detection and experimental treatment: a review. Front Neurol. 2020;11:701.
24. Gilron R, Little S, Perrone R, et al. Long-term wireless streaming of neural recording for circuit discovery and adaptive stimulation in individuals with Parkinson's disease. Nat Biotechnol. 2021;39:1078–85.
25. Krauss JK, Lipsman N, Aziz T, et al. Technology of deep brain stimulation: current status and future directions. Nat Rev Neurol. 2020;17:75–87.

26. Haridy R. Experimental brain implant instantly detects and relieves pain. San Francisco: New Atlas; 2021. p. 1–3.
27. Rao RPN. Brain co-processors: using AI to restore and augment brain function. In: Handbook of neuroengineering. Singapore: Springer; 2020. p. 1–36.
28. Musk E, Neuralink. An integrated brain-machine interface platform with thousands of channels. J Med Internet Res. 2019;21(10):e16194.
29. Patch K. Neural dust swept up in latest leap for bioelectronic medicine. Nat Biotechnol. 2021;39:255–6.
30. Lee J, Leung V, Lee AH, Huang J, Asbeck P, Mercier PP, Shellhammer S, Larson L, Laiwalla F, Nurmikko A. Neural recording and stimulation using wireless networks of microimplants. Nat Electron. 2021;4:604–14.
31. Optica. Tiny, injectable sensors could monitor brain activity without surgery or implants. Washington, DC: Optica; 2021. p. 1–2.
32. Hardy N, Habib A, Ivanov T, Yanik AA. Neuro-SWARM^3: system-on-a-nanoparticle for wireless recording of brain activity. IEEE Photon Technol Lett. 2021;33:900–3.
33. Hughes J. Using neurotechnologies to develop virtues: a Buddhism approach to cognitive enhancement. Account Res. 2013;20(1):27–41.
34. Hughes JJ. Moral enhancement requires multiple virtues: toward a posthuman model of character development. Camb Q Healthc Ethics. 2014;24:86–95.
35. Hughes JJ. Empathy is just one component of moral character. AJOB Neurosci. 2015;6:49–55.
36. Oakes P, Loukas M, Oskouian RJ, Tubbs RS. The neuroanatomy of depression: a review. Clin Anat. 2017;30:44–9.
37. Riggs W. Open-mindedness. In: The Routledge handbook of virtue epistemology. Oxfordshire: Routledge; 2018. p. 141–54.
38. Cremaldi A, Kwong JMC. Is open-mindedness a moral virtue? Ratio (Oxf). 2017;30:343–58.
39. Alsharif H, Symons J. Open-mindedness as a corrective virtue. Philosophy. 2021;96:73–97.
40. Deyoung CG. Openness/intellect: a dimension of personality reflecting cognitive exploration. In: Mikulincer M, Shaver PR, Cooper ML, Larsen RJ, editors. APA handbook of personality and social psychology: personality processes and individual differences. Washington, DC: American Psychological Association; 2015. p. 369–99.
41. Fabiano J. Virtue theory for moral enhancement. AJOB Neurosci. 2021;12:89–102.
42. Johnson JA. A closer look at the adequacy of proposed frameworks for a "virtue theory for moral enhancement". AJOB Neurosci. 2021;12:103–5.
43. Persson I, Savulescu J. Unfit for the future? Human nature, scientific progress, and the need for moral enhancement. In: Savulescu J, ter Meulen R, Kahane G, editors. Enhancing human capacities. Oxford: Blackwell; 2011. p. 486–500.
44. Menon V, D'esposito M. The role of PFC networks in cognitive control and executive function. Neuropsychopharmacology. 2022;47:90–103.
45. Kahneman D. Thinking, fast and slow. New York: Farrar, Straus and Giroux; 2011.
46. Damasio A. Descartes' error: emotion, reason, and the human brain. New York: Putnam; 1994.
47. Baumgartner T, Knoch D, Hotz P, Eisenegger C, Fehr E. Dorsolateral and ventromedial prefrontal cortex orchestrate normative choice. Nat Neurosci. 2011;14:1468–74.
48. Greene JD, Nystrom LE, Engell AD, Darley JM, Cohen JD. The neural bases of cognitive conflict and control in moral judgment. Neuron. 2004;44:389–400.
49. May J, Workman CI, Haas J, Han H. The neuroscience of moral judgment: empirical and philosophical developments. In: de Brigard F, Sinnott-Armstrong W, editors. Neuroscience and philosophy. Cambridge: MIT Press; 2021. p. 1–23.
50. Trémolière B, Maheux-Caron V, Lepage JF, Blanchette I. tDCS stimulation of the dlPFC selectively moderates the detrimental impact of emotion on analytical reasoning. Front Psychol. 2018;9:568–73.
51. Feeser M, Prehn K, Kazzer P, Mungee A, Bajbouj M. Transcranial direct current stimulation enhances cognitive control during emotion regulation. Brain Stimul. 2014;7:105–12.

52. Bernacer J, Murillo JI. The Aristotelian conception of habit and its contribution to human neuroscience. Front Hum Neurosci. 2014;8:1–10.
53. Jebari K. Brain machine interface and human enhancement—an ethical review. Neuroethics. 2013;6:617–25.
54. Bolloni C, Badas P, Corona G, Diana M. Transcranial magnetic stimulation for the treatment of cocaine addiction: evidence to date. Subst Abuse Rehabil. 2018;9:11–21.
55. Antonelli M, Fattore L, Sestito L, di Giuda D, Diana M, Addolorato G. Transcranial magnetic stimulation: a review about its efficacy in the treatment of alcohol, tobacco and cocaine addiction. Addict Behav. 2021;114:106760.
56. Lapenta OM, Marques LM, Rego GG, Comfort WE, Boggio PS. tDCS in addiction and impulse control disorders. J ECT. 2018;34:182–92.
57. Ma T, Sun Y, Ku Y. Effects of non-invasive brain stimulation on stimulant craving in users of cocaine, amphetamine, or methamphetamine: a systematic review and meta-analysis. Front Neurosci. 2019;13:1–10.
58. Zhao Y, Sallie SN, Cui H, Zeng N, Du J, Yuan T, Li D, de Ridder D, Zhang C. Anterior cingulate cortex in addiction: new insights for neuromodulation. Neuromodul Technol Neural Interface. 2021;24:187–96.
59. Ho AL, Salib AMN, Pendharkar AV, Sussman ES, Giardino WJ, Halpern CH. The nucleus accumbens and alcoholism: a target for deep brain stimulation. Neurosurg Focus. 2018;45:E12.
60. Ranjan M, Ranjan N, Deogaonkar M, Rezai A. Deep brain stimulation for refractory depression, obsessive-compulsive disorder and addiction. Neurol India. 2020;68:282–7.
61. Niu L, Guo Y, Lin Z, Shi Z, Bian T, Qi L, Meng L, Grace AA, Zheng H, Yuan TF. Noninvasive ultrasound deep brain stimulation of nucleus accumbens induces behavioral avoidance. Sci China Life Sci. 2020;63:1328–36.
62. Grant A, Schwartz B. Too much of a good thing: the challenge and opportunity of the inverted U. Perspect Psychol Sci. 2011;6:61–76.
63. Ciullo V, Spalletta G, Caltagirone C, Banaj N, Vecchio D, Piras F, Piras F. Transcranial direct current stimulation and cognition in neuropsychiatric disorders: systematic review of the evidence and future directions. Neuroscientist. 2020;27:285–309. https://doi.org/10.1177/1073858420936167.
64. Narita Z, Stickley A, DeVylder J, et al. Effect of multi-session prefrontal transcranial direct current stimulation on cognition in schizophrenia: a systematic review and meta-analysis. Schizophr Res. 2020;216:367–73.
65. Huo L, Zhu X, Zheng Z, Ma J, Ma Z, Gui W, Li J. Effects of transcranial direct current stimulation on episodic memory in older adults: a meta-analysis. J Gerontol Ser B. 2021;76:692–702.
66. Galli G, Vadillo MA, Sirota M, Feurra M, Medvedeva A. A systematic review and meta-analysis of the effects of transcranial direct current stimulation (tDCS) on episodic memory. Brain Stimul. 2019;12:231–41.
67. Farhat LC, Carvalho AF, Solmi M, Brunoni AR. Evidence-based umbrella review of cognitive effects of prefrontal tDCS. Soc Cogn Affect Neurosci. 2020;2020:1–18.
68. Lee JH, Lee TL, Kang N. Transcranial direct current stimulation decreased cognition-related reaction time in older adults: a systematic review and meta-analysis. Ageing Res Rev. 2021;70:101377.
69. Goldthorpe RA, Rapley JM, Violante IR. A systematic review of non-invasive brain stimulation applications to memory in healthy aging. Front Neurol. 2020;11:575075. https://doi.org/10.3389/fneur.2020.575075.
70. Salehinejad MA, Nejati V, Mosayebi-Samani M, Mohammadi A, Wischnewski M, Kuo MF, Avenanti A, Vicario CM, Nitsche MA. Transcranial direct current stimulation in ADHD: a systematic review of efficacy, safety, and protocol-induced electrical field modeling results. Neurosci Bull. 2020;36:1191–212.
71. Brunoni AR, Amadera J, Berbel B, Volz MS, Rizzerio BG, Fregni F. A systematic review on reporting and assessment of adverse effects associated with transcranial direct current stimulation. Int J Neuropsychopharmacol. 2011;14:1133–45.

72. Matsumoto H, Ugawa Y. Adverse events of tDCS and tACS: a review. Clin Neurophysiol Pract. 2017;2:19–25.
73. Reich MM, Hsu J, Ferguson M, et al. A brain network for deep brain stimulation induced cognitive decline in Parkinson's disease. Brain. 2022;145:1410–21.
74. Urban KR, Gao WJ. Performance enhancement at the cost of potential brain plasticity: neural ramifications of nootropic drugs in the healthy developing brain. Front Syst Neurosci. 2014;8:38. https://doi.org/10.3389/fnsys.2014.00038.
75. Wood S, Sage JR, Shuman T, Anagnostaras SG. Psychostimulants and cognition: a continuum of behavioral and cognitive activation. Pharmacol Rev. 2014;66:193–221.
76. Nihonsugi T, Ihara A, Haruno M. Selective increase of intention-based economic decisions by noninvasive brain stimulation to the dorsolateral prefrontal cortex. J Neurosci. 2015;35:3412–9.
77. Knoch D, Nitsche MA, Fischbacher U, Eisenegger C, Pascual-Leone A, Fehr E. Studying the neurobiology of social interaction with transcranial direct current stimulation—the example of punishing unfairness. Cereb Cortex. 2008;18:1987–90.
78. Dambacher F, Schuhmann T, Lobbestael J, Arntz A, Brugman S, Sack AT. Reducing proactive aggression through non-invasive brain stimulation. Soc Cogn Affect Neurosci. 2015;10:1303–9.
79. Romero-Martínez Á, Bressanutti S, Moya-Albiol L. A systematic review of the effectiveness of non-invasive brain stimulation techniques to reduce violence proneness by interfering in anger and irritability. J Clin Med. 2020;9:882–902.
80. Yuan BO, Tolomeo S, Yang C, Wang Y, Yu R. tDCS effect on prosocial behavior: a meta-analytic review. Soc Cogn Affect Neurosci. 2021;2021:1–17.
81. Yang CC, Khalifa N, Völlm B. The effects of repetitive transcranial magnetic stimulation on empathy: a systematic review and meta-analysis. Psychol Med. 2018;48:737–50.
82. Leopold A, Krueger F, Dal Monte O, Pardini M, Pulaski SJ, Solomon J, Grafman J. Damage to the left ventromedial prefrontal cortex impacts affective theory of mind. Soc Cogn Affect Neurosci. 2012;7:871–80.
83. Hiser J, Koenigs M. The multifaceted role of the ventromedial prefrontal cortex in emotion, decision making, social cognition, and psychopathology. Biol Psychiatry. 2018;83:638–47.
84. Roesmann K, Kroker T, Hein S, Rehbein M, Winker C, Leehr EJ, Klucken T, Junghöfer M. Transcranial direct current stimulation of the ventromedial prefrontal cortex modulates perceptual and neural patterns of fear generalization. Biol Psychiatry Cogn Neurosci Neuroimaging. 2021;7:210–20.
85. Sergiou CS, Santarnecchi E, Franken IHA, van Dongen JDM. The effectiveness of transcranial direct current stimulation as an intervention to improve empathic abilities and reduce violent behavior: a literature review. Aggress Violent Behav. 2020;55:101463.
86. Sergiou CS, Santarnecchi E, Romanella SM, Wieser MJ, Franken IHA, Rassin EGC, van Dongen JDM. Transcranial direct current stimulation targeting the ventromedial prefrontal cortex reduces reactive aggression and modulates electrophysiological responses in a forensic population. Biol Psychiatry Cogn Neurosci Neuroimaging. 2021;7:95–107. https://doi.org/10.1016/J.BPSC.2021.05.007.
87. Bahji A, Forth E, Yang CC, Khalifa N. Transcranial direct current stimulation for empathy: a systematic review and meta-analysis. Soc Neurosci. 2021;16:232–55.
88. Ahmad N, Zorns S, Chavarria K, Brenya J, Janowska A, Keenan JP. Are we right about the right TPJ? A review of brain stimulation and social cognition in the right temporal parietal junction. Symmetry (Basel). 2021;13:2219.
89. Bloom P. The case against empathy: the New Yorker. New York: Ecco HarperCollins; 2013.
90. Zhang H, Gross J, de Dreu C, Ma Y. Oxytocin promotes coordinated out-group attack during intergroup conflict in humans. elife. 2019;8:1–19.
91. van IJzendoorn MH, Bakermans-Kranenburg MJ. A sniff of trust: meta-analysis of the effects of intranasal oxytocin administration on face recognition, trust to in-group, and trust to out-group. Psychoneuroendocrinology. 2012;37:438–43.

92. Bethlehem RAI, Baron-Cohen S, van Honk J, Auyeung B, Bos PA. The oxytocin paradox. Front Behav Neurosci. 2014;8:48.
93. de Dreu CKW, Greer LL, van Kleef GA, Shalvi S, Handgraaf MJJ. Oxytocin promotes human ethnocentrism. Proc Natl Acad Sci U S A. 2011;108:1262–6.
94. Morales J, Lau H, Fleming SM. Domain-general and domain-specific patterns of activity supporting metacognition in human prefrontal cortex. J Neurosci. 2018;38:3534–46.
95. Qiu L, Su J, Ni Y, Bai Y, Zhang X, Li X, Wan X. The neural system of metacognition accompanying decision-making in the prefrontal cortex. PLoS Biol. 2018;16:e2004037.
96. Hallsson BG, Siebner HR, Hulme OJ. Fairness, fast and slow: a review of dual process models of fairness. Neurosci Biobehav Rev. 2018;89:49–60.
97. Strang S, Gross J, Schuhmann T, Riedl A, Weber B, Sack AT. Be nice if you have to—the neurobiological roots of strategic fairness. Soc Cogn Affect Neurosci. 2015;10:790–6.
98. Edgcumbe DR, Thoma V, Rivolta D, Nitsche MA, Fu CHY. Anodal transcranial direct current stimulation over the right dorsolateral prefrontal cortex enhances reflective judgment and decision-making. Brain Stimul. 2019;12:652–8.
99. Li X, Xiong G, Dong Z, Cai S, Zhao J, She Z, Guo Y. Causal role of the right dorsolateral prefrontal cortex in organizational fairness perception: evidence from a transcranial direct current stimulation study. Front Behav Neurosci. 2020;14:134.
100. Luo J, Ye H, Zheng H, Chen S, Huang D. Modulating the activity of the dorsolateral prefrontal cortex by tDCS alters distributive decisions behind the veil of ignorance via risk preference. Behav Brain Res. 2017;328:70–80.
101. Tassy S, Oullier O, Duclos Y, Coulon O, Mancini J, Deruelle C, Attarian S, Felician O, Wicker B. Disrupting the right prefrontal cortex alters moral judgement. Soc Cogn Affect Neurosci. 2012;7:282–8.
102. Darby RR, Pascual-Leone A. Moral enhancement using non-invasive brain stimulation. Front Hum Neurosci. 2017;11:77–87.
103. Jeurissen D, Sack AT, Roebroeck A, Russ BE, Pascual-Leone A. TMS affects moral judgment, showing the role of DLPFC and TPJ in cognitive and emotional processing. Front Neurosci. 2014;8:18. https://doi.org/10.3389/fnins.2014.00018.
104. Liu Y, Lin W, Xu P, Zhang D, Luo Y. Neural basis of disgust perception in racial prejudice. Hum Brain Mapp. 2015;36:5275–86.
105. Salvo G, Provenzano S, di Bello M, D'Olimpio F, Ottaviani C, Mancini F. Filthiness of immorality: manipulating disgust and moral rigidity through noninvasive brain stimulation as a promising therapeutic tool for obsessive compulsive disorder. Clin Psychol Sci. 2021;10:127–40.
106. Ottaviani C, Mancini F, Provenzano S, Collazzoni A, D'Olimpio F. Deontological morality can be experimentally enhanced by increasing disgust: a transcranial direct current stimulation study. Neuropsychologia. 2018;119:474–81.
107. Zheng H, Lu X, Huang D. tDCS over DLPFC leads to less utilitarian response in moral-personal judgment. Front Neurosci. 2018;12:193.
108. Civai C, Miniussi C, Rumiati RI. Medial prefrontal cortex reacts to unfairness if this damages the self: a tDCS study. Soc Cogn Affect Neurosci. 2015;10:1054–60.
109. Bartels DM, Pizarro DA. The mismeasure of morals: antisocial personality traits predict utilitarian responses to moral dilemmas. Cognition. 2011;121:154–61.
110. Pleeging E, Burger M, van Exel J. The relations between hope and subjective well-being: a literature overview and empirical analysis. Appl Res Qual Life. 2019;16:1019–41.
111. Walker M. Happy-people-pills for all. Hoboken: Wiley-Blackwell; 2013.
112. Zhou H, Zhang Q, Martinez E, Dale J, Robinson E, Huang D, Wang J. A novel neuromodulation strategy to enhance the prefrontal control to treat pain. Mol Pain. 2019;15:1744806919845739. https://doi.org/10.1177/1744806919845739.
113. Zhang Q, Hu S, Talay R, et al. A prototype closed-loop brain–machine interface for the study and treatment of pain. Nat Biomed Eng. 2021;2021:736.
114. Miyauchi E, Kawasaki M. Behavioural effects of task-relevant neuromodulation by rTMS on giving-up. Sci Rep. 2021;11:1–8.

115. Scangos KW, Khambhati AN, Daly PM, et al. Closed-loop neuromodulation in an individual with treatment-resistant depression. Nat Med. 2021;27:1696–700.
116. Niven L. Ringworld engineers. New York: Phantasia Press; 1979.
117. Crichton M. The terminal man. New York: Knopf; 1972.
118. Newstead S, Young H, Benton D, Jiga-Boy G, Andrade Sienz ML, Clement RM, Boy F. Acute and repetitive fronto-cerebellar tDCS stimulation improves mood in non-depressed participants. Exp Brain Res. 2017;236:83–97.
119. Qin P, Northoff G. How is our self related to midline regions and the default-mode network? NeuroImage. 2011;57:1221–33.
120. van Elk M, Arciniegas Gomez MA, van der Zwaag W, van Schie HT, Sauter D. The neural correlates of the awe experience: reduced default mode network activity during feelings of awe. Hum Brain Mapp. 2019;40:3561–74.
121. Palhano-Fontes F, Andrade KC, Tofoli LF, Jose ACS, Crippa AS, Hallak JEC, Ribeiro S, de Araujo DB. The psychedelic state induced by Ayahuasca modulates the activity and connectivity of the default mode network. PLoS One. 2015;10:1–13.
122. Smigielski L, Scheidegger M, Kometer M, Vollenweider FX. Psilocybin-assisted mindfulness training modulates self-consciousness and brain default mode network connectivity with lasting effects. NeuroImage. 2019;196:207–15.
123. Hafsteinsson M. Is the sense of self a threat to well-being? The default mode network and self-related processing in depression and meditation. Skovde: University of Skovde; 2020.
124. Johnstone B, Cohen D. Neuroscience, selflessness, and spiritual experience: explaining the science of transcendence. Amsterdam: Elsevier Science; 2019.
125. Luberto CM, Shinday N, Song R, Philpotts LL, Park ER, Fricchione GL, Yeh GY. A systematic review and meta-analysis of the effects of meditation on empathy, compassion, and prosocial behaviors. Mindfulness (N Y). 2018;9:708–24.
126. Aday JS, Mitzkovitz CM, Bloesch EK, Davoli CC, Davis AK. Long-term effects of psychedelic drugs: a systematic review. Neurosci Biobehav Rev. 2020;113:179–89.
127. Tenke CE, Kayser J, Svob C, Miller L, Alvarenga JE, Abraham K, Warner V, Wickramaratne P, Weissman MM, Bruder GE. Association of posterior EEG alpha with prioritization of religion or spirituality: a replication and extension at 20-year follow-up. Biol Psychol. 2017;124:79–86.
128. Miller L, Balodis IM, McClintock CH, Xu J, Lacadie CM, Sinha R, Potenza MN. Neural correlates of personalized spiritual experiences. Cereb Cortex (New York, NY). 2019;29:2331.
129. Miller L, Balodis IM, McClintock CH, Xu J, Lacadie CM, Sinha R, Potenza MN. Neural correlates of personalized spiritual experiences. Cereb Cortex. 2019;29:2331–8.
130. Grafman J, Cristofori I, Zhong W, Bulbulia J. The neural basis of religious cognition. Curr Dir Psychol Sci. 2020;29:126–33. https://doi.org/10.1177/0963721419898183.
131. Urgesi C, Aglioti SM, Skrap M, Fabbro F. The spiritual brain: selective cortical lesions modulate human self-transcendence. Neuron. 2010;65:309–19.
132. Crescentini C, Aglioti SM, Fabbro F, Urgesi C. Virtual lesions of the inferior parietal cortex induce fast changes of implicit religiousness/spirituality. Cortex. 2014;54:1–15.
133. Orrù G, Bertelloni D, Cesari V, Conversano C, Gemignani A. Targeting temporal parietal junction for assessing and treating disembodiment phenomena: a systematic review of TMS effect on depersonalization and derealization disorders (DPD) and body illusions. AIMS Neurosci. 2021;8:181.
134. Martin AK, Huang J, Hunold A, Meinzer M. Dissociable roles within the social brain for self-other processing: a HD-tDCS study. Cereb Cortex. 2019;29:3642–54.
135. Bartolomei F, Lagarde S, Scavarda D, Carron R, Bénar CG, Picard F. The role of the dorsal anterior insula in ecstatic sensation revealed by direct electrical brain stimulation. Brain Stimul. 2019;12:1121–6.
136. Ulrich M, Niemann J, Boland M, Kammer T, Niemann F, Grön G. The neural correlates of flow experience explored with transcranial direct current stimulation. Exp Brain Res. 2018;236:3223–37.

137. Gold J, Ciorciari J. A transcranial stimulation intervention to support flow state induction. Front Hum Neurosci. 2019;13:274. https://doi.org/10.3389/FNHUM.2019.00274.
138. Hodge AT, Sukpraprut-Braaten S, Narlesky M, Strayhan RC. The use of psilocybin in the treatment of psychiatric disorders with attention to relative safety profile: a systematic review. J Psychoactive Drugs. 2022;2022:2044096. https://doi.org/10.1080/02791072.2022.2044096.
139. Anderson T, Petranker R, Rosenbaum D, Weissman CR, Dinh-Williams LA, Hui K, Hapke E, Farb NAS. Microdosing psychedelics: personality, mental health, and creativity differences in microdosers. Psychopharmacology. 2019;236:731–40.
140. Canavero S. Criminal minds: neuromodulation of the psychopathic brain. Front Hum Neurosci. 2014;8:1–3.
141. Koi P, Uusitalo S, Tuominen J. Self-control in responsibility enhancement and criminal rehabilitation. Crim Law Philos. 2018;12:227–44.
142. Jotterand F. "Virtue engineering" and moral agency: will post-humans still need the virtues? AJOB Neurosci. 2011;2:3–9.
143. Pugh J, Pycroft L, Sandberg A, Aziz T, Savulescu J. Brainjacking in deep brain stimulation and autonomy. Ethics Inf Technol. 2018;20:219–32.
144. Karanasiou A. On being transhuman: commercial BCIs and the quest for autonomy. Cambridge: Cambridge University Press; 2020. https://doi.org/10.2139/SSRN.3840840.
145. Bublitz JC, Merkel R. Autonomy and authenticity of enhanced personality traits. Bioethics. 2009;23:360–74.
146. Coin A, Dubljević V. The authenticity of machine-augmented human intelligence: therapy, enhancement, and the extended mind. Neuroethics. 2021;14:283–90.

Brain Co-processors: Ethical and Social Implications

Rajesh P. N. Rao and Andreas Schönau

1 Introduction

The ability to record the activities of populations of neurons in the brain and "decode" these activities using a computer to control robotic prostheses and cursors forms the basis for the field of brain–computer interfaces (BCIs) [1–4] (also called brain–machine interfaces or BMIs). Closely related are devices called computer–brain interfaces (CBIs) that "encode" external signals such as sound, images, or artificial tactile measurements and deliver that information to the brain through neural stimulation. Pioneers in this field, which originated in the 1960s and 1970s, were Fetz [5], Delgado [6], and Vidal [7]. More recently, researchers have used a variety of machine learning techniques to decode neural activity for controlling prosthetic arms [8–10], cursors [11–16], spellers [17, 18], and robots [19–22]. Artificial sensory information has been delivered via neurostimulation to the brain and other regions of the nervous system for auditory [23], visual [24], proprioceptive [25], and tactile [26–30] perception.

In this chapter, we discuss a new class of brain interfaces called *brain co-processors* [31, 32] which leverage artificial intelligence (AI) to determine the best neural stimulation patterns for current brain activity to achieve predetermined goals, such as rehabilitation after injury or steering brain activity away from undesirable states associated with depression or other neurological conditions.

R. P. N. Rao (✉)
Center for Neurotechnology and Paul G. Allen School of Computer Science and Engineering, University of Washington, Seattle, WA, USA
e-mail: rao@cs.washington.edu

A. Schönau
Center for Neurotechnology and Department of Philosophy, University of Washington, Seattle, WA, USA

V. Dubljević, A. Coin (eds.), *Policy, Identity, and Neurotechnology*, Advances in Neuroethics, https://doi.org/10.1007/978-3-031-26801-4_10

We first review different kinds of brain co-processors, from simple co-processors relying on rudimentary types of AI to more sophisticated "neural co-processors" [32] which rely on artificial neural networks (ANNs) to interact with biological neurons in the brain. We briefly review past applications such as controlling prosthetic devices, reanimating paralyzed limbs, restoring sensorimotor and cognitive function, enhancing memory, and augmenting brain function, e.g., direct brain-to-brain interaction. We then discuss a variety of potential future applications of brain co-processors. A major part of this chapter is devoted to the ethical issues [33, 34] that arise as researchers and commercial enterprises start exploring augmentative applications of brain co-processors.

2 Brain Co-processors

Figure 1a depicts the general architecture of a brain co-processor. A co-processor uses AI to transform neural activity *and/or* external inputs into stimulation patterns *and/or* external control signals for actuators. Inputs to the co-processor may include neural recordings, (e.g., spikes or local field potentials [LFPs] from microelectrodes, electrical activity from electrocorticography (ECoG) or electroencephalography (EEG), optical/optogenetic recordings, and blood flow changes from functional Magnetic Resonance Imaging (fMRI) or functional Near-InfraRed Spectroscopy (fNIRS) and external information sources (e.g., from sensors [infrared, ultrasonic, etc.], the Internet, local storage device providing data, a source of user input, or even another nervous system). Outputs of the co-processor can be multidimensional neural stimulation patterns (e.g., stimulation delivered electrically via microelectrode or ECoG arrays, optical stimulation via optogenetic techniques, focused ultrasound stimulation, and magnetic stimulation) and control signals for an external actuator (e.g., commands for a robot, a computing device, an internet search engine, messages to another nervous system, etc.).

The algorithms implemented on the co-processor for transforming its inputs into suitable outputs can range from simple mappings (e.g., each input spike results in a stimulation pulse) and rules based on medical or other domain knowledge (e.g., a fixed stimulation pattern when a seizure is detected) to sophisticated machine learning algorithms for classification, regression, or reinforcement learning that map complex multidimensional neural recordings and sensor inputs to appropriate multidimensional stimulation patterns and control signals for external devices.

A powerful type of brain co-processor that relies on ANNs as the basis for its AI is a *neural co-processor* (Fig. 1b) [32]. A neural co-processor, in its most general form, uses two ANNs: a co-processor network (CPN) and an emulator network (EN). The CPN is used to map input neural activity patterns in one set of areas to output stimulation patterns in the same or other areas. The CPN's weights are optimized to minimize brain-activity-based error (between stimulation patterns and target neural activity patterns *when known*), or more generally, to minimize behavioral/task error or maximize reward using the EN. The EN is designed or pre-trained (e.g., via the backpropagation algorithm) to learn the biological transformation

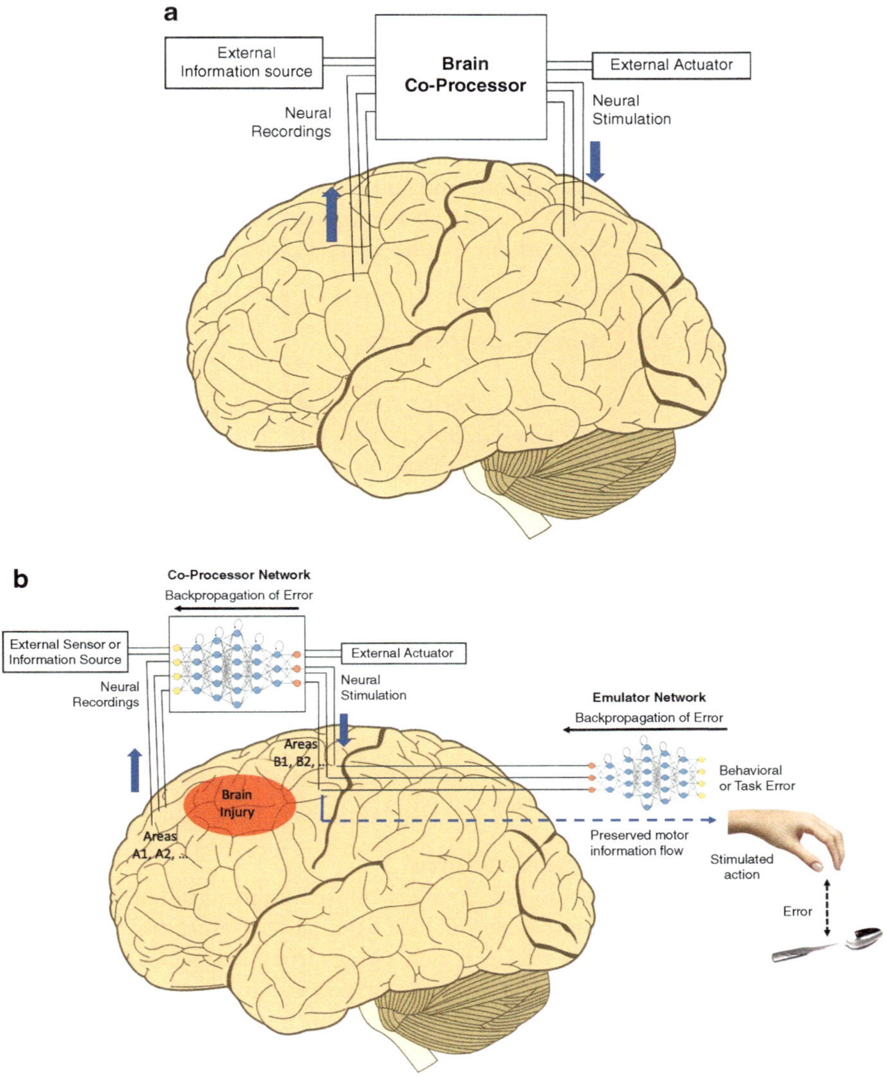

Fig. 1 (**a**) General architecture of a brain co-processor. The co-processor receives as input both external information from sensors, the internet or other information source as well as ongoing neural activity. The output of the co-processor includes commands to external actuators (robots, internet search, message to another brain, etc.) as well as multidimensional stimulation patterns delivered to one or more regions of the nervous system to achieve a desired goal. (**b**) Neural co-processor. In this type of brain co-processor, an ANN called the "Co-Processor Network" (CPN) is used to map input neural activity patterns in one set of areas A1, A2, … to output stimulation patterns in the same or other areas B1, B2, ... in order to achieve a neural or behavioral goal using another ANN, an "Emulator Network" (EN). The example here shows the CPN creating a new information processing pathway between prefrontal cortex and motor cortex, bypassing an intermediate area affected by brain injury (e.g., stroke)

from stimulation and/or neural activity patterns at the stimulation site to the resulting output behaviors.

Using a trained EN, the CPN is trained to produce optimal stimulation patterns, thereby creating a goal-directed artificial information processing pathway between the input and output areas. External information from artificial sensors or other information sources can be integrated into the CPN's information processing as additional inputs to the neural network, and outputs of the CPN can include control outputs for external actuators. CPNs can also be trained using a reward/cost-based *reinforcement learning* [35] algorithm. Further details can be found in [31].

2.1 Examples of Brain Co-processors

Brain co-processors can be used for restoring sensory or motor function, controlling a robotic arm, reanimating a paralyzed limb, modulating neural circuits for alleviating the symptoms of motor or cognitive disorders, and inducing neuroplasticity for targeted rehabilitation of the injured brain.

2.1.1 Sensory Restoration

The brain co-processor approach to designing sensory prostheses is depicted in Fig. 2a. The co-processor receives as input not only the sensor values (e.g., image pixel values from a camera) but also the current neural recordings from relevant regions. The co-processor's AI algorithm takes into account both the ongoing neural dynamics and the external sensory input to compute stimulation patterns appropriate for current brain state in order to achieve a reliable percept. The AI algorithm's parameters (e.g., weights of ANNs) can be tuned based on the subject's feedback to optimize the parameters for reliable perception.

2.1.2 Closed-Loop Prosthetic Control

The brain co-processor shown in Fig. 2b allows closed-loop control of a prosthetic device. The inputs to the co-processor include external measurements from tactile and proprioceptive sensors, as well as neural signals from both motor and sensory regions of the brain. The motor neural signals are decoded by an AI algorithm such as the Kalman filter [36, 37] to generate control signals for the prosthetic hand. The same algorithm or a different method is used to encode information from the artificial sensors on the prosthetic device, in conjunction with neural recordings in sensory areas, to appropriately stimulate somatosensory neurons for tactile and proprioceptive feedback.

Although the co-processor framework of Fig. 2b is yet to be fully tested, several research groups have explored versions of this co-processor framework for prosthetic control [38–41].

2.1.3 Reanimating Paralyzed Limbs

A brain co-processor can be used for reanimating a paralyzed limb by translating motor commands from the brain to stimulation patterns for spinal neurons (Fig. 2c) or muscles. As an example, Bouton et al. [42] showed that a quadriplegic man with

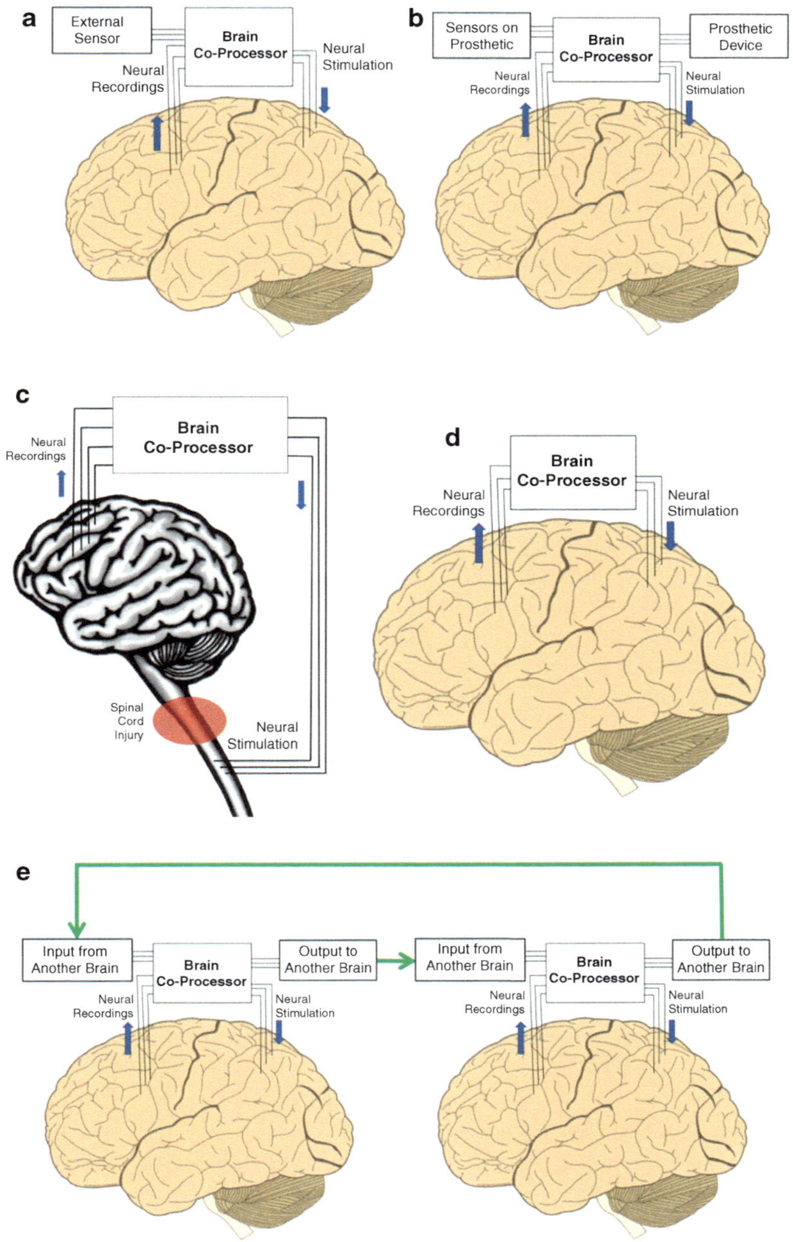

Fig. 2 (**a**) Brain co-processor for sensory restoration/augmentation. Co-processor takes into account ongoing dynamics of the brain to tailor its stimulation pattern for reliable perception. (**b**) Brain co-processor for closed-loop prosthetic control. (**c**) Brain co-processor for reanimation of paralyzed limbs. Motor commands from the brain are translated by AI to stimulation patterns delivered to the spinal cord to reanimate a paralyzed limb. (**d**) Brain co-processor for neuromodulation and plasticity induction. (**e**) Brain co-processors for direct brain-to-brain interaction

a 96-electrode array implanted in the hand area of the motor cortex could use cortical signals to electrically stimulate muscles in his paralyzed forearm and produce six different wrist and hand motions. These results were extended to multi-joint reaching and grasping movements by Ajiboye et al. [43]. Continued electrical stimulation of muscles may result in muscle fatigue, rendering the technique impractical for day-long use. The brain co-processor approach in Fig. 2c avoids this problem by using brain signals to stimulate the spinal cord rather than muscles.

2.1.4 Neuromodulation for Restoring Motor and Cognitive Function

Figure 2d shows how a brain co-processor can be used to translate neural recordings from one region of the brain to appropriate stimulation patterns delivered to the same or other region of the brain for (a) modulating ongoing neural dynamics to correct undesirable behaviors and symptoms such as tremors and (b) replacing lost function by emulating an injured neural circuit and conveying information from one brain region to another bypassing the injured region. One of the early pioneers in this area was Jose Delgado [6] who designed an implantable co-processor called the stimoceiver that detected neural activity patterns in one brain region of a monkey and triggered stimulation in another to make the monkey quiet and withdrawn, suggesting possible use of such co-processors for treating depression. Delgado's work anticipated later commercial brain implants such as Neuropace's RNS system for reducing seizures and deep brain stimulation (DBS) for reducing tremors in Parkinson's patients.

2.1.5 Inducing Plasticity and Rewiring the Brain

The co-processor in Fig. 2d can also be used for induction of plasticity and rewiring neural connections. Hebb's principle for plasticity states that if a group of neurons A consistently fires before another group of neurons B, connections from group A to group B should be strengthened since this indicates a causal relationship from A to B. Such plasticity can be artificially induced in the motor cortex of freely behaving primates using a simple "AI" algorithm mapping from each input spike to an output stimulation pulse [44].

This method could be useful for rewiring the brain after traumatic brain injury [45], stroke or neuropsychiatric disorders such as depression and post-traumatic stress disorder (PTSD).

2.1.6 Enhancing Memory

Besides rehabilitation and restoration of lost function, brain co-processors can also be used for augmentation of existing brain function. As an example, a co-processor such as the one depicted in Fig. 2d can be used to enhance short-term memory, as demonstrated by Berger and colleagues [46, 47]. They implanted a co-processor system in the hippocampus of monkeys and rats and demonstrated improved performance due to memory enhancement in delayed match-to-sample and nonmatch-to-sample tasks. The drawback of the approach, namely that we do not have training data from the time the brain was healthy to train the AI, is addressed by the neural co-processor framework above using emulator networks.

2.1.7 Brain-to-Brain Interfaces

Figure 2e depicts how brain co-processors can be used to augment human communication and collaboration capabilities by facilitating direct brain-to-brain interactions. Each person utilizes a co-processor to send information to one or more other brains and receive information from these brains. The co-processor is optimized to reliably interpret and encode the signals from another brain for stimulation and reliably decode information from one's own brain for transmission to another brain. The first such human brain-to-brain interface was demonstrated by Rao, Stocco, and colleagues utilizing noninvasive recording and stimulation technologies (EEG and transcranial magnetic stimulation [TMS], respectively) [48–50]. The researchers showed that tasks such as a video game [48] or "20 questions" [50] could be completed successfully through direct brain-to-brain collaboration in humans (for other examples in humans, see [51, 52] and in animals, see [53–55]). A more recent experiment [56] demonstrated a "BrainNet" that allows groups of humans to collaborate and solve tasks together via direct brain-to-brain interaction.

2.2 Applications of Brain Co-processors

The examples above mostly involved proof-of-concept demonstrations. Except for deep brain stimulators and Neuropace's RNS closed-loop stimulation system for controlling epilepsy, the vast majority of brain co-processors are still in their "laboratory testing" phase. However, given the entry of commercial entities in this space, a range of co-processor applications may start appearing on the market within the next few decades, if not sooner. We discuss some of these potential future applications of co-processors below (Fig. 3).

2.2.1 Medical Applications

The first co-processor applications to be commercialized will likely be in the space of medical devices, building on the track record of FDA-approved devices such as deep brain stimulators. These future medical applications will include restoring

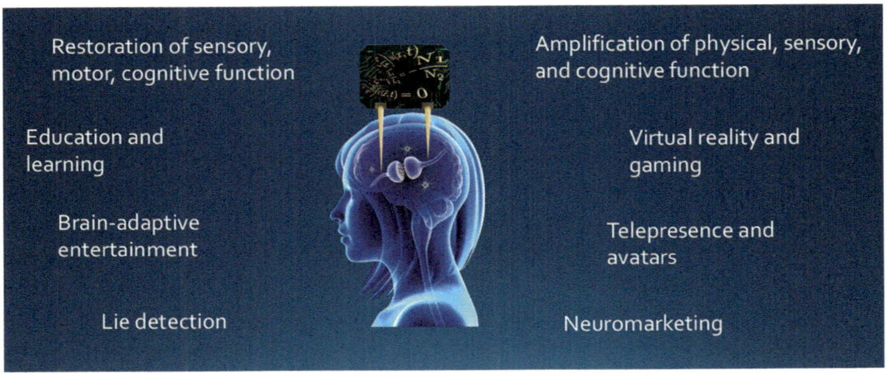

Fig. 3 Medical and non-medical applications of brain co-processors

motor, sensory or cognitive function via closed-loop control of prosthetic devices, closed-loop encoding for sensory prostheses, and neuromodulation and plasticity induction for cognitive restoration and rehabilitation after stroke or other injury, as discussed above in the context of laboratory experiments. Given that the medical community has prior experience with risk-to-benefit analysis and regulatory aspects for medical applications, the path to commercialization for medical co-processors may have fewer hurdles and unknowns compared to the non-medical augmentative applications discussed below.

2.2.2 Non-medical Applications: Augmenting Human Function

Beyond medical applications, brain co-processors in the future could potentially be used to augment the capacity of the human brain in variety of ways:

- **Amplification of physical, sensory, and cognitive function:** Co-processors could be used to amplify the physical capacity of a human through external actuators such as exoskeletons, robotic arms, or even "Iron Man"-style body armor, with applications in firefighting, nuclear inspections, and maintaining law and order. Sensory amplification may include augmenting the brain with the ability to sense beyond the visible spectrum by providing as input to the brain co-processor measurements from infrared, hyperspectral, ultrasonic, laser-based, or other types of sensors. Cognitive amplification may be achieved in a variety of ways, e.g., by allowing the co-processor to augment the brain's knowledge and information processing capacity by rapidly accessing and integrating information from the internet.
- **Education and learning:** A brain co-processor could serve as an assistive device during knowledge acquisition by monitoring a student's progress, tracking attention, delivering lessons tailored to the student's optimal pace of learning, etc. Furthermore, the ability of co-processors to induce plasticity through closed-loop stimulation could potentially be used to accelerate the acquisition of knowledge and skills, or even transfer knowledge from an expert brain to another via brain-to-brain interfacing [49]. Eventually, the ability to verify whether a student has grasped the concepts in a course directly by monitoring corresponding changes in their brain activity may obviate the need for examinations and tests.
- **Virtual reality and gaming:** An obvious application is using brain co-processors for closed-loop brain stimulation for high fidelity virtual reality/augmented reality (VR/AR) and gaming. A proof-of-concept brain-stimulation-based VR game was described in [57]. Unlike today's VR/AR headsets that are limited to providing visual and auditory inputs, brain-stimulation-based VR systems would potentially allow a complete sensory experience including artificial smell, taste, proprioception, hunger, thirst, and somatic senses such as touch, heat, pressure, and pain. Generating realistic sensations through stimulation will however require significant advances in our understanding of the neural basis of these sensations under natural circumstances.

- **Brain-adaptive entertainment:** Brain co-processors may open the door to personalized entertainment where the content may adapt not only to a person's overall preferences but also to current brain activity.
- **Telepresence and avatars:** The ability to touch, see, and sense using sensor-rich robotic "avatars" in remote locations coupled to a co-processor conveying sensations through stimulation opens up the possibility of ultra-realistic telepresence, posing a potential threat to the air travel industry.
- **Lie detection and biometrics:** Methods for lie detection and "brain fingerprinting" for identification based on EEG and fMRI have already been proposed [58, 59], but the insufficient accuracy of these methods has prevented their adoption in law and policing. Closed-loop methods based on co-processors may eventually increase the accuracy of brain-based lie detection and biometrics to an acceptable level for real-world use, assuming the ethical issues can be satisfactorily addressed (see below).
- **Neuromarketing:** Marketing professionals may be interested in gauging a person's response to an advertisement and in tracking a person's interest in a product by monitoring and studying their brain signals. A co-processor could potentially learn to track and even predict a person's interests over time as it interacts with the person's brain. While such an application is currently not feasible, it nevertheless raises the important issue of ethics of these types of applications.

3 Ethical, Moral, and Social Justice Implications of Brain Co-processors

The possibility of augmenting human abilities with brain co-processors, as discussed in the previous section, brings to the forefront the urgent need to identify and address the ethical, moral, and social justice issues (Fig. 4) before these technologies become feasible enough to be commercialized.

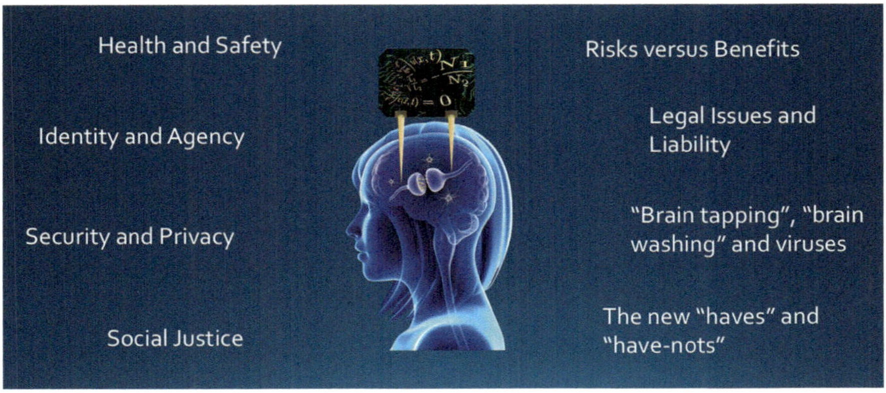

Fig. 4 Summary of ethical, moral, and social justice issues associated with the use of brain co-processors

3.1 Health, Safety, and Full Life Cycle Design

The most powerful co-processors will likely be invasive and require implantation, requiring the user to weigh the risks to health versus benefits of the technology. Those risks include, but are not limited to, the individual assessment of short- and long-term safety, cognitive and communicative impairment, inappropriate expectations, involuntariness, affective impairment, as well as privacy and security issues [60]. Early trial studies especially need to attend to those concerns by adapting best practices into the informed consent process [61], for instance, through a user-centered design approach that builds upon the experience voiced by end users to inform future and current participants about the challenges that might lay ahead of them [62].

Additional ethical issues arise in the context of commercial co-processors: What happens when the company that manufactured and sold a co-processor goes out of business? How are patients who have become dependent on co-processor technology supported when the technology becomes obsolete? Such a situation arose recently when more than 350 blind people with retinal implants found out that the company that manufactured the implants has stopped making and supporting them [63, 64]. To prevent such situations in the future, several options could be explored:

1. Full life cycle design: Neurotechnologists and companies could adopt a patient-centric design strategy with the full life cycle of their device in mind. This could involve building the device using, as much as possible, easily available commercial and industry-standard components instead of customized parts, facilitating easier replacement of faulty parts and upgrades. The design team should include ethicists from the outset to ensure a patient-centric design process.
2. Right to repair laws and open sourcing: In the case of consumer electronics, right to repair laws have been adopted in some countries that allow consumers to repair products they buy or choose their own service providers instead of going through the manufacturer. Right to repair laws do not currently exist for neurotechnologies given that the devices are highly specialized and there are typically no other service providers besides the manufacturing company itself. However, with further growth of the industry, such laws and the possibility of open sourcing a company's neurotechnologies should be seriously considered to ensure continued support for patients.
3. Industry standards and compatibility: The chances of co-processor technologies becoming obsolete can be reduced if industry standards are established, enabling compatibility of device parts across companies. The neurotechnology field can look to the cardiac pacemakers industry for guidance: voluntary standards established by the industry in the 1980s have helped promote compatibility of parts and enhanced patient safety during routine part replacements. Similarly, a company that develops cochlear implants for the hearing-impaired has adopted the standard that each next-generation sound processor the company develops is compatible with all of the old implants the company has sold.

3.2 Identity and Agency

The use of a co-processor has the potential to change a user's behavior in the long term, thereby affecting their sense of identity [65–68]. Furthermore, in some cases, the user may feel that they have lost their sense of agency and ceded control to the co-processor [69]. Here, it is crucial to recognize that the loss of agency has multi-faceted effects that influence the way in which end users perceive themselves across the ethical domains of authenticity, privacy, trust, and responsibility [70]. Addressing and alleviating these potential threats to our notions of being human and having agency will be critical requirements for future co-processors.

3.3 Security and Privacy

Like most new technologies, there is a significant risk of brain co-processors being abused. Wireless communication from or to a brain could be intercepted ("brain tapping") and exploited by criminals, terrorists, commercial enterprises, or spy agencies as well as legal, law enforcement, and military entities. Brain stimulation opens up the dangerous possibility that an unsecure device may be hijacked and used to coerce a person to perform objectionable acts (e.g., commit a crime or sign a document such as a will). A device with access to memory-related regions in the brain could potentially be subverted to selectively erase memories [71] or write in false memories ("brainwashing"). Malicious entities could send a "virus" to a device, resulting in cognitive impairment or cognitive manipulation. "Brain spy-ware" could add or replace legitimate components of a BCI system to extract the users' cognitive and behavioral processes without their permission [72].

Given the potential for unprecedented abuse and malicious attacks, it is imperative that strong legal and technological safeguards are put in place before wide-spread deployment of any co-processor, for instance, by considering neurosecurity during the design process itself [73]. Activities that violate a co-processor's security and privacy should be made illegal, with stringent punishments for breaking the law. Encryption techniques and security methods will need to have much stronger guarantees against attacks than current techniques and methods. However, before we can provide those safeguards, we need to develop an understanding of the kind of information end users might want (or need) to share and the kind of information they might want to keep private. Security principles that are at least partly informed by their perspectives are crucial in order to develop the type of regulations that respect their autonomy and privacy.

3.4 Legal Issues

Given the scope for abuse, lawmakers will need to pass sufficiently nuanced legislation to regulate what type of co-processors are legal to use and what are not—this may vary from country to country similar to how laws governing controlled substances today are different in different countries. Additionally, liability laws may need to change—courts will need to decide who is responsible for accidents or unlawful acts committed using a co-processor [74]. Since co-processors use AI to adapt and learn, it may not be clear if the law was broken due to a volitional command issued by the human user or due to the action of the AI [75]. In those scenarios, the difficulty of determining who caused the action leads to a responsibility gap [76, 77]. Specifically, it is unclear who is responsible for the unintended outcome—the end user, the device, the manufacturer, or the software developer? In order to find an answer to this intricate question, it is critical to map out the types of control end users experience when using neurotechnological devices [78]. After a careful assessment of the actions end users can potentially carry out and the ways in which AI might interfere to create unintended outcomes, agreements can be formulated that regulate who is at fault for certain actions, e.g., placing full responsibility only on trained users for actions we can reasonably expect them to carry out (similar to how we assign full responsibility only to licensed drivers). This would free the co-processor company from liability except for manufacturing defects. In addition, courts could maintain panels of AI experts charged with investigating whether a company is at fault due to the behavior of the company's AI algorithm.

3.5 Moral and Social Justice Issues

The use of co-processors as an integral part of the human brain has the potential to fundamentally redefine what it means to be human by enhancing agentive or cognitive capacities beyond what is considered normal. Some authors argue that the ethical issue is not *whether* neurotechnologies should be used but *how* widely they should be used [79]. While there is consensus that the use of neurotechnologies for the treatment of disorders or diseases is ethical, it is debatable whether implanting a device into a healthy person is actually desirable. Among other topics discussed in the literature, enhancement touches upon ethical issues such as how we perceive peak performances in sport [80], assess legal consequences [81], or develop standards for education and employment [82]. Will some humans forego the advantages of augmenting their physical and mental capabilities and choose to live a co-processor-free existence? This could divide human society into a new type of "haves" and "have-nots."

Furthermore, the rich might have their children implanted at an early age to give them an edge in mental and/or physical capabilities, leaving the poor behind, with potentially drastic social consequences. One potential solution is for governments to subsidize certain basic types of co-processors for those who otherwise would not be able to afford them, similar to free public education and healthcare in certain countries. Another moral dilemma arises from parents having to decide whether or not to implant their child to augment the child's *future* mental and/or physical capabilities. Is it ethical for parents to decide what type of augmentation a child should have? Is it ethical for them to opt out of such augmentation, potentially leaving the child at a significant disadvantage in the future compared to augmented children? Should there be different schools for students with and without cognitive enhancement? These questions challenge our current conceptions of what it means to be human and point to the need for a comprehensive discussion of these issues among all stakeholders.

We hope that the ethical and moral issues raised above will help in the formulation of an internationally accepted code of regulations and ethics for co-processor development and future use.

4 Conclusion

Brain co-processors use AI to simultaneously decode neural activity from one brain region and deliver information to the same or another region, thereby "closing-the-loop" between an AI and the brain. This chapter provided a high-level overview of brain co-processors, including neural co-processors that rely on the interaction between artificial and biological neural networks to restore or augment brain function. For example, a co-processor could be used to map inputs from one memory-related area to another to facilitate or restore access to particular memories (e.g., in memory loss) or to unlearn traumatic memories (e.g., in PTSD). A related application is using a co-processor to unlearn unwanted behaviors (e.g., in obsessive compulsive disorders [OCD] or addiction) or retrain the brain in schizophrenia.

Non-medical augmentative applications of co-processors include mapping inputs from novel external sensors (e.g., infrared, ultrasonic, etc.) to augment sensation and using brain signals to control external actuators to augment motor capabilities. More generally, co-processors open up the possibility of augmenting the brain's knowledge, skills, information processing, and learning capabilities with the computational power of AI such as deep ANNs for harnessing external information and guidance from sensors, the internet, and other brains.

Although we are in the early stages of co-processor design and development, there is an urgent need to identify and address the ethical, moral, and social justice concerns associated with this new technology before it leaps too far ahead.

In this chapter, we identified health and safety, identity and agency, security and privacy, and moral issues pertaining to the use of brain co-processors. The neural engineering community will need to work closely with ethicists, medical care providers, end users, policy makers, legal experts, and the general public in formulating appropriate guidelines and best practices for the development of safe, secure, ethically informed and morally grounded brain co-processors.

Acknowledgements This work was supported by a Weill Neurohub Investigator grant and a CJ and Elizabeth Hwang endowed professorship (RPNR), National Institutes of Health grant no. MH117800-01 (AS) and National Science Foundation grant no. EEC-1028725.

Industry Funding This research was not supported by funding from industry.

References

1. Rao RPN. Brain–computer interfacing: an introduction. Cambridge: Cambridge University Press; 2013.
2. Wolpaw J, Wolpaw EW, editors. Brain–computer interfaces: principles and practice. Oxford: Oxford University Press; 2012.
3. Moritz CT, Ruther P, Goering S, Stett A, Ball T, Burgard W, Chudler EH, Rao RP. New perspectives on neuroengineering and neurotechnologies: NSF-DFG workshop report. IEEE Trans Biomed Eng. 2016;63(7):1354–67.
4. Lebedev MA, Nicolelis MA. Brain–machine interfaces: from basic science to neuroprostheses and neurorehabilitation. Physiol Rev. 2017;97(2):767–837.
5. Fetz EE. Operant conditioning of cortical unit activity. Science. 1969;163(870):955–8.
6. Delgado J. Physical control of the mind: toward a psychocivilized society. New York: Harper and Row; 1969.
7. Vidal JJ. Toward direct brain–computer communication. Annu Rev Biophys Bioeng. 1973;2:157–80.
8. Chapin JK, Moxon KA, Markowitz RS, Nicolelis MA. Real-time control of a robot arm using simultaneously recorded neurons in the motor cortex. Nat Neurosci. 1999;2(7):664–70.
9. Velliste M, Perel S, Spalding MC, Whitford AS, Schwartz AB. Cortical control of a prosthetic arm for self-feeding. Nature. 2008;453:1098–101.
10. Hochberg LR, Bacher D, Jarosiewicz B, Masse NY, Simeral JD, Vogel J, Haddadin S, Liu J, Cash SS, van der Smagt P, Donoghue JP. Reach and grasp by people with tetraplegia using a neurally controlled robotic arm. Nature. 2012;485(7398):372–5.
11. Wolpaw JR, McFarland DJ, Neat GW, Forneris CA. An EEG-based brain–computer interface for cursor control. Electroencephalogr Clin Neurophysiol. 1991;78(3):252–9.
12. Serruya MD, Hatsopoulos NG, Paninski L, Fellows MR, Donoghue JP. Instant neural control of a movement signal. Nature. 2002;416(6877):141–2.
13. Wolpaw JR, McFarland DJ. Control of a two-dimensional movement signal by a noninvasive brain–computer interface in humans. Proc Natl Acad Sci U S A. 2004;101(51):17849–54.
14. Li Z, O'Doherty JE, Hanson TL, Lebedev MA, Henriquez CS, Nicolelis MA. Unscented Kalman filter for brain–machine interfaces. PLoS One. 2009;4(7):e6243.
15. Gilja V, Pandarinath C, Blabe CH, Nuyujukian P, Simeral JD, Sarma AA, Sorice BL, Perge JA, Jarosiewicz B, Hochberg LR, Shenoy KV, Henderson JM. Clinical translation of a high-performance neural prosthesis. Nat Med. 2015;21(10):1142–5.
16. Pandarinath C, Nuyujukian P, Blabe CH, Sorice BL, Saab J, Willett FR, Hochberg LR, Shenoy KV, Henderson JM. High performance communication by people with paralysis using an intracortical brain–computer interface. elife. 2017;6:e18554.
17. Farwell LA, Donchin E. Talking off the top of your head: toward a mental prosthesis utilizing event-related brain potentials. Electroencephalogr Clin Neurophysiol. 1988;70(6):510–23.
18. Sellers EW, Kübler A, Donchin E. Brain–computer interface research at the University of South Florida Cognitive Psychophysiology Laboratory: the P300 speller. IEEE Trans Neural Syst Rehabil Eng. 2006;14(2):221–4.
19. Bell CJ, Shenoy P, Chalodhorn R, Rao RPN. Control of a humanoid robot by a noninvasive brain–computer interface in humans. J Neural Eng. 2008;5(2):214–20.
20. Galán F, Nuttin M, Lew E, Ferrez PW, Vanacker G, Philips J, del Millán JR. A brain-actuated wheelchair: asynchronous and non-invasive brain–computer interfaces for continuous control of robots. Clin Neurophysiol. 2008;119(9):2159–69.
21. del Millán JJR, Galán F, Vanhooydonck D, Lew E, Philips J, Nuttin M. Asynchronous non-invasive brain-actuated control of an intelligent wheelchair. Conf Proc IEEE Eng Med Biol Soc. 2009;2009:3361–4.
22. Bryan M, Nicoll G, Thomas V, Chung M, Smith JR, Rao RPN. Automatic extraction of command hierarchies for adaptive brain-robot interfacing. Proc ICRA. 2012;2012:5–12.

23. Niparko J, editor. Cochlear implants: principles and practices. 2nd ed. Philadelphia: Lippincott; 2009.
24. Weiland JD, Liu W, Humayun MS. Retinal prosthesis. Annu Rev Biomed Eng. 2005;7:361–401.
25. Tomlinson T, Miller LE. Toward a proprioceptive neural interface that mimics natural cortical activity. Adv Exp Med Biol. 2016;957:367–88.
26. Tabot GA, Dammann JF, Berg JA, Tenore FV, Boback JL, Vogelstein RJ, Bensmaia SJ. Restoring the sense of touch with a prosthetic hand through a brain interface. Proc Natl Acad Sci U S A. 2013;110(45):18279–84.
27. Tyler DJ. Neural interfaces for somatosensory feedback: bringing life to a prosthesis. Curr Opin Neurol. 2015;28(6):574–81.
28. Dadarlat MC, O'Doherty JE, Sabes PN. A learning-based approach to artificial sensory feedback leads to optimal integration. Nat Neurosci. 2015;18(1):138–44.
29. Flesher SN, Collinger JL, Foldes ST, Weiss JM, Downey JE, Tyler-Kabara EC, Bensmaia SJ, Schwartz AB, Boninger ML, Gaunt RA. Intracortical microstimulation of human somatosensory cortex. Sci Transl Med. 2016;8(361):361ra141.
30. Cronin JA, Wu J, Collins KL, Sarma D, Rao RP, Ojemann JG, Olson JD. Task-specific somatosensory feedback via cortical stimulation in humans. IEEE Trans Haptics. 2016;9(4):515–22.
31. Rao RPN. Brain co-processors: using AI to restore and augment brain function. In: Thakor NV, Garg A, Nargund S, Nuan C, editors. Handbook of neuroengineering. Cham: Springer; 2022.
32. Rao RPN. Towards neural co-processors for the brain: combining decoding and encoding in brain–computer interfaces. Curr Opin Neurobiol. 2019;55:142–51.
33. Coin A, Mulder M, Dubljević V. Ethical aspects of BCI technology: what is the state of the art? Philosophies. 2020;5(4):31.
34. Coin A, Dubljević V. The authenticity of machine-augmented human intelligence: therapy, enhancement, and the extended mind. Neuroethics. 2021;14(2):283–90.
35. Sutton RS, Barto A. Reinforcement learning: an introduction. 2nd ed. Cambridge: MIT Press; 2018.
36. Kalman RE. A new approach to linear filtering and prediction problems. J Basic Eng. 1960;82:35–45.
37. Bryson AE, Ho Y. Applied optimal control. New York: Halsted Press; 1975.
38. O'Doherty JE, Lebedev MA, Hanson TL, Fitzsimmons NA, Nicolelis MA. A brain–machine interface instructed by direct intracortical microstimulation. Front Integr Neurosci. 2009;3:20.
39. O'Doherty JE, Lebedev MA, Ifft PJ, Zhuang KZ, Shokur S, Bleuler H, Nicolelis MA. Active tactile exploration using a brain–machine–brain interface. Nature. 2011;479(7372):228–31.
40. Klaes C, Shi Y, Kellis S, Minxha J, Revechkis B, Andersen RA. A cognitive neuroprosthetic that uses cortical stimulation for somatosensory feedback. J Neural Eng. 2014;11(5):056024.
41. Flesher S, et al. Intracortical microstimulation as a feedback source for brain–computer interface users. Brain Comput Interface Res. 2017;6:43–54.
42. Bouton CE, Shaikhouni A, Annetta NV, Bockbrader MA, Friedenberg DA, Nielson DM, Sharma G, Sederberg PB, Glenn BC, Mysiw WJ, Morgan AG, Deogaonkar M, Rezai AR. Restoring cortical control of functional movement in a human with quadriplegia. Nature. 2016;533(7602):247–50.
43. Ajiboye AB, Willett FR, Young DR, Memberg WD, Murphy BA, Miller JP, Walter BL, Sweet JA, Hoyen HA, Keith MW, Peckham PH, Simeral JD, Donoghue JP, Hochberg LR, Kirsch RF. Restoration of reaching and grasping movements through brain-controlled muscle stimulation in a person with tetraplegia: a proof-of-concept demonstration. Lancet. 2017;389(10081):1821–30.
44. Jackson A, Mavoori J, Fetz EE. Long-term motor cortex plasticity induced by an electronic neural implant. Nature. 2006;444(7115):56–60.
45. Guggenmos DJ, Azin M, Barbay S, Mahnken JD, Dunham C, Mohseni P, Nudo RJ. Restoration of function after brain damage using a neural prosthesis. Proc Natl Acad Sci U S A. 2013;110(52):21177–82.
46. Berger T, Hampson R, Song D, Goonawardena A, Marmarelis V, Deadwyler S. A cortical neural prosthesis for restoring and enhancing memory. J Neural Eng. 2011;8(4):046017.

47. Deadwyler SA, Hampson RE, Song D, Opris I, Gerhardt GA, Marmarelis VZ, Berger TW. A cognitive prosthesis for memory facilitation by closed-loop functional ensemble stimulation of hippocampal neurons in primate brain. Exp Neurol. 2017;287(Pt 4):452–60.
48. Rao RP, Stocco A, Bryan M, Sarma D, Youngquist TM, Wu J, Prat CS. A direct brain-to-brain interface in humans. PLoS One. 2014;9(11):e111332.
49. Rao RPN, Stocco A. When two brains connect. Sci Am Mind. 2014;25:36–9.
50. Stocco A, Prat CS, Losey DM, Cronin JA, Wu J, Abernethy JA, Rao RPN. Playing 20 questions with the mind: collaborative problem solving by humans using a brain-to-brain Interface. PLoS One. 2015;10(9):e0137303.
51. Grau C, Ginhoux R, Riera A, Nguyen TL, Chauvat H, Berg M, Amengual JL, Pascual-Leone A, Ruffini G. Conscious brain-to-brain communication in humans using non-invasive technologies. PLoS One. 2014;9(8):e105225.
52. Lee W, Kim S, Kim B, Lee C, Chung YA, Kim L, Yoo SS. Non-invasive transmission of sensorimotor information in humans using an EEG/focused ultrasound brain-to-brain interface. PLoS One. 2017;12(6):e0178476.
53. Nicolelis MAL. Beyond boundaries. New York: Macmillan; 2011.
54. Pais-Vieira M, Lebedev M, Kunicki C, Wang J, Nicolelis MA. A brain-to-brain interface for real-time sharing of sensorimotor information. Nat Sci Rep. 2013;3:1319.
55. Pais-Vieira M, Chiuffa G, Lebedev M, Yadav A, Nicolelis MA. Building an organic computing device with multiple interconnected brains. Nat Sci Rep. 2015;9(5):11869.
56. Jiang L, Stocco A, Losey DM, Abernethy JA, Prat CS, Rao RPN. BrainNet: a multi-person brain-to-brain interface for direct collaboration between brains. Nat Sci Rep. 2019;9(1):6115.
57. Losey DM, Stocco A, Abernethy JA, Rao RPN. Navigating a 2D virtual world using direct brain stimulation. Front Robot AI. 2016;3:72.
58. Farwell LA, Donchin E. The truth will out: interrogative polygraphy ("lie detection") with event-related brain potentials. Psychophysiology. 1991;28(5):531–47.
59. Kozel FA, Johnson KA, Mu Q, et al. Detecting deception using functional magnetic resonance imaging. Biol Psychiatry. 2005;58:605–13.
60. Klein E. Informed consent in implantable BCI research: identifying risks and exploring meaning. Sci Eng Ethics. 2016;22:1299–317.
61. Klein E, Ojemann J. Informed consent in implantable BCI research: identification of research risks and recommendations for development of best practices. J Neural Eng. 2016;13:043001.
62. Klein E. Informed consent for next-generation deep brain stimulation psychiatric research: engaging end users to understand risks and improve practice. Research involving participants with cognitive disability and difference. Oxford: Oxford University Press; 2019. p. 149–60.
63. Strickland E, Harris M. Their bionic eyes are now obsolete and unsupported. IEEE Spectr. 2022. https://spectrum.ieee.org/bionic-eye-obsolete
64. Strickland E, Harris M. Should right-to-repair laws extend to bionic body parts? IEEE Spectr. 2022. https://spectrum.ieee.org/bionic-right-to-repair
65. Goering S, Yuste R. On the necessity of ethical guidelines for novel neurotechnologies. Cell. 2016;167(4):882–5.
66. Baylis F. "I am who I am": on the perceived threats to personal identity from deep brain stimulation. Neuroethics. 2013;6:513–26.
67. Bluhm R, Cabrera L, McKenzie R. What we (should) talk about when we talk about deep brain stimulation and personal identity. Neuroethics. 2019;13:289–301.
68. Mackenzie C, Walker M. Neurotechnologies, personal identity, and the ethics of authenticity. In: Clausen J, Levy N, editors. Handbook of neuroethics. Dordrecht: Springer; 2015. p. 373–92.
69. Yuste R, Goering S, et al. Four ethical priorities for neurotechnologies and AI. Nature. 2017;551(7679):159–63.
70. Schönau A, Dasgupta I, Brown T, Versalovic E, Klein E, Goering S. Mapping the dimensions of agency. AJOB Neurosci. 2021;12:172–86.
71. Rich MT, Huang YH, Torregrossa MM. Plasticity at thalamo-amygdala synapses regulates cocaine-cue memory formation and extinction. Cell Rep. 2019;26(4):1010–1020.e5.

72. Bonaci T, Calo R, Chizeck HJ. App stores for the brain: privacy and security in brain–computer interfaces. IEEE Technol Soc Mag. 2015;34:32–9.
73. Denning T, Matsuoka Y, Kohno T. Neurosecurity: security and privacy for neural devices. Neurosurg Focus. 2009;27:E7.
74. Grübler G. Beyond the responsibility gap. Discussion note on responsibility and liability in the use of brain–computer interfaces. AI Soc. 2011;26:377–82.
75. Gurney D. Killer robot arms: a case-study in brain–computer interfaces and intentional acts. Mind Mach. 2018;28:775–85.
76. Porter Z, Habli I, Monkhouse H, Bragg J. The moral responsibility gap and the increasing autonomy of systems. In: Hoshi M, Seki S, editors. Developments in language theory. Cham: Springer International Publishing; 2018. p. 487–93.
77. Rainey S, Maslen H, Savulescu J. When thinking is doing: responsibility for BCI-mediated action. AJOB Neurosci. 2020;11:46–58.
78. Schönau A. The spectrum of responsibility ascription for end users of neurotechnologies. Neuroethics. 2021;14:423–35.
79. Demetriades AK, Demetriades CK, Watts C, Ashkan K. Brain–machine interface: the challenge of neuroethics. Surgeon. 2010;8:267–9.
80. Fronda G, Crivelli D, Balconi M. Neurocognitive enhancement: applications and ethical issues. NeuroRegulation. 2019;6:161–8.
81. Errigo MC. Neuroenhancement and law. In: D'Aloia A, Errigo M, editors. Neuroscience and law. Cham: Springer; 2020. https://doi.org/10.1007/978-3-030-38840-9_10.
82. Bard I, Gaskell G, Allansdottir A, et al. Bottom up ethics—neuroenhancement in education and employment. Neuroethics. 2018;11:309–22. https://doi.org/10.1007/s12152-018-9366-7.

Part III

Legal and Policy Implications

United States Policy on BCIs: Funding Research, Regulating Therapies, and Commercializing Consumer Technology

Robert H. Blank

1 Introduction

The range of BCIs is broad, and there remains some disagreement as to what qualifies. However, BCIs promise many applications in medicine, communication, rehabilitation, the military, and education. Although preliminary results of BCI research are promising and generate considerable media coverage, there are many barriers to medical uses of BCIs, and most BCI systems remain as prototypes [1, 2]. In order to succeed, medical BCIs must be cost-effective to be reimbursed by health insurance and designed to fit in the needs of users. Not surprisingly, rehabilitation professionals are less optimistic about the state-of-the-art of BCI technology than the BCI developers [3].

Although most BCIs are still in the early research stages, recently attention and apprehension have grown over their development. Considerable media fanfare and concerns followed Elon Musk's announcement of Neuralink's plans for invasive BCIs. Moreover, the announcement by Facebook in 2017 that it was working on a wearable device that would allow users to think up to 100 words per minute to then be converted into text triggered considerable media attention. In the same year, Neurable launched the world's first brain-controlled virtual reality (VR) game. In 2020, NextMind introduced a wearable EEG-based device to record the brain's electrical activity using machine learning to translate it into commands.

As discussed in detail elsewhere [4, 5], there are three main approaches for recording brain signals with BCIs. The distinction is crucial for public policy for reasons discussed below. Non-invasive methods record signals from sensors applied on the scalp to track and record brain activity. Until now, the most common method is EEG, but use of fMRI, MEG, PET, and fNIRS has increased [6]. Non-invasive BCIs can be placed and removed easily but record inferior signals. In contrast,

R. H. Blank (✉)
Political Science Department, University of Canterbury, Christchurch, New Zealand

V. Dubljević, A. Coin (eds.), *Policy, Identity, and Neurotechnology*, Advances in Neuroethics, https://doi.org/10.1007/978-3-031-26801-4_11

invasive BCIs require surgery to implant electronic devices directly into the cortex. Invasive methods provide much clearer and accurate signals between the brain and the device, but the implant procedures come with many risks including pulmonary embolisms, hemorrhages, seizures, and infection as well as post-surgery complications [7, 8]. For this reason, partially invasive BCI devices are now being implanted inside the skull but outside the brain. For instance, electrocorticography (ECoG) is a promising intermediary method because it offers more accurate results than non-invasive BCIs and is less risky than invasive BCIs [6].

A commonly used BCI is the cochlear implant that bypasses the auditory apparatus and allows the person to hear sounds from outside by converting them into electrical signals that directly reach the brain. Over 200,000 adults and children in the USA have benefitted from this device [9]. Similarly, there have been many experimental measurements of brain activity for human control commands of wheelchairs [10]. Investigational BCI systems are being developed in rehabilitation both as neuroprostheses to replace lost function and as potential plasticity-enhancing therapy aimed at aiding neurorecovery [11, 12]. BCIs also offer much promise to restore the ability to communicate in stroke and paralyzed persons who cannot speak due to aphasia [13] or anarthria [14] as well as ALS [15]. An estimated 40 million Americans have communication disorders [16]. With mixed results, researchers from several VA medical centers tested how a BCI system designed to aid with communication worked when ALS patients used it independently without a technician on hand [17]. It also might be used to treat elderly patients suffering from physical impairments and people suffering from psychiatric disorders such as schizophrenia and depression [18]. Although BCIs were initially conceived for biomedical applications, much current research and investment have been extended to develop BCIs for healthy persons which raises numerous other issues and to gaming and VR. To date, most BCI applications available to the public are non-invasive.

2 Ethical Issues of BCIs

Although ethics discussions are largely absent in BCI studies published in technical journals [19], recently there has been an increase in works that address social and ethical issues raised by BCI research [20–22]. Some have focused on specific issues such as personhood [23], authenticity [24], and treatment of BCI-mediated action [25]. Others have reviewed a range of ethical issues found in this emerging literature. Useful reviews by Burwell et al. [6] and Coin et al. [26] confirm that informed consent is a frequently mentioned ethical issue followed by concern over research ethics and safety, privacy and security, and issues of stigma and justice. Informed consent is especially problematic because many potential BCI patients suffer from diseases that can be comorbid with dementia and other cognitive deficiencies that develop over the course of treatment, rendering them unable to provide continuous consent [20]. Ensuring full decision-making capacity and confirming consent at each stage of the BCI treatment process is critical for preserving users' agency and autonomy.

Moreover, while BCIs promise to increase quality of life and contribute to a higher level of independence for persons with physical impairments, they can also lead to impingements on human autonomy, a creation of dependency, and confusion over self-perception. Implanting any device in the brain has the potential to disrupt an individual's identity or sense of self in unpredictable ways [27]. Until now, the qualitative self-experience of BCI users including aspects related to personal identity, agency, and responsibility has seldom been examined [20].

Although ethical analyses have increased, Coin et al. [26] note that questions of BCI policy are rare in the literature on BCI ethics. There has been little discussion of what would be the best public policies to regulate the development and use of BCI technologies, how governments might regulate them, and who will have access and who will pay. Moreover, since most commercial activity occurs in the international sphere, what implications do policies in one country have on the diffusion of BCIs? Some important issues related to BCIs that require policy attention are:

- Guaranteeing informed consent
- Questions of research ethics and oversight
- Risk/benefit analysis
- Anticipated and unintended consequences
- Licensing of practitioners
- Liability issues
- Fair practice and protection of consumers
- Use/protection of brain data
- Equity and social stratification
- Resource allocation

Clearly the breadth of the issues means that no single government entity can address them all.

3 Public Policy Context

Although some scientific and bioethics works and occasional media stories note the political and policy dimensions of BCIs, with few exceptions, they are not highlighted. The eventual move of the BCI debate to the policy domain will alter the context by bringing to the forefront political considerations and divisions and placing the resolution of these issues in the milieu of interest group politics. Given the significant economic, social, and personal stakes surrounding BCIs, this is unavoidable. The emerging policy issues are framed by the frequent announcements of new technologies by a mass media that tends to dramatize them and heighten expectations. As Pham and Gilbert [21] note: "positively biased narratives surrounding BCIs in the media, that make speculative promises about their uses while failing to address risks and ethical issues, can create serious problems related to informed consent, among other things."

BCIs raise challenging policy issues and trade-offs that reveal a need for more systematic and anticipatory analysis of the social consequences of these innovations. The gap between the rapid advancement of BCI technologies and sluggish development of the legal and social frameworks poses significant challenges for policy makers. Moreover, a dependence on technological solutions to health and social problems makes it difficult to slow diffusion of new technologies. Media hype and active marketing and publicity often promote their use long before their risks are fully understood. Although BCI usage is currently limited, research is moving quickly. Rather than reacting retrospectively to the inevitable issues that the proliferation of BCIs will engender, now is the time for anticipatory policy making.

Although many of the issues raised by BCIs are distinctive, fundamentally, the policy dimensions are similar to other areas of biomedical research. At their base, there are three relevant dimensions. First, decisions must be made concerning the research and development of the technologies. Because a considerable proportion of this research is funded either directly or indirectly with public funds, civilian and military, it is important that public input be included at this stage. Despite a growing prominence of forecasting and assessing the social as well as technical consequences of technologies early in the process, it remains problematic as to how to best design assessment processes to evaluate efficacy, short-and long-term safety, and the social impact of BCIs, especially when there is a growing market and demand for them.

The second policy dimension relates to the individual access to and use of technologies. Although direct governmental intrusion into individual decision-making in the medical arena is, by its nature, restrained, governments have at their disposal an array of strategies to encourage or discourage individual use, including tax incentives or disincentives, the provision of services, licensing, and education programs. Although conventional regulatory mechanisms might be utilized to protect potential users or targets of BCIs, at a minimum the government has a responsibility of ensuring safety and quality control standards as well as consumer protection and fair market practices.

The third dimension of BCI policy centers on the aggregate consequences of widespread use, particularly for non-medical purposes. For instance, what impact might enhancement applications have on society? Will they aggrandize social inequalities or break down social barriers? Policy making here requires a clear conception of goals, data to predict the consequences of each possible course of action, and an accurate means of monitoring these consequences. Furthermore, some forms of BCI are likely to be expensive, posing questions of affordability and coverage under health care plans. They could add considerably to the costs of health care without proportionate benefit. Is investment in BCI technology the best use of limited medical research resources, and should it be a high priority for public funding? Also, if the regulatory issues over medical devices make it more financially feasible for companies to focus on consumer devices, this could limit the ability of people with severe disabilities to access BCI as an assistive technology [6].

Government involvement can occur at many points from basic research and innovation stages to placing a technology on the market. Basically, BCI policy can be permissive, affirmative, regulatory, or prohibitive. Theoretically, a government can opt to take no action, thus allowing unfettered activity by the private sector. Or, it can make affirmative policies that promote or encourage certain activities, for example, public funding of research or provision of services to facilitate its use. The question of whether the government ought to be providing such encouragement, and if so by what means, will be a matter of debate. Should public funds be used to pay for BCI interventions? Should private insurers be required to cover these expenses? Should we distinguish between medical treatment and uses by healthy individuals for enhancement?

Prohibitive policies could be implemented that would reduce the options available at each stage of BCIs. The most straightforward form would be to impose criminal sanctions on a particular research activity or application. A softer type of prohibitive policy is to preclude public funding of specific areas of research and development (as with certain types of embryo/fetal research) or specific services. Not surprisingly, these policies often reflect political motives or a response to the demands of opposing interest groups. Since it is unlikely that any BCI methods will be banned, attention here focuses on regulation. Although regulatory policy can apply solely to government-supported activities, it normally consists of sweeping rules governing activities in both the public and private sectors.

Throughout the policy process, governments have many mechanisms for facilitating expert input. Permanent mechanisms include the use of internal bureaucratic expertise, science advisors, offices of science, and technology and science advisory councils. Temporary mechanisms comprise task forces, ad hoc committees, commissions, consultants, conferences, hearings, and issues papers. Their remit can be specific to a particular application such as non-invasive BCIs or wider in scope.

At the broadest level, the policy controversy over BCIs will center on a clash between public and private regulation. Although a government has ultimate responsibility for the health of its population, the dominance of the medical model and the power of the private sector have meant that a significant proportion of medical care in the USA has remained the domain of non-public interests. Therefore, the range of regulatory options is more complicated than the public–private distinction suggests. Figure 1 illustrates the scope of options available for BCIs. Given its complexity, it is likely that a workable approach must involve some combination of these mechanisms.

Regulatory policies are problematic because rapidly advancing technologies and alterations in social values raise the prospects of obsolescence of any regulation no matter how carefully drafted. Legislation, in particular, risks freezing technology in place and the inability to offer the flexibility needed to adapt to new applications. Furthermore, the moral underpinning of the debate over BCIs means that legislation

Fig. 1 Regulatory mechanisms

could be made based on emotions rather than rational choice. There is no guarantee that government involvement will be objective, nor helpful, in resolving the social issues, and it could even exacerbate them. Attempts to fit medical BCIs into models used for other areas of public policy also fail to account for several unique features of medicine. First, traditionally the conduct of medical decision-making has been based on professional judgments monitored primarily by professional standards of care. A second special feature of BCIs is their focus on the human brain and any intervention risks compromising constitutionally based liberties and the common law principle of self-determination.

4 BCI Policy Making in USA

Policy making in the USA is complicated by federalism where both the nation and the states have potential roles in framing BCI policy and by the separation of powers among branches. In part because of this fragmented authority, policy making in the USA is a measured process, not manifested in quick, decisive action. Thus, any policy on BCIs is likely to come in fits and starts in an unsystematic manner. Moreover, the crisis-oriented emphasis is on ills to be remedied rather than on positive goals to be met, and the analysis of future consequences is scant. Since BCIs are viewed as a future issue, they are unlikely to engender much attention by policy makers.

The USA is the prototype of an individualistic society. Although individual rights have a role in all democratic countries, in the USA they enjoy supremacy over collective interests. This cultural tenet helps explain why the USA expends substantially more of its GDP on health care than any other country without providing universal access. The U.S. culture also exhibits an unrealistic dependence on technology to fix health problems at the expense of public health. This demand for technological solutions is bolstered by the dominance of medical specialists who expand the indications for use of innovations and a robust liability system that encourages overuse, thus leading an early and wide diffusion of new diagnostics and treatment modalities as compared to other countries [28]. It will be critical for BCI developers as to what FDA rules apply to BCI software and whether they will be protected from lawsuits. If not, BCI companies might develop their devices in places where U.S. rules and lawsuits cannot reach.

Another critical policy issue involves questions of distributive justice. While it is premature to speculate about the relative costs and benefits of yet undeveloped BCI treatments, cumulatively their cost could be significant. If access is to be equitable, how will BCIs be funded? Assuming they will be used largely by the elderly, those on disability, or veterans, all BCIs will have to get approval for reimbursement by either the Centers for Medicare and Medicaid Services (CMS) or the Veterans' Administration (VA). Although it is possible that some private insurers would pay for BCI procedures in the absence of CMS approval, it is unlikely.

Although BCIs currently are not high on the public's radar, the Pew Research Center conducted two surveys of U.S. adults on potential uses of BCIs that offer valuable insights [29, 30]. Although both found large majorities in favor of medical applications, by a 56–13% margin, respondents felt that the use of BCIs to enhance cognitive function in healthy individuals would be bad for society. Seventy-eight percent would *not* want an implant to better process information. Moreover, a large majority worry that BCIs will increase the wealth inequality that exists between "haves" and "have nots." Overall, 57% say that the widespread use would increase the wealth gap, while just 10% think it would decrease it.

When asked which statement better describes their views about the widespread use of BCIs on healthy people, 63% say this idea "is meddling with nature and crosses a line we should not cross." Far fewer (35%) say that "as humans we are always trying to better ourselves and this is no different." Not surprisingly, there are sizable differences by religious commitment and across religious groups. An overwhelming majority (81%) of highly religious Americans say that the widespread use of computer chip implants is meddling with nature, while those with low religious commitment are closely divided. Despite their concerns, six-in-ten think that if BCI use became widespread, people would feel pressure to get an implant. Furthermore, 78% say BCIs will be used before we fully understand how they affect health. As a result, 83% think these implants should be tested using a higher standard than is used for medical devices. As noted by Chan "Without competent and smart regulation from the very start, negative public reaction may conceivingly lead to a moratorium or outright ban on neuroelectronics... A laissez-faire approach is not the solution" [31].

5 Reports on BCIs

The USA would benefit from authoritative reports such as that of UK Royal Society [32] which called for the creation of a public-driven, flexible regulatory framework for BCI technology. It warned against the dangers of commercializing BCIs, especially if Big Tech manages to obtain monopolistic access to human thoughts and ideas for financial gain. Among its recommendations, it urged the government to launch a national investigation to clarify the ethical issues behind BCIs, create a UK Neural Interface Ecosystem to promote greater sharing of technology and increased output of new ideas, and create regulatory frameworks that offer the best methods for ensuring innovation while curbing the tendency of big companies to capture the field. Previously, the European Commission funded a coordination and support action for the BCI community called "BNCI Horizon 2020" [33]. Major goals included developing a roadmap for the next decade and beyond, encouraging discussion within the BCI community, fostering communication with the public, and the creation of an official BCI Society.

A recent promising sign is that the US Government Accounting Office (GAO) launched a Science, Technology Assessment, and Analytics (STAA) team to help Congress understand and address critical trends that profoundly affect the nation. They identified BCIs/augmented reality of one of five emerging technologies that will potentially transform society [34]. Since its inception, the STAA team has expanded its network of experts to increase the depth, breadth, and diversity of its knowledge. In October 2020, the team hosted an inaugural meeting of the Polaris Council, a group of science, technology, and policy leaders and experts from many fields, established to advise it on emerging science and technology issues facing Congress. Hopefully, BCIs will soon be a target for concentrated analysis by the GAO.

6 Public Funding of BCI Research in the USA

The National Institutes of Health (NIH) comprised of 27 Institutes and Centers is the primary federal agency conducting and supporting basic, clinical, and translational medical research, and investigating the causes, treatments, and cures for both common and rare diseases. Key institutes for BCIs include the National Institute of Neurological Disorders and Stroke (NINDS), the major funder of research on the brain and nervous system, and the National Institute on Deafness and Other Communication Disorders (NIDCD) that supports and conducts research on the processes of hearing, balance, taste, smell, voice, speech, and language.

The NIH's Brain Research through Advancing Innovative Neurotechnologies (BRAIN) Initiative began in 2013 as a comprehensive effort to accelerate neuroscience research [35]. It is collaboratively managed by ten institutes, including the NIDCD. The BRAIN Initiative is supported by Congress through the regular appropriations process and with funds from the Twenty-first Century Cures Act. To date, more than 900 awards, totaling approximately $1.8 billion, have been made by the Initiative, with anticipated $5.8 billion in total funding through 2026. Congress allocated $560 million to the Initiative for the 2021 fiscal year, a $60 million increase from 2020 [36]. A primary aim of the BRAIN Initiative is to build tools and knowledge resources across diverse fields for understanding how neural circuits function. For instance, one Initiative-funded project, BrainGate2, focuses on restoration of the capacity to communicate in patients with spinal cord injuries and neurological disorders [37].

The BRAIN Multi-Council Working Group [38] includes representatives from each of the ten institutes and centers that contribute to the BRAIN Initiative, ex officio members from DARPA, FDA, IARPA, and NSF which comprise NIH's federal partners in the Initiative, and at-large members appointed to supplement the Group's expertise. It provides oversight of the long-term scientific vision of the Initiative and serves as a forum for initial "concept clearance" of ideas for new enterprises before they become funding announcements. In addition, the Working Group ensures that each of the BRAIN IC Advisory Councils is informed about proposals and awards as well as assessment of the progress of current projects and programs supported by the Initiative.

In 2017, NIH began funding neuroethics research as part of the BRAIN Initiative and in 2018 it announced support for embedding ethicists into Initiative-supported research and encouraged incorporating neuroethics into existing awards [39]. In 2019, the Advisory Committee to the Director (ACD) Working Group on BRAIN

2.0 Neuroethics Subgroup (BNS) was formed to consider the ethical implications of ongoing research and forecast what the future advancements might entail by crafting a neuroethics "roadmap" for the Initiative [40]. The BNS conducted a review, held a workshop on neuroethical issues posed by such research, and presented a draft report in Spring 2019. The final report, *The BRAIN Initiative and Neuroethics: Enabling and Enhancing Neuroscience Advances for Society,* was released in Fall 2019.

The Defense Advanced Research Projects Agency (DARPA), an arm of the US Department of Defense, has funded research and development on BCIs since the 1970s. Two broad categories of DARPA programs for BCIs include efforts aimed at restoring neural and/or behavioral function of injured veterans and those intended to improve human training and performance of active duty members. As to the second category, it should be noted that the use of BCIs to improve the performance of military personnel and warfare capabilities raises unique ethical questions. A key issue is whether soldiers are able to give free informed consent. Moreover, there is the possibility that they might receive invasive forms of BCIs that would be problematic when they return to civilian life.

In 1999, the Brain Machine Interface program was launched with the goal of enabling service members to communicate by thought alone. Since then, at least eight DARPA programs have funded research to restore memory and treat psychiatric disorders. Meanwhile, DARPA has funded six groups, mostly in academia, to develop a device capable of instantly sensing and stimulating the brain. As part of a larger Neurally Enhanced Operations (NEO) project, researchers from Johns Hopkins University Applied Physics Laboratory have demonstrated the ability to "feel" virtual objects by integrating neural stimulation in a mixed-reality environment. Founded in 1942, the Applied Physics Laboratory is the nation's largest university-affiliated research center with 7200 staff and a budget of $1.52 billion [41].

DARPA recently awarded funding to six organizations to support the Next-Generation Nonsurgical Neurotechnology (N3) program. Teams selected for DARPA's N3 pursue a mix of approaches to developing wearable interfaces for communicating with the brain. By removing the need for surgery, N3 systems promise to expand the pool of patients to the civilian population to manage neurological illnesses. Currently, federal regulators are cooperating with DARPA to help the teams better understand human-use clearance. As the work progresses, these regulators will help guide strategies for submitting applications for Investigational Device Exemptions and Investigational New Drugs to enable human trials of N3 systems. Importantly, once the N3 program participants prove a technology's feasibility, it will be up to the commercial world to market it [42].

A recent comprehensive report by RAND Corporation examined military applications of BCI and their potential risks as part of its Security 2040 initiative which explores new technologies that are shaping the future of global security [7]. The report assessed current and potential BCI applications to ensure that they respond to actual needs, practical realities, and legal and ethical considerations. Among its recommendations, it called for planning ahead for BCI technology's implications including ethical and policy issues. Based on their analysis of current BCI development and the types of tasks that future tactical military units might face, the RAND team created a toolbox to catalog how BCIs might be useful. While noting that some BCI functions may be available within a relatively short time (within a couple of decades or so) and others could take much longer, the report concludes that it is crucial now to begin analyzing emerging technologies from a policy perspective. "We have an opportunity to get ahead of the game. This is something we should be thinking about now, before BCI technologies become a reality in the everyday world" [7].

7 Regulating BCIs in the USA

The Office for Human Research Protections (OHRP) provides leadership in the protection of the rights and well-being of subjects involved in research conducted or supported by the DHHS. As in most countries, institutional review boards (IRBs) are charged with providing an independent evaluation that proposed research is ethically acceptable and reviewing compliance with regulations and laws designed to protect human subjects [43]. Review is required for research funded in any part by all federal agencies, as well as for research testing interventions, including devices under the jurisdiction of the FDA. Private research institutions often extend federal regulatory requirements to all human subjects research [44]. Research conducted outside the USA but funded by the US government is subject to the same regulations or equivalent protections. To be ethically acceptable and comply with regulatory requirements, the IRB determines that risks to subjects are minimized and reasonable in relation to the importance of the knowledge the study is expected to produce, that the process and outcomes of subject selection are fair, and that there are acceptable plans for obtaining informed consent. Furthermore, the Portability and Privacy Act (HIPAA) would be appropriate to regulate the data gathered in BCI use.

Statutory authority to promote the safety, effectiveness, and ethical marketing of BCIs in the USA rests with the Consumer Product Safety Commission (CPSC), the FDA, and the Federal Trade Commission (FTC). They use a variety of formal and informal mechanisms to oversee and monitor technological developments, novel uses, and marketing of new products. In 2012, the Neuromarketing Science and Business Association (NMSBA) adopted a code of ethics for its members, though it is not clear how stringently and uniformly this code is enforced [45]. Also, OpenBCI is an organization dedicated to creating open standards for BCI devices.

All invasive and some non-invasive BCIs will be regulated by the FDA, an agency within the DHHS. The FDA is mandated to protect the public health by ensuring the safety, effectiveness, and security of drugs, vaccines and other biological products, and medical devices. Traditionally, the FDA approval process has been extremely slow which has led to complaints from stakeholders. In response to political pressures, the FDA introduced the Breakthrough Devices Program (BDP) in 2016 to expedite the development and review of certain medical devices and device-led combination products for effective treatment or diagnosis of life-threatening or irreversibly debilitating diseases or conditions. The aim is to provide timely access to these devices and lower the burden required for Medicare reimbursement while preserving the statutory standards for premarket approval, clearance, and de novo marketing authorization. The BDP offers manufacturers an opportunity to interact with FDA experts to address topics arising during the premarket review phase and receive timely feedback.

In August 2020, the FDA granted Breakthrough Device Designation (BDD) to Stentrode, the first implantable device delivered to the brain through blood vessels rather than open brain surgery [46]. Manufactured by Synchron, it has been implanted in patients with upper-limb paralysis. Safety and efficacy data from the clinical trial will be used to finalize the protocol for a FDA-enabling study to guide evaluation for marketing approval. Future research is planned to assess its use in patients with paralysis due to spinal cord injury, ALS, stroke, and muscular dystrophy.

In April 2021, IpsiHand became the first BCI device to receive FDA market approval for clinical use in the USA [47]. The IpsiHand device consists of two separate parts—a wireless exoskeleton that is positioned over the wrist and a small headpiece that records brain activity with EEG electrodes. Neurolutions has begun commercialization of the device with the goal to make it clinically accessible later in 2021. Earlier, in March 2021, NeuroPace, Inc. received BDD status for the use of its RNS System to treat idiopathic generalized epilepsy. Under the program, the FDA provides NeuroPace with priority review for clinical trial protocols and commercialization decisions. The BDD can also facilitate Medicare reimbursement following FDA approval of the technology. On April 14, NeuroPace filed proposed terms for its $85 million IPO to market the RNS system.

Another FDA initiative that is relevant for some non-invasive BCIs is the de novo classification pathway for low- to moderate-risk devices. Traditionally, these devices were summarily classified as class III devices requiring full review, but since 2012 the FDA considers the de novo classification appropriate for devices that do not fit into any particular class or have no equivalent device that is currently marketed. In addition, the device has to be low-to moderate-risk and meet all the requirements for classification as a class I or II device. Normally, the manufacturer submits a pre-submission to the FDA which then determines whether the de novo process is appropriate. If so, the FDA provides information on the documentation necessary to submit the application. Approval of a device depends on whether its manufacturer has conducted an effective search for an equivalent currently marketed device, determined the risks and identified mechanisms to decrease such risks, and collected enough data for the FDA to determine its safety and efficacy.

In 2021, the FDA issued leapfrog guidance for non-clinical testing and study design for implanted BCIs for patients with paralysis or amputation [48]. Leapfrog guidance allows the agency to share its initial thoughts on emerging technologies early in development. The guidance covers considerations for Investigational Device Exemptions (IDEs) feasibility and provides non-clinical testing and clinical study design recommendations for implanted BCI devices. In addition, in order to explore future innovations, the FDA established the Emerging Sciences Working Group, a team of 15 FDA experts representing various specialties and FDA Centers. The Group is charged with leveraging scientific expertise and resources to conduct long-range forecasting and advising FDA Center leadership on how emerging issues and cross-cutting scientific advances may affect the FDA's activities.

Finally, the FDA's pilot Pre-Cert program focuses regulatory attention not on specific products but on the companies and developers making them. Once the FDA deems that a company is responsible and using safe practices to develop software, approval for each product is not required. In one option, BCI companies could go through clinical trials for the safety of the physical device, while the software was addressed through more flexible programs like Pre-Cert. It is unclear what FDA rules would apply to BCI software and whether developers would be protected from lawsuits as with other implantable device designers.

The FDA Center for Devices and Radiological Health (CDRH) facilitates medical device innovation by helping stakeholders navigate the regulatory landscape. In 2015, CDRH held an open workshop on design, implementation, and evaluation considerations for physiological closed-loop controlled (PCLC) devices used in critical care environments. CDRH is currently developing regulatory recommendations and guidelines to facilitate innovation for them. Although PCLC devices have been available in parts of Europe for over a decade, regulatory obstacles imposed by the FDA as well as medical liability concerns have barred this technology from entering American markets [49].

Even with these new FDA initiatives to speed up the process for BCIs and other innovative technologies, the Big Tech sector lacks the patience needed to comply with FDA regulations. According to Tournes and Johnson [50], the clash between the FDA's legal lag and an overzealous BCI industry could lead to real patient harm and damage the Agency's ability to oversee cutting-edge technologies. Technology companies that are used to moving quickly seem unprepared for the vast regulatory oversight, approval scheme, and interdisciplinary approach required for successful health care projects. Much to Elon Musk's antipathy, any invasive BCI like Neuralink must go through the most stringent FDA approval protocols to satisfy the rigorous premarket approval process. Prior to that they will need to get an investigational device exemption to test their device, but the company's secrecy makes it difficult to know when it might be ready to apply for testing approval [51].

8 Commercialization of BCIs

Despite lingering ethical and policy concerns, the march to market BCIs is accelerating. The BCI market was valued at $1.6 billion in 2020, and it is expected to reach $3.2 billion by 2026 [52]. It is likely to demonstrate rapid growth due to the increase of neurodegenerative disorders in an aging population, escalating research and development activities to improve the BCI technology, and technological advancements such as miniaturization of devices. However, the commercial non-invasive BCI market is still in its infancy, and creation of easy-to-use and safe BCIs that can give high accuracy remains a major challenge. Although North America continues to dominate the market for BCIs, the Asia-Pacific BCI industry is anticipated to grow significantly, particularly Japan and China. Low-cost manufacturing sites plus favorable regulatory and taxation policies have attracted foreign players [53].

Non-invasive BCI currently represents the largest share of the BCI market in part due to its less stringent regulatory context. Although health care dominated the BCI market in 2019 owing to its use in the treatment of sleep disorders and neurological diseases, applications in mobile and virtual gaming, home control systems, and communication is fueling its growth. Furthermore, the development of non-invasive BCI devices based on an EEG is expected to increase its accessibility and marketing potential for healthy individuals. However, invasive BCI is also expected to enjoy substantial growth in the next decade in medical and rehabilitation applications. Partially invasive BCI is also expected to register significant growth because of its rapid technological improvement and easy adaptability [52].

In 2016, Bryan Johnson invested $100 million to establish startup Kernel to develop BCIs to enhance human intelligence and extend cognition with neural chip implants. In October 2019, neurotech startup Cognixion launched a non-invasive (AI) BCI that enables the speech impaired to communicate their thoughts. They expect their products to soon be covered by both Medicare and Medicaid reimbursement [16]. Neural signals is also developing BCIs to restore speech to disabled people. In 2021, Neuralink raised $205 million from investors, including Google Ventures, and total funding now stands at $363 million [54]. Cyberkinetics Neurotechnology Systems is marketing BrainGate, a neural interface system that allows disabled people to control a wheelchair, robotic prosthesis, or computer cursor. BitBrain is marketing a range of products for restoration and rehabilitation. In 2020, Naxon Labs launched the Brain to Computer Interface Solution and Neurotechnology and announced the BrainAccess Development Kit for BCI applications. Moreover, in October 2018, Advanced Brain Monitoring Inc. received a grant from the National Institute on Aging to use brain activity biomarkers to prevent cognitive decline associated with aging and dementia.

BCI technology is increasingly used in mobile and virtual gaming industries by integrating BCI with VR headsets [55]. In 2017, Neurable invented the world's first brain-controlled VR game. It recently raised $six million to move beyond its role as a VR game developer and work on building a next-generation BCI with a variety of real-world applications. Valve, Tobii, and OpenBCI are currently collaborating on the Galea hardware and software platform [56]. Similarly, Oculus would replace touch controllers with neural interfaces in a wristband that picks up electrical impulses and turns them into digital inputs for use in VR games. It will give the sensation of being able to interact with digital objects [57]. Currently available consumer BCI devices include the game Mindball; the Epoc Neuro-headset; Neurosky Mindwave; Mindmaze Mind Motion PRO; Neurable Enten; The XWave Headset and gTec Nautilus; MindX; Paradromics; Meltin MMI MELTANT-α cyborg, and NextMind.

Some observers believe that these kinds of BCIs, such as haptic feedback [58] or the use of BCIs to navigate through virtual environments and shape avatars' body language and facial expressions, could make virtual/augmented applications the biggest market for BCIs. However, others express concern over the vast data they can collect on personal likes, dislikes, and other interests or tracking nerve signals while typing out documents on virtual keyboards. As noted by digital expert Ray Walsh, dangers posed by these technologies include the exploitation of people's eye movements and nerve impulses to detect whether they are interested in certain content. "Legislators should look closely at the legality of this new data collection and subsequent usage to ensure that consumers are adequately protected" [59].

Among the major players in the BCI market are Advanced Brain Monitoring, Inc., Emotiv, Inc., Guger Technologies OEG, Mind Solutions, Inc., Neurosky, Inc., Nihon Kohden Corporation, OpenBCI, Quantum Applied Science and Research, Inc., Brain Products GmbH and Natus Medical, Inc. To date most effort is in the research stages and there have been many false starts and pivots. For instance, after 4 years and widespread publicity following announcement of a project to build a "silent speech" interface using optical technology to read thoughts, Facebook abandoned the project, noting that consumer brain-reading remains far off. It will instead focus on an experimental wrist controller for VR that reads muscle signals in the arm [60].

9 Summary

Although BCIs are still low on the policy agenda in the USA, both NIH and DARPA have increased funding of basic research for medical and military uses, respectively. Moreover, the GAO has targeted BCIs for study that includes ethical dimensions. Fifteen years after Eric Chan challenged the FDA to "foster and encourage device development" [31], it has worked to speed up the approval process for BCIs. However, there remains a need to differentiate the policy approaches between invasive and non-invasive BCIs and among BCIs for medical/rehabilitative uses, their use on healthy individuals for enhancement, and gaming and VR applications [61]. We must recognize that policy issues vary significantly by type of BCI application. Whereas safety and efficacy concerns are critical for medical and other invasive BCIs, questions of privacy and potential misuse of data are important for consumer applications of non-invasive techniques.

As emphasized by Coin et al. [26], it is crucial to discern which ethical and policy concerns are less urgent for highly speculative future developments from those that are more pressing based on the current state of BCI technology. Although many blogs and the mass media will continue to highlight dramatic, but decidedly hypothetical, future BCI scenarios, we need a more focused analysis on the present and near future developments. We must also remember that many of the highly optimistic promises of gene therapy and stem cell therapies to revolutionize medicine in the 1980s and 1990s have as yet failed to materialize.

Unfortunately, there are presently few scholars trained in political science or policy analysis with an interest in the issues surrounding BCIs, or science and technology in general. Therefore, it is imperative that concerned ethicists take up the challenge to emphasize the urgent need to clarify the relevant policy options for targeted regulation throughout the stages of development and application of BCIs and the emerging challenges they raise.

References

1. Tiwari N, Edla DR, Dodia S, Bablani A. Brain computer interface: a comprehensive survey. Biol Inspir Cogn Archit. 2018;26:118–29.
2. Yadav D, Yadav S, Veer K. A comprehensive assessment of brain computer interfaces: recent trends and challenges. J Neurosci Methods. 2020;346:108918. https://doi.org/10.1016/j.jneumeth.2020.108918.
3. Nijboer F. Technology transfer of brain–computer interfaces as assistive technology: barriers and opportunities. Ann Phys Rehabil Med. 2015;58:35–8.
4. Mudgal SK, Sharma SK, Chaturvedi J, Sharma A. Brain computer interface advancement in neurosciences: applications and issues. Interdiscip Neurosurg. 2020;20:10069.
5. Wolpaw JR, del Millán JR, Ramsey NF. Brain–computer interfaces: definitions and principles. Handb Clin Neurol. 2020;168:15–23.
6. Burwell S, Sample M, Racine E. Ethical aspects of brain computer interfaces: a scoping review. BMC Med Ethics. 2017;18(1):60. https://doi.org/10.1186/s12910-017-0220-y.
7. Binnendijk A, Marler T, Bartels EM. Brain–computer interfaces: U.S. In: Military applications and implications, an initial assessment. Santa Monica: RAND Corporation; 2020. https://www.rand.org/pubs/research_reports/RR2996.html.
8. Gonfalonieri A. Business applications of brain–computer interfaces and importance of brain data: challenges of non-invasive commercial brain–computer interfaces and brain data. 2020. https://towardsdatascience.com/business-applications-of-brain-computer-interfaces-importance-of-brain-data-615c230bb930.
9. National Institute on Deafness and Other Communication Disorders. What are cochlear implants for hearing? 2022. NIDCD (nih.gov).
10. Abiyev RH, Akkaya N, Aytac E, Günsel I, Çalman A. Brain–computer interface for control of wheelchair using fuzzy neural networks. Biomed Res Int. 2016;2016:9359868. https://doi.org/10.1155/2016/935986.
11. Bockbrader MA, Francisco G, Lee R, Olson J, et al. Brain computer interfaces in rehabilitation medicine. Phys Med Rehabil. 2018;10(9 Suppl 2):S233–43.
12. Leeb R, Pérez-Marcos D. Brain–computer interfaces and virtual reality for neurorehabilitation. Handb Clin Neurol. 2020;168:183–97.
13. Bocquelet F, Hueber T, Girin L, Chabardès S, Yvert B. Key considerations in designing a speech brain–computer interface. J Physiol. 2017;110:392–400.
14. Moses DA, Metzger SL, Liu JR, Anumanchipalli GK, et al. Neuroprosthesis for decoding speech in a paralyzed person with anarthria. N Engl J Med. 2021;385:217–27.
15. Shahriari Y, Vaughan TM, McCane LM, Allison BZ, et al. An exploration of BCI performance variations in people with amyotrophic lateral sclerosis using longitudinal EEG data. J Neural Eng. 2019;16(5):056031. https://doi.org/10.1088/1741-2552/ab22ea.
16. Forsland A. New AI brain computer interface for healthcare launched: Cognixion's brain computer interface enables the speech-impaired to communicate. 2019. https://www.psychologytoday.com/us/blog/the-future-brain/201910/new-ai-brain-computer-interface-healthcare-launched.
17. Horrom T. Home use test of brain–computer interface for ALS patients yields 'complicated' results. 2018. https://www.research.va.gov/currents/0818-Home-use-test-of-brain-computer-interface-for-ALS-patients-yields-complicated-results.cfm.
18. Papanastasiou G, Drigas A, Skianis C, Lytras M. Brain computer interface-based applications for training and rehabilitation of students with neurodevelopmental disorders. A literature review. Heliyon. 2020;6(9):e04250.

19. Specker Sullivan L, Illes J. Ethics in published brain–computer interface research. J Neural Eng. 2018;15(1):013001. https://doi.org/10.1088/1741-2552/aa8e05.
20. Davidoff EJ. Agency and accountability: ethical considerations for brain–computer interfaces. Rutgers J Bioethics. 2020;11:9–20.
21. Pham C, Gilbert F. Predicting the future of brain–computer interface technologies: the risky business of irresponsible speculation in news media. Bioethica Forum. 2021;12(1/2):15–28.
22. Kögel J, Schmid JR, Jox RJ, Friedrich O. Using brain–computer interfaces: a scoping review of studies employing social research methods. BMC Med Ethics. 2019;20:18. https://doi.org/10.1186/s12910-019-0354-1.
23. Sample M, Aunos M, Blain-Moraes S, Bublitz C, et al. Brain–computer interfaces and personhood: interdisciplinary deliberations on neural technology. J Neural Eng. 2019;16(6):063001
24. Coin A, Dubljević V. The authenticity of machine-augmented human intelligence: therapy, enhancement, and the extended mind. Neuroethics. 2020;14(2):283–90. https://doi.org/10.1007/s12152-020-09453-5.
25. Rainey S, Maslen H, Savulescu J. When thinking is doing: responsibility for BCI-mediated action. AJOB Neurosci. 2020;11(1):46–58.
26. Coin A, Mulder M, Dubljevic V. Ethical aspects of BCI technology: what is the state of the art? Philosophies. 2020;5(4):31. https://doi.org/10.3390/philosophies5040031.
27. Klein E, Ojemann J. Informed consent in implantable BCI research: identification of research risks and recommendations for development of best practices. J Neurol Eng. 2016;13:043001
28. Blank RH, Burau V, Kuhlmann E. Comparative health policy. London: Palgrave; 2018.
29. Funk C, Kennedy B, Sciupac EP. Public opinion on the future use of brain implants. 2016. https://www.pewresearch.org/science/2016/07/26/public-opinion-on-the-future-use-of-brain-implants/.
30. Rainie L, Funk C, Anderson M, Tyson A. Public cautious about enhancing cognitive function using computer chip implants in the brain. 2022. https://www.pewresearch.org/internet/2022/03/17/public-cautious-about-enhancing-cognitive-function-using-computer-chip-implants-in-the-brain/.
31. Chan E. The FDA and the future of the brain–computer interface: adapting FDA device law to the challenges of human–machine enhancement. John Marshall J Inf Technol Priv Law. 2007;25:117–64.
32. The Royal Society. IHuman: blurring lines between mind and machine. 2019. https://royalsociety.org/topics-policy/projects/ihuman-perspective/.
33. Brunner C, Birbaumer N, Blankertz B, Guger C, et al. BNCI Horizon 2020: towards a roadmap for the BCI community. Brain Comput Interfaces. 2015;2(1):1–10.
34. GAO. GAO Science, Technology Assessment, and Analytics Team: initial plan and considerations moving forward. 2019. https://www.gao.gov/assets/2020-02/GAOScienceTechPlan-2019-04-10_0.pdf.
35. NIH. Brain 2025: a scientific vision. 2014. https://braininitiative.nih.gov/strategic-planning/brain-2025-report.
36. Tucci DL. An update on the NIH BRAIN initiative and on the NIDCD's January Advisory Council meeting. 2021. https://www.nidcd.nih.gov/about/nidcd-director-message/update-nih-brain-initiative-nidcds-january-advisory-council-meeting.
37. Willett FR, Avansino DT, Hochberg LR, Henderson JM, Shenoy KV. High-performance brain-to-text communication via handwriting. Nature. 2021;593:249–54. https://doi.org/10.1038/s41586-021-03506-2.
38. NIH. Multi-Council Working Group. 2014. https://braininitiative.nih.gov/about/multi-council-working-group.
39. Eberwine J, Kahn J. The BRAIN initiative and neuroethics: enabling and enhancing neuroscience advances for society. AJOB Neurosci. 2020;11(3):135–9.
40. NIH. The BRAIN Initiative® and neuroethics: enabling and enhancing neuroscience advances for society. 2019. https://acd.od.nih.gov/working-groups/brain2.0-subgroup.html.
41. Tullis P. The brain–computer interface is coming and we are so not ready for it. 2020. https://thebulletin.org/2020/09/the-brain-computer-interface-is-coming-and-we-are-so-not-ready-for-it/.

42. Miranda RA, Casebeer WD, Hein AM, et al. DARPA-funded efforts in the development of novel brain–computer interface technologies. J Neurosci Methods. 2015;244:52–67.
43. US Department of Health and Human Services. The office for Human Research Protections (OHRP). 2021. https://www.hhs.gov/ohrp/index.html.
44. Hyun I, Scharf-Deering JC, Lunshof JE. Ethical issues related to brain organoid research. Brain Res. 2020;1732:146653. https://doi.org/10.1016/j.brainres.2020.146653.
45. Thomas AR, Pop NA, Iorga A, Ducu C. Ethics and neuromarketing. London: Springer; 2017.
46. Business Wire. Stentrode brain–computer interface receives breakthrough device designation from FDA. 2020. https://www.businesswire.com//home/news/20200827005748/en/.
47. Hardy R. First ever FDA-approved brain–computer interface targets stroke rehab. 2021. https://newatlas.com/medical/first-fda-approved-brain-computer-interface-ipsihand-stroke/#:~:text=Called%20IpsiHand%2C%20the%20system%20is%20the%20first%20brain-computer,records%20brain%20activity%20using%20non-invasive%20electroencephalography%20%28EEG%29%20electrodes.
48. FDA. Implanted brain–computer interface (BCI) devices for patients with paralysis or amputation. 2021. https://www.fda.gov/regulatory-information/search-fda-guidance-documents/implanted-brain-computer-interface-bci-devices-patients-paralysis-or-amputation-non-clinical-testing.
49. Parvinian B, Scully C, Wiyor H, Kumar A, et al. Regulatory considerations for physiological closed-loop controlled medical devices used for automated critical care: Food and Drug Administration workshop discussion topics. Anesth Analg. 2018;126(6):1916–25.
50. Tournas LM, Johnson WG. Elon Musk wants to hack your brain: how will the FDA manage that? 2019. https://slate.com/technology/2019/08/elon-musk-neuralink-facebook-brain-computer-interface-fda.html.
51. Mordor Intelligence. Brain–computer interface market—growth, trends, COVID-19 impact, and forecasts (2021–2026). 2020. https://www.mordorintelligence.com/industry-reports/brain-computer-interface-market.
52. Grand View Research. Brain computer interface market size, share and trends analysis report by product (invasive, partially invasive, non-invasive), by application (healthcare, communication and control), by end-use, and segment forecasts, 2020–2027. 2020. https://www.grandviewresearch.com/industry-analysis/brain-computer-interfaces-market.
53. Marketwatch. Brain computer interface market to witness astonishing growth with key players, analysis with impact of COVID-19, analysis, demand, forecast 2027. 2021. https://www.marketwatch.com/press-release/brain-computer-interface-market-to-witness-astonishing-growth-with-key-players-analysis-with-impact-of-covid-19-analysis-demand-forecast-2027-2021-07-08.
54. Shead SL. Elon Musk's brain computer start-up raised $205 million from Google Ventures and others. 2021. https://www.msn.com/en-us/money/companies/elon-musks-brain-computer-start-up-raises-24205-million-from-google-ventures-and-others/ar-AAMJVj1.
55. Porter J. Gabe Newell has big plans for brain–computer interfaces in gaming. 2021. https://www.theverge.com/2021/1/25/22248202/gabe-newell-valve-brain-computer-interface-bci-meat-peripherals.
56. Tobii GP. Valve and OpenBCI collaborate on 'Galea' VR brain–computer interface. CI's could be the future of immersive experiences. 2021. https://www.gmw3.com/2021/02/tobii-valve-openbci-collaborate-on-galea-vr-brain-computer-interface/.
57. Benson J. Oculus is making a virtual reality headset you control with your brain. 2019. https://www.gamingbible.co.uk/news/games-oculus-is-making-a-virtual-reality-headset-you-control-with-your-mind-20190925.
58. Fleury M, Lioi G, Barillot C, Lécuyer A. A survey on the use of haptic feedback for brain–computer interfaces and neurofeedback. Front Neurosci. 2020;14:528. https://doi.org/10.3389/fnins.2020.00528.

59. Donovan I. Facebook's new VR wristband is powered by brain signals, apparently. 2021. https://www.gamingbible.co.uk/news/games-facebooks-vr-wristband-will-predict-your-likes-and-dislikes-20210323.
60. Regalado A. Facebook is ditching plans to make an interface that reads the brain. 2021. https://www.technologyreview.com/2021/07/14/1028447/facebook-brain-reading-interface-stops-funding.
61. Blank RH. Cognitive enhancement: social and public policy issues. London: Palgrave Macmillan; 2016.

Memory Enhancement and Brain–Computer Interface Devices: Technological Possibilities and Constitutional Challenges

Marc Jonathan Blitz and Woodrow Barfield

1 Introduction

Laws are built around certain background assumptions about how the world works. For example, consider how the law of privacy assumes certain facts about the world. Generally, we often have privacy protections within our own homes: The Fourth Amendment of the U.S. Constitution, for example, bars law enforcement from searching a home unless it convinces a court it has "probable cause" to believe there is evidence of a crime there [1, 2]. By contrast, law enforcement is allowed to investigate more vigorously, free from such constitutional constraints, in public space: A police officer or other official is free to walk on the street outside of a house and look around and can do so without permission from a judge (or the occupant of the house). As the U.S. Supreme Court put this point in one case, "Fourth Amendment protection of the home has never been extended to require law enforcement officers to shield their eyes when passing by a home on public thoroughfares" [3].

However, this constitutional doctrine is built around an assumption about the natural world: that our privacy and intimate sphere isn't as threatened by observation from a public street as it is when police enter our homes. As the legal scholar, Lawrence Lessig, observed in 1996, "nature helps protect my privacy" because "police, unlike Superman, don't have X-ray vision, so they can't simply look through my walls to see what sorts of stuff I have on the other side" [4]. If they could, then the Fourth Amendment's "probable cause" requirements for entering homes wouldn't have much force, because police could simply circumvent the protections to citizens offered by the Fourth Amendment by gathering the same

M. J. Blitz (✉)
Oklahoma City University School of Law, Oklahoma City, OK, USA
e-mail: mblitz@okcu.edu

W. Barfield
Visiting, School of Law, University of Turin, Turin, Italy

V. Dubljević, A. Coin (eds.), *Policy, Identity, and Neurotechnology*, Advances in Neuroethics, https://doi.org/10.1007/978-3-031-26801-4_12

information from outside that they would obtain from entering. But technology, of course, can unsettle the background assumptions that underlie these legal doctrines. While police don't have X-ray vision, for example, they have used infrared scanners (exploiting another form of electromagnetic energy) to see through walls in a sense, and the US Supreme Court responded by interpreting Fourth Amendment law to ban this technological entry to the same extent as it does physical entry [5].

This chapter asks whether and how BCI technology might unsettle law in a similar way: such technological advances can reshape the world to which our existing law is designed to apply, and force courts to respond. Perhaps with a simple adjustment (like extending existing legal categories to these new technologies) [6]. Or perhaps with a more radical transformation that comes with new legal frameworks for addressing unfamiliar challenges [7, 8]. More specifically, in this chapter, we are interested in how certain laws—the constitutional law of the US, Europe, and other jurisdictions—might have to respond to, and change with, a certain type of BCI, namely that which can be used to repair, enhance, or augment human memory.

It may, at first, seem less obvious why technological transformation of human memory might alter our world—and unsettle law—in the same way as the X-ray vision-equivalent made possible by infrared scanners. The latter devices allow us to enhance our vision so as to perceive *other peoples'* private environments. The most obvious use of technologically aided memory enhancement would let us enhance only *our own internal* recollections.[1] While we don't have unlimited freedom of action, laws—in the United States, the European Union, international law, and other jurisdictions—do give us a right to "freedom of thought" that some judges and commentators have treated as nearly absolute.[2] But alteration of our memory can have consequences for others. As we discuss at more length below, this notion has already been explored in other scholarship about the ethical and legal implications of memory dampening or elimination [12]. When criminals or others erase their own memories about a crime (and who committed it), this may prevent the judicial system from learning critical facts about a crime—and may even count as obstruction of justice (a crime itself) [12, 13].

Moreover, government is not entirely barred from engaging in paternalistic action to prevent individuals from harming themselves. That is, apart from protecting us from using drugs or medical devices that might harm us physically, governments might plausibly argue they are justified—in using what is often called their "police power" in American law [14]—to protect us in other ways. They can argue they have power to protect us from taking certain risks when we wish to trade an existing natural form of memory and psychological processing—refined over eons by evolution to allow us to function—for an uncertain BCI-enabled substitute, at least if such a substitute causes changes more radical than those already caused by

[1] As we note later in the chapter, some argue that BCI may one day allow us to share mental operations such that many people can share in the same perceptual experience at the same time or experience the memories of others.

[2] See Jones v. Opelika, 316 U.S. 584 (1942), which describes the "privileges of thought" as "illimitable" [9]; see also [10, 11].

more familiar uses of notebooks, smartphones, and other computers. Arguments for that kind of paternalism may focus not only BCI technology itself—but on the ways individuals might, in using it, unwisely invite others (individuals, corporations, government actors) to access and shape their memories, sometimes, without appreciating the risks that such sharing could carry for their privacy and autonomy.

If BCI brings more radical changes in the future to the way memory works, this can destabilize our existing legal doctrine even more significantly. Imagine, for example, that we do not obtain another person's memory the way we might currently obtain a diary entry of theirs—by asking them to share it with us voluntarily—but rather that we participate in a system where there is a strong default (if not a requirement) that everyone's recorded memory of certain experiences will be automatically available to everyone else, with the aid of BCI and perhaps other technology. Such a radical transformation in how human beings interact with the world would undermine numerous key features of the background that law assumes. The privacy and liberty rights we have are generally rights that belong to individuals. So too are many property rights. Such rights enable an individual to exclude others from their homes or other spaces they have property in and to shield their personal affairs from others' observation. Liberty rights, and the autonomy they protect, are secure only if an individual can control key aspects of that person's own life. It is hard to imagine what might become of these rights in a world where memories and other raw materials of thoughts are collective property (for example, in the public domain), and where creating and accessing them is (by default at least) a collective act. Such a world is likely too starkly different from our own to easily imagine how a legal framework built for individuals with their own mental autonomy and private memory stores can be adapted to it. But to the extent that the BCI of the future enables such a world, and allows individuals to seek it voluntarily, this raises the question of whether, and to what extent, society might (through government) require that the technology be designed in a way that largely preserves the boundaries between selves that are generally a condition of modern liberty and privacy.

Addressing these issues necessarily involves speculation. We can't state with certainty how courts in the United States, the European Union, or other jurisdictions will apply existing constitutional and human rights law to emerging technologies when we remain unsure what form these technologies will take and how they will change our lives—and particularly what form they will take in the distant future. Still, given the rapid pace at which BCI and other technologies are developing, it is helpful to *begin* to think about the nature of the challenges BCI technologies may raise in altering memory processing. We will do so in two parts. First, we will briefly discuss BCI generally, and then look at efforts to develop BCI and related technologies to restore or enhance memory. We will also discuss more ambitious plans to use BCI to bring us to a science fiction like world where memories can be recorded onto and replayed from computer chips, shared with people who didn't experience them, and perhaps even forged to make us clearly remember events that never occurred. There have already been simple forms of memory transfer between animals in laboratory experiments [15, 16]—but, as we will note below, there is debate about the extent to which this means that human beings could feasibly experience each others' memories in future decades or centuries. Second, we will reflect on some of the legal challenges which might arise if BCIs allow for two types of memory transformation—either by (1) transforming our memory *capacity* and (2) allowing us to custom-design memory *content*—or have others design or create it for us. At the extreme, the latter could allow us to begin to *dissolve the boundaries* that divide individuals into separate selves, or otherwise leave us uncertain about how use of the technology will transform fundamental elements of our mental processing.

2 BCI Technology: An Overview

It is useful to discuss BCI—and some claims writers have made about how it might develop in future debates—before considering the questions it raises for thinking about constitutional rights. We will also discuss "neural prostheses" or "neuroprostheses" that might use BCI—but may also use other, different kinds of technologies—to replace brain functions that have been damaged by injury or illness. And we will also briefly touch on some technologies (such as AI or virtual, augmented, or other "extended" reality technologies) that can be used in conjunction with BCI.

A "brain computer interface"—also sometimes called a "brain–machine interface," "mind–machine interface," "direct neural interface," or "neural control interface" [17]—is a device that uses a computer(s) to provide inputs to, or receive outputs from, our neuronal processing. Perhaps the most well-known form of BCI is that which captures outputs from our brain and translates them into some kind of action in the world. For example, a BCI device might consist of sensors that measure brain signals, an amplifier to boost the magnitude of these brain signals, and a computer that translates the signals into commands to control computer programs and/or devices attached to, or external to, the body. Such BCIs operate using a closed-loop control structure, are upgradeable, and increasingly allow technology to be controlled by thought. In widely-reported demonstrations of brain–computer interface technology, animals or human subjects are able to do something that resembles using "telekinetic" powers [18]: They can move an item external to their body—be it a cursor on a computer screen or a robot arm—merely by thinking a command rather than by pressing a button or pulling a lever or physically interacting with the object. The neural command that allows a person to take action in the external world, in other words, is sent not to some muscle in their body (in their arm or hand, for example) but rather directly to a computer or other machine.

Other types of BCI technology alter the way the brain receives inputs *from* the world rather than send outputs *to* it: A computer device that replaces parts of our natural biology for vision and auditory perception, for example, can partially replace—with artificial substitutes—the natural processes by which individuals see and hear. As Jeremiah D. Wander and Rajesh P. N. Rao point out, although "the model of a BCI that often comes to mind" involves transferring signals from "the motor regions of the brain... to an output device, either a cursor on a screen or a robotic arm," BCI can also replace sensory mechanisms—for example, by "providing artificial sensory inputs directly to the auditory system" [19]. As Allen Coin and Veljko Dubljevic likewise write, BCI technology might not only restore "mobility," for example, by sending outputs from the brain to a device that provides movement capacity no longer provided by muscles—it might also restore "perceptive sense" with devices such as "cochlear and visual cortical implants" [20]. A recent article in MIT Technology Review discusses one such device developed for restoring sight to blind individuals "by feeding signals directly to the brain." It consists of a "modified pair of glasses... fitted with a tiny camera" that in turn connects to a computer which transforms the "live video feed" from the camera "into electronic signals" which are in turn sent to "a port embedded in the... skull that is wired to a 100-electrode implant in the visual cortex" of the brain [21].

This kind of BCI technology is most commonly developed as a kind of *prosthetic*. That is, it is intended not to endow individuals with any new capacity but to *replace* a capacity they have lost because of damage to a part of their nervous system, or because they were born without certain biological capacities that, for example, enable vision or hearing. Such a prosthesis could also consist of technology other than BCI—for example, in a device that enables movement or hearing without connecting a computer to the nervous system.

Whether it receives outputs from, or provide inputs to, neuronal processes, BCI might also be used for the purpose of enhancement rather than medical treatment: Instead of repairing damage to biological processes that involve the nervous system, they might instead augment their function—for example, with infrared vision or hearing ranges of a kind humans don't naturally have. As Coin and Dubljevic write, some of the same types of BCI technology that can restore lost powers of movement or sensation can also give individuals the capacity to "accomplish tasks they were previously unable to." And "one can imagine a future scenario in which BCI technology could potentially allow users access to above average capabilities" [20].

BCI technology might not only repair or enhance the natural ways our brains link to the outside world in order to sense it or to act on - it might also transform aspects of our cognition, such as memory and in some cases allow third parties access to an individual's internally stored memories. As with existing artificial retinas and cochlear implants, many of the devices currently being proposed or tested are intended to *treat* injuries or diseases not to enhance or augment human functions by giving individuals capacities they never had. They involve, in other words, use of BCI as a "neural prostheses"—that is, a device implanted in the brain meant to replace missing biological functionality that underlies memory or some other aspects of cognition [22].

Consider one kind of design and use of BCI technology being funded by the US government through the Defendant Advanced Research Projects Agency (DARPA). Among the programs DARPA supports is the Restoring Active Memory Program (RAM) whose purpose is to mitigate the effects of traumatic brain injury (TBI) by developing neurotechnologies to facilitate memory formation and recall in injured brains [23]. One goal of this program that is of particular interest for our chapter is the aim of developing and testing a wireless, fully implantable neural interface for human clinical use. Projects in and supported by the European Union also aim to use BCIs to modify and strengthen memory: the Human Brain Project (HBP), for example, is one of the three FET (Future and Emerging Technology) Flagship initiatives [24]. The purpose of the HBP is to create a research infrastructure to help advance neuroscience, medicine, computing, and brain interface technologies within the European Union.

These proposed designs are far from the only means of using BCI to restore memory or of using technology in general to repair memory function. In 2011, Theodore Berger and colleagues gained significant attention for developing a prototype "artificial hippocampus" which led to the implementation of a proof-of-concept system for restoring memory function for those suffering from Alzheimer's disease [25]. Since the 1960s, scientists have known that the hippocampus—a structure within the temporal lobe of the brain—plays a crucial role in the consolidation of long-term memory. The hippocampus takes the sensory experiences the brain generates as we sense the world and organizes them into stored neural patterns that allow us to recall such experiences. In Berger's artificial hippocampus—or "hippocampus cognitive prosthesis"—a computer (implanted within the hippocampus) rather than a biological structure performs a similar translation of mental inputs to outputs that can be stored in the brain's "biological memory." And other scientists' work suggests the possibility of other kinds of memory prosthesis using computers to replace, or add to, memory processing: The European Union's Coronet project, for example, is "developing a biohybrid interface between biological and artificial neural networks" [26].

Like the forms of BCI discussed above, use of BCI to alter memory processing or other cognition can be used not only as a prosthetic that restores lost biological function—it can also potentially be a means of augmenting or enhancing cognitive function. Coin and Dubljevic, for example, consider one kind of cognition-enhancing BCI when they discuss a device that might not only restore—but "augment"—"the user's ability to perform one specific kind of purely cognitive activity: namely, mental mathematical calculations"—and could conceivably provide "enhancement beyond the normal abilities of a human being to perform mental mathematical calculations" [19]. The same might be true of memory-enhancing BCI: Not only can it allow individuals to regain memory formation or retrieval capacities they *once* had—it can give them new memory capacities they *never* had, including, perhaps, memory capacity that human beings have not previously had.

To be sure, an analysis that goes beyond the brief discussion of BCI devices in this chapter would look more closely at how different forms of BCI might enhance different forms or aspects of memory. As O. Carter Sneads notes, current understandings of memory conceive of it as "including an array of distinctive processes and systems, involving different neural structures" [27]. They distinguish between "non-declarative" or "implicit" memory, which includes "procedural memory" of "skills and habits," and "declarative memory" or "'explicit' or 'conscious' memory," which is "present in the individual's conscious and can be intentionally called to mind" [26]. And the latter can in turn be either "short-term" or "long-term," with "long-term" memory being either "semantic" memory of facts or "episodic" memory of life experiences. BCI that enhances one of these types of memory might not enhance other types and it is conceivable that some BCI enhancement of memory could be even more specific, enhancing only some kinds of episodic memory. For purposes of this discussion, however, we will generally bracket these nuances and ask questions that might pertain to enhancement of either semantic or episodic long-term memory (and perhaps, to some other forms).

The type of memory-enhancing BCI that has received the most attention in the popular press is a type focused on episodic memory that is, for now, still in the realm of futuristic reverie: that which would enable individuals to record and replay memories from implanted computer chips, and, perhaps, allow people to directly *experience* each other's memories, or rapidly inject tremendous amounts of factual knowledge into their mental banks with computer technology. This is a familiar theme in science fiction. The movie, *Strange Days*, depicts an underground market in the memories of individuals who have engaged in dangerous (and often, illegal) activity. Buyers in this market can feed themselves (or "jack into") the experiences of others who have lived more exciting lives [28]. The video game, Remember Me, is likewise set, as one review describes it, in a "Paris of the future, where technology has allowed us to exchange and purchase memories" and where "happy memories can be bought and abused like drugs"[3] [30]. In the television show, The Prisoner, individuals use "speedlearning" to "imprint" onto their minds—in only a few minutes—knowledge that would take months to learn in the traditional fashion [31]. In *3001: The Final Odyssey,* Arthur C. Clarke dedicates a significant amount of his plot to "Braincaps" which can likewise instantaneously import knowledge into the memory bank of the person wearing it—and also allow expressionless sharing of thoughts with one's neighbors [32]. BCI has made only modest strides in this direction. But it is interesting to note that some lab experiments appear to involve very limited demonstrations of sharing or replaying of memories in animals [15].

Some writers are predicting that those strides will become more significant in the coming decades. The efforts of Elon Musk's company, Neuralink, have received the most attention from journalists reporting on its goal of allowing individuals to store their memories on computers integrated with their brains and download them into a new body or share them with others (who can experience the memories as their own rather than simply listen to someone else try to convey them) [33, 34].

To be sure, these extraordinary predictions have been met with strong skepticism: For example, a June 2021 Endgadget article quotes BCI-researcher, Miguel Nicolelis, as saying: BCI technology "will never make people download their emotions or their deep cognitive functions, and they'll never make people learn French by uploading French grammar to a brain–machine interface" [35]. One basis for such skepticism is that—while popular imagination sometimes analogizes a memory to a stored video or image—it is in many respects quite different. As Jane Campbell Moriarty has pointed out, the brain's memory systems do "not record and recall information like a video recorder." It instead "layers memory over memory, changes, loses, restructures, and adapts to continual addition of new information.

[3]Another video game, Cyberpunk 2077, similarly explores the concept of chips that can transfer someone else's consciousness into one's mind [29].

Every time a memory is recalled, it is altered" [36]. One might thus ask how replaying data recorded from one person would generate anything like the same memory in a different person, which may well generate a markedly different perceptual experience from those inputs, as it weaves them together with the distinctive past experiences of the recipient.

But it may be the case that even if BCI will not be able to augment or transform memory in quite the way that some futurists envision it, it will do so in other ways. It is possible, for example, that individuals in the future could use virtual reality to make 3D recordings of all that they see and hear—a futuristic version of the "life-logging" some individuals already engage in when they record each moment of their lives over long stretches of time [37]. BCI might supplement and enhance these visual and auditory records—which allow one to recreate another's experience from outside the brain (with senses) rather than inside—perhaps by generating certain emotional or visceral reactions it is hard to capture in a video, or perhaps by allowing individuals to access the recorded memories in others' lifelogging videos from a device storage implanted within their brain rather than from a device outside of it [38]. In any event, it seems likely that the BCI of the future will in some ways give us means to reshape our mental functioning.

It is worth noting that different writers tend to use the term "brain–computer interface" or "BCI" somewhat differently—and might disagree about whether these categories include *every* technological means of making *any kind of* computing device respond to an input from brain activity (unmediated by other human action, such as typing into a keyboard). For example, the examples we have focused on above all involve implanting a computer, or a device that connects with a computer, into the brain, skull, or some other part of the body. But some devices extract information about neural activity from a device that is worn *outside of* the body—like a helmet of some kind. This is true, for example, of certain "mind-controlled" video games or neurofeedback devices: Computers generate video game activity, or display brain activity so that a person may exert greater control over it, in response to commands or signals that are extracted from electroencephalography (EEG) readings taken by a band or helmet that sits over a person's head (for the duration of the game or neurofeedback session)—not from a device implanted into a body or from electrodes fixed with adhesive to the outside of a person's scalp [39, 40]. Like the devices we have discussed earlier, such devices forge a link between computers and neural activity that by-passes the sensory or muscular paths.

But one might argue that, if writers define terms such as "brain–computer interface" device or "BCI" to include any and all devices that link brains and computers in this way, it will be difficult to make any useful generalizations about this technology—and the ethical and legal challenges it raises. Certain problems raised by implanted BCI, for example, don't arise for machines that are linked to the brain in a more temporary fashion and without any intrusion into the body. On the other hand, some problems might encompass many different methods of linking brains to computers. As a general matter, when we use the term "BCI" or "brain–computer interface" by itself below, we will use it to refer to devices in which a computer is implanted, or otherwise permanently attached to, the body in such a way that the computer can receive information from neuronal activity and translate it into action, input information to the brain (to be translated into sense perception), or alter neuronal activity by processing information transferred from one part of the brain to another. Some technology that links computers to brain processes—in what is sometimes called "wearable BCI"—does so without such a permanent attachment, and we will describe that technology with other language. We will also note when some of the legal claims we make might apply to this kind of "wearable BCI" or other technologies that link brains to computers.

3 Constitutional and Other Legal Implications

What predictions, if any, can one make about how constitutional law in the United States, Europe, and perhaps other jurisdictions—or how international law—will apply to such possible uses of BCI technology? Clearly, significant questions of law will be raised by technology that interfaces with the brain. For example, in the near future, will individuals have a right to repair their own memory with artificial hippocampuses or other neuroprosthetic devices? Will they have a right to alter their own memory with such technology? They would likely have a right *against others'* use of BCI technology to manipulate their memory without their consent or even knowledge that it is happening. But to what extent would this interest (against being subject to mental manipulation) justify government not only in protecting them from others' hostile entries into their memories and minds, but also from their own, arguably unwise decision to *invite* others to share in, and possibly reshape their memories? To what extent, in other words, can government exercise a kind of paternalism in how individuals use BCI technology?

Farther into the future, if a BCI device gives individuals a way to dissolve some of the boundaries that separate themselves from the world—to erase the dividing line between data stored in their natural memories and that stored in computers or the "cloud," or erase the boundaries that separate their own minds from others—what rights, if any, might individuals have to transform themselves in this way? Moreover, what rights may they retain against manipulation of their minds from the outside (by governments, corporations, or other external actors) in a world where individuals have used wirelessly connected BCIs to push the doors of their memory and mental space wide open to such outsiders? How, in other words, might they invoke rights against mental trespass and other egregious violations of the mind in a world where others have the technological ability to infuse their minds with memories, knowledge, or other mental content?

This section of the chapter will briefly analyze how such challenges might arise if BCIs allow for two types of memory transformation—either by (1) transforming our memory *capacity* or (2) allowing us to custom-design memory *content* or have others design or create it for us. At the extreme, the latter could allow us to begin to *dissolve the boundaries* that divide individuals into—separate selves, or otherwise leave us uncertain about how use of the technology will transform fundamental elements of our mental processing. The latter of these (modifying memory content), we will suggest as a general matter, presents a more difficult challenge for courts trying to adapt long-standing protections for privacy and liberty to a new technological landscape. Extraordinary changes in memory capacity may unsettle our expectations of what others can remember about us—but there has long been some variation in how well different individuals can remember the details of particular interactions, and we've long been able to enhance our memory with mnemonic and other learning strategies, and with outside supplements to it—from old-fashioned diaries to iPhone apps and photo libraries. We are less familiar with a world where an AI-equipped machine (or any third party for that matter) might insert a semantic memory of facts we never learned or episodic memory of an event we never experienced directly within our brain, perhaps in conjunction with other technologies such as virtual reality simulations of others' experiences or of fictional events. These novel ways of generating memory may unsettle not only our familiar sense of what we can keep private (since observations of us to which only one person was privy might be "remembered" by countless others who can access a library of memories)—it might also raise the risk that people will remember, as being real, custom-designed memories that have no grounding in reality. Where such modification of memory content results in a world where the experience of calling memories to mind is a collective phenomenon rather than something where each individual summons a memory in a private "theater" of sorts, it will leave us in a world markedly different from the one for which our constitutional and other law has been constructed. Thus, courts may have to let society reconstruct, within the design of BCI technology or otherwise, some of the features of our world that the BCI technology of the future could erode or destabilize.

3.1 A Framework for Beginning to Think About the Legal Implications of BCI

We began the chapter not with a discussion of BCI—but rather of another technology that has unsettled American constitutional privacy law in recent years: thermal imaging. The problem raised by that technology is that it undermines a boundary line that has been significant under the Fourth Amendment of the U.S. Constitution. Under this system, police need special permission (from a court) to enter into the enclosed *private* space of the home but are free to look at, and scrutinize, what is visible in the *public* space outside of it [2]. Thermal imaging undermines this division of space by giving police a kind of "X-ray" vision[4] that enables them to make the private home life—previously shielded from outside observation by walls—visible from public space [4].

[4] Thermal imaging works by letting police or others construct images from infrared radiation with greater wavelengths than that of visible, not X-ray, light with shorter wavelengths. But the term "X-ray vision," as Lessig's previous example illustrations, has become shorthand for any kind of vision in which one can see through physical barriers.

One might use a similar framework to begin asking about the constitutional status of different forms of memory-enhancing BCI: To the extent it is used by a person only to change their *internal* mental operations, one might assume, it is not the government's business to monitor or restrict such purely inward-facing activity—except, perhaps, to assure (as is true for any other medical treatment) that the technology involved in it is physically safe for individuals to use. Our "freedom of thought," one might observe, must entail a right to solidify a memory by silently repeating information inside of our heads, or learning a particular mnemonic device. If so, one might argue this freedom should *also* entail a right to do so with tools to supplement our natural thinking, including BCI. Individuals in the US have a right to "freedom of thought" under the First Amendment [41, 42]. Article 10 of the European Union Charter of Fundamental Rights and Article 9 of the European Convention on Human Rights similarly guarantee the freedom of thought, as does Article 18 of the International Covenant on Civil and Political Rights [43]. These rights to freedom of thought might give individuals a right to alter their cognitive functions using a BCI, including their mental processing and memory. Moreover, all of these legal sources give individuals a "right to receive information and ideas" from others' speech (and arguably, from other sources) and, in some circumstances, use of a BCI may provide a crucial conduit for receiving information, or altering the way we receive and process the information we obtain through our senses [44].

By contrast, if BCI is exploited by some individuals (or organizations) not simply to enhance *their own* mental operations, but rather to manipulate or spy upon those *of others*, [45] then this is something that can and should be legally restricted, and—especially where the manipulator is a state actor—perhaps even treated as a constitutional or international human right violation. Hacking into other people's private computer files, for example, is generally a crime in the United States [46, 47], Canada [48], and in many other jurisdictions. Using BCI to hack into computers within (or linked to) their brain operations or other body processes is intuitively an even greater violation of privacy and autonomy (and, possibly, bodily integrity) than hacking into their property. Far from an exercise of freedom of thought this invasion of others' mental operations seems like a violation of it.[5] As long as we can

[5] Consider, for example, Article 3 of the EU Charter of Fundamental Rights, which provides a right to "mental integrity", which is a right that not only protects a person from interventions that undermine mental health, but also from illicit and harmful manipulations of people's neural activity through the misuse of neurotechnology [49]. The European Union's recently enacted General Data Protection Regulation (GDPR) has set out requirements for companies and organizations which collect, store, and manage personal data, including organizations which process the personal data of individuals [50]. And 2021 the European Commission released the proposed Regulation Laying Down Harmonized Rules on Artificial Intelligence. Article 5 (5.2.2) states: "The Regulation prohibits certain AI practices that are deemed to pose an unacceptable level of risk and contravene EU values. These practices include the provision or use of AI systems that either deploy subliminal techniques (beyond a person's consciousness) to materially distort a person's behavior" [51]. These laws may place limits on the powers that individuals may have to shape someone else's mental processes using BCI technology, and possibly even the power individuals have to reshape their own memory or mental processing—especially where the outcomes of certain changes they willingly undergo have uncertain (and possibly harmful) consequences.

easily place any given use of BCI memory-enhancement technology on either side of this line—as long as we can classify it as either (1) a voluntary alteration of one's own thoughts or (2) an intrusion into someone else's—it may well be the case that this novel technology can be regulated using traditional and familiar legal categories.

But just as thermal imaging blurs the line between private and public space, so might certain uses of memory-enhancing BCI. In fact, some of the challenges it raises have already been explored by scholars analyzing the ethical and legal implications of *other* technologies for altering one's memory or other mental processes. Consider the debates that have occurred in the past two decades about drugs for *dampening or erasing* memories. As Adam Kolber notes, drugs such as propranolol are already being used by physicians to dull traumatic memories of patients suffering from PTSD—and scientists have found other methods that might be used to dull, or even erase, memories. Kolber has suggested that an individual's decision to dull a painful memory might often come within a right to "freedom of memory" [12].

But in making this point, he has taken note of—and responded to—other arguments, for example, those made by members of former President George W. Bush's Council on Bioethics and others, that use of technology to alter our memories might be subject to ethical objections or legal restrictions. One such argument is rooted in a familiar kind of paternalism: Individuals may decide to use new technologies for altering their memory or other mental operations in ways that bring immediate satisfaction but only at the cost of long-term harm to their lives. The Council, for example, argues that "by 'rewriting' memories pharmacologically we might succeed in easing real suffering at the risk of falsifying our perception of the world and undermining our true identity" [13]. This argument is similar to the concern that some ethicists have raised about other forms of cognitive enhancement: That it is at odds with "authenticity." Carl Elliott, for example, has expressed skepticism about the benefits of cognitive enhancement drugs—noting that even if my taking a drug like Prozac or another SSRI can be described as giving me a "better personality" it "isn't my personality" [52]. Kolber expresses skepticism toward this authenticity-based objection to memory dampening, noting that it is hard to say altered memory makes life less genuine without a more specific account of what makes life genuine and that those with PTSD might feel "by preventing people from being overtaken by trauma," memory dampening drugs "may actually make them more genuine, more true to what they take their lives to be, than they would be if they were gripped by upsetting memories." In any event, this is one ground on which some might argue for legal limits on BCI-based memory enhancement: Even if individuals should presumptively be left free to exercise their "freedom of thought" to reshape their mental operations by using their natural capacities (for example, by meditating or silently using a technique to remember facts or events), one might argue that matters are different when they use drugs or BCI to alter these operations in ways that arguably bring unintended side effects.

Another focus of such paternalism might be individuals' decisions not merely to use BCI to change their mental operations—for example, by learning more and consolidating more information about history or mathematics—but to *invite* other individuals (individuals, organizations, or state actors) to technologically exercise control over their minds. Consider again a futuristic world where information is automatically downloaded each day into our memory banks with the aid of BCI that links our brains to some outside data source via the Internet. Imagine further that this downloading occurs with the consent of the person who receives the BCI-transferred information. Arguably, even after such consent is given, BCI-alteration of memory might still circumvent our autonomy in a way that violates freedom of thought. Thus, in recent work, J. Adam Carter considers various circumstances in which—even if a person consents to be connected to a Neuralink system to supplement memory, in a "pre-arranged BCI" system—automatic updates to memory through this system may nonetheless violate a "right not to have one's thoughts or opinions manipulated." Even with consent, he argues, using hypothetical scenarios, there may be freedom of thought concerns raised when a person has "beliefs or desires 'implanted' in a clandestine fashion." One might argue that such a violation may not only be a matter of ethics, but of law—and that the state should thus be able to restrict the way individuals might use BCI to voluntarily give others access to, or power over, their memories [53].

Moreover, sometimes our memories are *not* exclusively our own business. For example, we might be obligated to remember events related to a crime in order to testify about it as a witness at a trial. If we intentionally erase a memory in order to deprive the justice system of the information in it, this might—some have argued—count as obstruction of justice [13, 14]. Altering our memory in this case is not just a matter of our own interest in freedom of thought. It can undermine key social interests in solving crime. Also, generating false memories can also have social consequences: For example, if a person technologically generates in their mind a vivid but false memory that someone else was at fault for a wrong that they are responsible for, this mental alteration could conceivably impact the person scapegoated in one's memory as well as the person who alters it (particularly if the person who alters their memory in this way then testifies honestly, but falsely, about the subject of their false memory). The self-transformation enabled by BCI then can raise difficult legal challenges about what memory-enhancing use of BCI technology is insulated against state restriction by our "freedom of thought" and which can be regulated—on paternalistic or other grounds.

3.2 Transforming Memory Capacity

It is helpful, we think, to consider separately two ways that BCI could alter how we remember: by increasing memory capacity and by changing the ways we add memory content.

First, as we noted previously, the use of a BCI device might in the future enhance our memory capacity. The "artificial hippocampus" discussed earlier is a device that concerns memory capacity: it might help restore damaged memory processing in amnesiacs. BCI technology that alters memory could also conceivably allow us to convert more of our perceptions into long-term memories. Brain–computer interfaces might, more modestly, allow us to engage in forms of training of our psychology in a way that enhances memory. Some experiments have also indicated that BCI could enhance the success of memory coding by inducing theta or alpha oscillations in certain parts of the brain [54]. Of course, memory capacity might be radically enhanced if the storage for long-term memory is not limited to our own neurons' functionality.

But apart from the benefits it brings, enhancing memory capacity could potentially destabilize law and especially the law of privacy if the enhancement is significant enough. We briefly discussed above the way that societal interests could be threatened by *memory erasure* (which could amount to obstruction of justice), or by the creation of *false memory* (which could allow for spreading defamatory information without the knowledge or other mental state that subjects the speaker to legal liability under American law[6]). It is also possible that societal interests—and particularly interests in privacy—could be threatened by BCI-enabled enhancements to *accurate* memory.

In 1971, Justice Harlan of the US Supreme Court noted the important difference between a situation where one's words are addressed to a natural person, and where one's words are recorded for future use by some kind of technology (whether it is a written record or an audio or visual recording). The latter, he stressed, ensures "full and accurate" capture of "all that is said, free of the possibility of error and oversight that inheres in human reporting" [56]. Much off-hand exchange, he continued, "is easily forgotten, and one may count on the obscurity of his remarks, protected by the very fact of a limited audience, and the likelihood that the listener will either overlook or forget what is said, as well as the listener's inability to reformulate a conversation without having to contend with a documented record." But "[a]ll these values," said Harlan, "are sacrificed by a rule of law that permits official monitoring of private discourse" with some system of recording what was said [56].

Harlan was worrying here about special efforts by the government to preserve the contents of a person's words in a way that natural memory cannot. The US Fourth Amendment, he argued, should bar the government from making such efforts unless they could first show that they had probable cause to believe the recording would pick up evidence of a crime.

But it is harder for constitutional law to make such a demand in a world where "full and accurate" capture of what someone says is *routinely* made by enhanced memory, perhaps supplemented by technology (like a "brain chip" or a permanent connection between brains and computer storage). In such a situation, Fourth Amendment law could *not* protect privacy simply by maintaining a status quo where words are easily forgotten—since that wouldn't be the status quo. It would rather

[6] In order for a person to be liable for defamation of a public figure in the United States, they must have "'actual malice'—that is, with knowledge that it was false or with reckless disregard of whether it was false or not" [55]). But where the statement stems from a false memory, such actual malice will not be present at the time they make the statement: They will be saying something they sincerely believe to be true. New York Times Co. v. Sullivan, 376 U.S. 254, 280, 84 S. Ct. 710, 726, 11 L. Ed. 2d 686 (1964).

have to *force* informants or officers to somehow forget what they naturally absorb and retain.

To be sure, this kind of situation would only present a problem if BCI were able to substantially change our memory capacity—for example, by giving us all something closer to "photographic" memory. And one might argue that even if technology makes this possible, other considerations would make it unlikely: Photographic or eidetic memory often has a downside for those who have it. As memory researchers have written, remembering details irrelevant to one's survival and success often interferes with focusing on what is important. It also interferes with the process evolution has generated for distilling details into a more general, and practically useful, framework for moving in the world. But the problems created by photographic memory may be avoidable when individuals can store more information—for example, in a brain-connected computer memory bank—and recall it selectively. As noted earlier, some BCI designed to restore vision works by transmitting visual images, from cameras mounted in front of a person's eyes, to computers implanted in or near the visual cortex [21]. A modified version of such a design could conceivably retain a library of videos or photographic images for individuals to rewatch with a mental command, and perhaps with the aid of BCI that not only retrieves but also enhances memory function.

This problem is already being raised to some degree by more familiar technology: The cameras and computer storage of Smartphones already allow for more pervasive recording than was possible in the past—and technologies where recording was made constantly by Google Glass or other visors would exacerbate this problem. BCI can make the threat to privacy more serious—by creating a situation where courts may no longer be able to counter it except by interfering in activity that seems to be integral to someone's ordinary mental functioning.

Enhancement of memory capacity with BCI devices would similarly raise problems for the "right to be forgotten" that is now recognized in the law of the European Union and can be invoked there to force Google or other search engines to remove search results relating to stories about long ago events in a person's life [49]. The logic of this right is that the public nature of such episodes has a "shelf life" after which it is no longer relevant and may be consigned back to obscurity by removal from computer searches. That logic will be undermined to a certain degree when information removed from searches is likely to remain in the future in most individuals' ordinary equipment for generating and storing memories.

Now imagine, for example, that someone is involved in a sex scandal in a particular year (say it is 2050), and that Elon Musk's vision of Neuralink-enhanced memory has come to fruition: Individuals store and retrieve memories not only from their neurons, but from devices within their brains that are connected to computers [51]. In such a world, enforcing the right to be forgotten will require not merely removing information from a set of search results, or a database external to human action. It will require an *intervention* into readers' personal memory itself or, more specifically, into what personal memory will have become in the future. The sex scandal in which that individual was involved, once known by numerous people, will not simply fade naturally from human memory—and be recoverable only with new Google searches. It may remain in a new hybrid of human and computer memory until it is forcibly erased—and in a world where forcible erasure of it can only occur with government intervention.

Given the uncertain consequences of enhanced memory capacity, might individuals nonetheless claim a right—enshrined in a national constitution or an international legal document—to alter their mental processing in this way? As noted earlier, basic sources of rights in the US, Europe, and international law have all given recognition to a right for "freedom of thought." This right has sometimes been viewed as a synonym for the right to express thoughts—in speech or in some kind of religious practice [57, 58], or as the underlying justification for free speech rights [59, 60]. Others have seen it as a guarantor against criminal punishment of thought [61].

But assuming such a right to freedom of thought can have an independent effect, can it give us a right to modify our thinking *capacity*? Each of us has previously argued in separate works that, at least in US law, it should be understood to do so [57, 58], Chaps. 3 and 4 in [59, 62, 63]. Other scholars have argued that a right to freedom of thought entails protection for autonomous decisions not merely to generate or adhere to certain thoughts, but to change *how* one thinks. In the American context, this form of freedom of thought is less familiar in the First Amendment context than is protection for the right to adhere to, or express, particular beliefs (for example, about which policy deserves support or about what religious doctrine to hold) [57, 60].

Still, there are at least two reasons to think that when individuals alter their own memory capacity, they are covered by at least some versions of a right to freedom of thought. First, in US law, the First Amendment's right to freedom of speech entails a right to access and use the resources necessary to create that speech: to spend money on speech [64], have spaces to speak [65], or use video-recorders or computers to create visual or written expression [66], for example. A First Amendment right to think may likewise then entail a right to develop means to engage in thought and memory capacity is a resource central to the exercise of our mental powers [57]. A right to think may in part be a right to think with computers[7] [69, 70]. And scholars have considered particular examples: Kolber, for example,

[7] One might also argue that even if generating mental capacity—with a BCI device that restores or enables it—is not a part of the same right that gives us a right to think discrete thoughts, it may still possibly be covered by a different constitutional or basic right: The Fifth and Fourteenth Amendment's "due process" right to liberty has been found by US courts to include a right to education—to self-formation free from state control, and the Court has said this includes a right to "acquire useful knowledge" [67]. This could be said to include a right to develop certain intellectual capacities and skills—including memory capacity—and to do so with technology. To be sure, a US Supreme Court decision issued in 2022 seemingly makes it harder to find that such rights cover new and unfamiliar uses of cognitive enhancement technology: In Dobbs v. Jackson Women's Health Organization, the Court not only found there was no US constitutional right to obtain an abortion—it also stressed that that courts should only find that these constitutional provisions give individuals protection for specific liberties that are "deeply-rooted in the nation's history or traditions"—description courts are unlikely to find to use of emerging technologies to generate new thinking capacities [68].

has asked whether freedom of thought might not only protect—from legal restrictions—the calculations "card counters" do in their heads while in a casino, [71] but also their use of card-counting devices to supplement their thinking [72].

Second, there are strong intuitions supporting the idea that enhancing memory capacity through use of a BCI is covered. As Andy Clark and David Chalmers point out, when our thinking is inextricably linked to, and supported by, technologies that lie outside of our natural biomemory, then officials or others might interfere with our freedom of thought if they interfere with those supports rather than the biomemory itself[8] [73]. Although the above discussion has focused on US law, there is little reason to think these considerations would not have some force in elaborating freedom of thought protections in other jurisdictions.

This does not mean, however, that use of BCI to enhance memory would simply be tightly insulated against government regulation or limitation. As we have noted, there is little doubt that government would be left with leeway to regulate BCI safety—and BCIs of all types are already subject to regulation to assure medical safety (and protect patient interests) by the FDA, for example, in the US.[9] What is more uncertain is whether our freedom of memory may nonetheless lead courts to impose limits, or exhibit skepticism, when agencies limit individuals' voluntary decisions to enhance memory capacity—for example, by requiring and closely evaluating a safety-based justification. What is also uncertain is how much leeway such a jurisprudence would give government to protect individuals not only from physical harms or threats of external manipulation that might accompany BCI devices—but *also* from the way memory-enhancing BCI could destabilize general assumptions about individuals' privacy. As Woodrow Hartzog and Frederic Stutzman have emphasized, privacy in many circumstances depends heavily on obscurity [77] and Hartzog has proposed further that the design of new technologies (especially Internet technologies) has features that safeguard this condition of privacy [78]. The

[8] Coin and Dubljevic use Clark and Chalmer's "extended mind" hypothesis to argue that, if the mathematical ability-enhancing BCI they consider can be viewed as an part of our extended mind, it may be viewed as an "authentic" part of a person's own identity (unlike certain other enhancement) [20]. A similar argument might be made that such a BCI would be encompassed by the individual's freedom of thought.

[9] In the U.S. the Food and Drug Administration (FDA) regulates medical devices (such as implantable technologies which are Class III devices and thus require the highest scrutiny) to assure their safety and efficacy [74]. Moreover, inserting a computerized device inside of, and connected to, a person raises cybersecurity risks: The FDA shares this responsibility with device manufacturers, hospitals, health care providers, patients, security researchers, and other government agencies, including the U.S. Department of Homeland Security's Cybersecurity and Infrastructure Security Agency (CISA) and U.S. Department of Commerce. The FDA urges manufacturers to monitor and assess cybersecurity vulnerability risks, and to be proactive about disclosing vulnerabilities and solutions to address them [75]. If a vulnerability or weakness in software, hardware, or other factor that could pose a risk is identified, the FDA may issue what is called a "safety communication" [76]. These messages contain information about the vulnerability of and recommended actions for patients, providers, and manufacturers. It seems unlikely that, as courts develop a jurisprudence of freedom of thought or some other right that entails "freedom of memory," they will do so in a way that disables these agencies from protecting patient and consumer safety as they have long done.

same could be required of a BCI device: Rather than wait for a world where BCI enhancement of memory makes obscurity harder to obtain, courts and regulatory bodies could proactively take steps to prevent it from arising—for example, by trying to thwart wide, unrestricted adoption, and the use of BCI devices that enhance memory in this way, and to assure that the technology is designed so as to limit its privacy-eroding potential.[10] In US First Amendment law, courts have generally been very skeptical of arguments that certain kinds of speech (or speech capacities) should be subject to regulation because they will otherwise lead to negative social effects. If they were to view First Amendment freedom of thought claims, or other constitutional claims to mental autonomy, in the same way, they may be inclined to reject arguments that the long-term effects that BCI technology on privacy justify limits on individuals' decisions about when and how to augment their own memory. Given their endorsement of a right to be forgotten [49]. European courts may perhaps be more open to this type of argument.

3.3 Transforming Memory Content

Almost since its beginnings, science fiction has envisioned a world where technology might not only strengthen memories—but also transfer or create them—giving individuals memories of events they have never experienced personally (and perhaps events that have never even occurred), and of facts they have never learned. In an 1886 story called "The Memory Clearinghouse," by Israel Zangwill, a man gathers memories from individuals, stores them, and publishes a "Memories for Sale Catalogue" from which numerous buyers select memories to experience [80]. A. E. Van Vogt's stories from 1944 and 1945, published in The Changeling [81] and The World of Null-A [82], similarly imagine a world where memories—including false memories—are implanted in people. More recently, author Gene Wolfe in the *Book of the New Sun* discusses memory and how it may be distorted, forcing the reader to untangle which strands of memory have been changed or corrupted [83, 84]. In recent years, scientists have made very modest strides toward memory implantation: They implanted in mice, for example, memories of sensory experiences the mice never had, and paired it with a reaction of fear [15]. More specifically, they have used the technique of optogenetics to activate patterns of neurons, the activation of which causes a particular experience—of an odor, for example, or of fear associated with an external stimulus—and having had this experience generated from within the brain rather than from the outside world, the mouse forms a memory of it [16].

While it seems unlikely that, in the near term, scientists will be able to use a BCI device to activate neurons to generate specific memories in human brains, the possibility that they will do so raises interesting questions. Moreover, even if a BCI cannot do so by itself, it might still generate artificial memories in combination with other techniques and technologies, such as virtual reality recordings of what others

[10] This kind of safeguard is proposed by Andrea Lavazza. See [79].

have perceived, or of fictional experiences. As noted above, one can imagine a futuristic, 3D version of "life-logging" wherein someone plays back their own virtual reality experience—or that of another person. BCI could conceivably be used to enhance those memories—perhaps by regenerating, in someone's brain, some of the neural patterns that were present when it first occurred, or by letting individuals enter into and interact with such a virtual world through mental commands rather than by pressing buttons to operate a machine.

If human beings can rely on freedom of thought, or perhaps, more specifically, "freedom of memory" to argue that technologically enhancing mental capacity is something they should be free to pursue without government interference, why not also the freedom to pursue measures that generate specific content? If, like the protagonist of the movie, *Total Recall*—based on a Philip K. Dick story—a person wishes to give themselves vivid memories of experiences they never had, would this be something they have a constitutional right to do? [85].

On the surface, it may seem there is an even stronger case that we have a right to give ourselves a particular memory—or simpler access to a specific episodic memory we already have—than there is to alter our memory capacity as a general matter. While there is uncertainty, as noted earlier, about how and to what extent constitutional rights give us a right to reshape core aspects of our mental functioning, in the US, Europe, and other jurisdictions, we clearly have a right to form and express particular beliefs or thoughts. It is hard to see how a state measure would not run afoul of such a right if it interfered with the way we form or recall a particular memory. Of course, the kind of "memory implants" that discussions of Neuralink and other BCI technology envision is in some respects starkly different from natural memory formation—possibly opening the door, for example, to vastly more memory that is entirely false. But government has not had free rein in controlling the way memory is reshaped by storytelling, or by forcing people to record diary memories with accuracy, for example, so one might argue that courts should not allow restriction of BCI-enabled memory implants unless it is clear how and why that merits different treatment.

There are, however, difficulties raised by a potential right to generate artificial memory. In fact, for reasons discussed below, technological alterations in memory content likely raise a greater challenge for law as a general matter than enhancement of memory capacity. One concern has already been discussed. Our judicial system and many other institutions currently rely on people to have memories that are in many respects accurate reflections of experiences they have actually had. When witnesses take the stand in a trial and testify from their "personal knowledge," judges and juries know that their memory (and the perception it is based on) can be imperfect, or their reports of it can be dishonest. But there often must be *enough* accuracy in these reports of witnesses to give the justice system some way to reconstruct the facts of the case and apply the law to these facts. The same is true in numerous less formal investigatory processes: school and workplace investigations, for example, or the institutional learning that countless organizations and other groups engage in when they draw lessons from past experience. Artificial memories—to the extent they are inaccurate or false—could severely undermine these processes. At one extreme, as we have already noted, they could count as obstruction of justice—when someone wishing to hide evidence intentionally erases it from the mind of a witness, or someone who wishes to manufacture evidence inserts false memories into witnesses' minds [13]. The research of psychologist Elizabeth Loftus has already revealed how false memories can be created (even without the aid of technology) and provide a basis for false testimony: Therapists, according to Loftus, have sometimes unwittingly planted fake memories about crimes in their patients [83].

The possibility of creating memories through artificial means can generate at least three other potentially significant harms [86]. First, it can make the modern-day problem of disinformation far worse. If an individual's basis for believing a lie they have heard comes not merely from false Tweets or other statements (which others know to distrust) but from vivid memories of what they think they have seen and heard, it will be far harder to combat: Others might have some chance of convincing them that they have been lied to by a social media user, but will not likely have a chance of convincing them to reject what they've seen (or are sure they've seen) with their own eyes.

Second, if allowing artificial memory is an exercise of freedom of thought and individual autonomy, it is an exercise with tremendous potential to undermine autonomy. If I consent (e.g., through a contract) or let someone else form my memories for me, I am giving them potentially extraordinary power over my beliefs. Such power would be especially worrisome if in the hands of the government or a corporate entity.

In fact, there is a danger in a world where I experience others' memories—even if I have some control over this process. As one Scientific American article notes, "[w]e learn from our personal interaction with the world, and our memories of those experiences help guide our behaviors. Experience and memory are inexorably linked" [87]. Breaking this link, as the article notes, destabilizes a system where an organism's memory is a guide to its future behavior: *Another* organism's memory might not be a good guide. An animal that remembers classifying a certain area as safe does in part because it is safe for *that* animal. Problems could likewise arise if transformations to our memory content don't preserve the features of our memory processing that let us use memory as accurate guides for our own purposes.

Moreover, it is worth noting that these technological advances in memory may be adopted even when they have uncertain effects. This is an issue that could be raised with the use of BCIs in general. For example, Mark Gasson was the first human to be infected by a computer virus when a chip implanted in Gasson's hand was purposely given a virus to see how simple radio-frequency identification (RFID) chips could host and spread "technological diseases" [88]. Similarly, to what extent, legal thinkers might ask, does a right to freedom of thought include a right to "roll the dice" and alter our psychological processing or to open our mind to the possibility of a computer virus without confidence about the result? For example, do individuals have a right to connect their brain's memory processing functions to computer storage *without knowing* exactly what use the brain will be able to make of these connections (or what effects might follow)? One might argue society has already (though not intentionally) conducted such an experiment on itself by rapidly adopting pervasive use of smartphones and various types of social media without being able to predict their (perhaps very powerful) psychological effects and they should be able to do the same with BCI technologies. Or some might take a contrary position and argue that unpredictable self-alterations that may result from using a BCI (especially a BCI that is implanted within the brain) may constitute a more fundamental—and less easily reversible - change to our psyches than self-experimentation one might conduct with external devices, or even with enhancement drugs.

4 Conclusion

Our focus in this chapter has been on how a particular use of BCI (to enhance memory) may raise challenges for courts seeking to understand whether individuals are constitutionally shielded—in certain ways—from state regulation as they use BCI to modify their own memories and processing of memories. This is likely *not* the form of BCI that legal thinkers will worry about first when they try to predict what voluntary uses of BCI will unsettle our laws. Some BCI has been used to restore individuals' visual capacities and could conceivably be used to enhance it: Consider again the thermal imaging technology that has unsettled Fourth Amendment law by letting police officers outside a home see through its walls, and into its interior. One can imagine BCI that incorporates such thermal imaging technology and allows individuals with such BCI to see through walls with their own eyes. That could conceivably present new challenges for courts—since protecting the privacy of the home would then require not only asking people to refrain from using a thermal imager, but to refrain from seeing the way they normally do (after BCI-enhancement of vision).

BCI that alters my own private internal memory processes—not my perception of the external world—may seem at first to be less likely to be the business of law and government. First, private recollection does not threaten others' interests in the same way surveillance does. Second, our control over our own thought processes is considered by legal scholars to be central to freedom of thought. Our analysis in this chapter has largely accepted the idea that use of BCI to enhance our memory implicated our right to freedom of thought.

But we have also argued that even if use of memory-enhancing BCI constitutes an exercise of an individual right (to freedom of thought), it may also constitute a kind of conduct that government sometimes is justified in regulating and may even have a responsibility to regulate. This is most clearly true in cases where use of BCI to alter one's memory can make one's mind vulnerable to external hacking and other kinds of manipulation. However, it is also arguably true even when individuals voluntarily use BCI, and use it as intended. Even such voluntary use of BCI, by altering our memory capacity, and ability to generate or share memory content, might undermine the foundations of our privacy, and legal and social institutions that rely on memory that is relatively insulated against technological manipulation.

We have not predicted how courts might reconcile the freedom of thought entailed by use of such memory-enhancing BCI with the justifications for potential legal limits on it. Nor have we taken a firm position on how courts should address this challenge. How they should and will do so will likely depend on the specific form this technology takes. As courts and legal thinkers grapple with this challenge more deeply in the future, they will have to decide whether they should respond by prohibiting or tightly restricting memory-enhancing BCI until they are sure it is safe for the fundamental conditions of our privacy and autonomy. This is something they may be inclined to do if freedom of thought is something we can exercise even without such BCI, and the risks it generates. Alternatively, they might find that we must be insulated from such restriction as we use BCI to reshape, and use, our memories—at least so long as those who develop the technology build certain safeguards for privacy and autonomy into its design.

References

1. Kerr O. The case for third-party doctrine. Mich Law Rev. 2009;107:1–15.
2. City of Los Angeles, Calif. v. Patel, 576 U.S. 409, 419. 2015.
3. California v. Ciraolo, 476 U.S. 207, 213. 1986.
4. Lessig L. Constitution and code. Cumberl Law Rev. 1997;27:1–15.
5. Kyllo v. United States, 533 U.S. 27. 2001.
6. Kerr O. An equilibrium-adjustment theory of the fourth amendment. Harv Law Rev. 2011;125:476–543.
7. Han DS. Constitutional rights and technological change. UC Davis Law Rev. 2020;54:71–5.
8. Blitz MJ. Video surveillance and the constitution of public space: fitting the fourth amendment to a world that tracks image and identity. Tex Law Rev. 2004;82:1349–52.
9. Jones v. Opelika, 316 U.S. 584. 1942.
10. Richards NM. Intellectual privacy. Tex Law Rev. 2008;87:387–445.
11. Lavazza A. Freedom of thought and mental integrity: the moral requirements for any neural prosthesis. Front Neurosci. 2018;12:82. https://doi.org/10.3389/fnins.2018.00082.
12. Kolber AJ. Therapeutic forgetting: the legal and ethical implications of memory dampening. Vanderbilt Law Rev. 2006;59(5):1559–626.
13. President's Council on Bioethics. Beyond therapy: biotechnology and the pursuit of happiness. New York: ReganBooks; 2003. p. 209.
14. Rivera-Corraliza v. Morales, 794 F.3d 208, 220 (1st Cir. 2015). 2015.
15. Liu X, Ramirez S, Pang P, Puryear CB, Govindarajan K, Deisseroth K, Tonegawa J. Optogenetic stimulation of a hippocampal engram activates fear memory recall. Nature. 2012;484:381–5. https://doi.org/10.1038/nature11028.
16. Lloreda CL. Memories can be surgically implanted into brains now (yes, in mice). Massive Science. 2019. https://massivesci.com/notes/implanted-memories-memory-mice-neuroscience-science-fiction-brains-mind-control/.
17. Binnendijk A, Marler T, Bartels EM. Brain–computer interfaces: U.S. military applications and implications: an initial assessment. Santa Monica: RAND; 2020. p. 1–43.
18. Reddy GSR, Lingaraju GM. A brain–computer interface and augmented reality neurofeedback to treat ADHD: a virtual telekinesis approach. In: 2020 IEEE International symposium on mixed and augmented reality adjunct (ISMAR–Adjunct). 2020; 2020:123–128.
19. Wander JD, Rao RP. Brain–computer interfaces: a powerful tool for scientific inquiry. Curr Opin Neurobiol. 2014;25:70–5. https://doi.org/10.1016/j.conb.2013.11.013.
20. Coin A, Dubljevic V. The authenticity of machine–augmented human intelligence: therapy, enhancement, and the extended mind. Neuroethics. 2021;14:283–90.

21. Juskalian R. A new implant for blind people jacks directly into the brain. MIT Technology Review. 2020. https://www.technologyreview.com/2020/02/06/844908/a-new-implant-for-blind-people-jacks-directly-into-the-brain/.
22. Prochazka A, Mushahwar VK, McCreery DB. Neural prostheses. J Physiol. 2001;533(Pt 1):99–109. https://doi.org/10.1111/j.1469-7793.2001.0099b.x.
23. Miranda RA, Casebeen WD, Hein AM, et al. DARPA-funded efforts in the development of novel brain–computer interface technologies. J Neurosci Methods. 2015;244:52–67.
24. Rose N. The human brain project: social and ethical challenges. Neuron. 2014;82(6):1212–5.
25. Berger TW, Hampson RE, Song D, Goonawardena A, Masmarelis Z, Deadwyler SA. A cortical neural prosthesis for restoring and enhancing memory. J Neural Eng. 2011;8(4):046017.
26. Keren H, Partzsch J, Marmon S, Mayr CG. A biohybrid setup for coupling biological and neuromorphic neural networks. Front Neurosci. 2019;13:432. https://doi.org/10.3389/fnins.2019.00432/full.
27. Snead OC. Memory and punishment. Vanderbilt Law Rev. 2011;64:1195–203.
28. Bigelow K. Strange days. Santa Monica: Lightstorm Entertainment; 1995.
29. Agida T. Cyberpunk 2077 review-great story, too many bugs. Inquirer.net. Accessed 24 Sept 2021.
30. Van Ord K. Remember me review. Gamespot. www.gamespot.com. Accessed 29 Sept 2013.
31. Carraze A, Oswald H. The prisoner: a televisionary masterpiece (trans. Christine Donougher). London: Virgin; 1996.
32. Clarke AC. 3001: The final odyssey. New York: Del Rey Books; 1997.
33. Hinchliffe T. The sociable, the future's gonna be weird': Musk's says memories could be downloaded into a new body or robot. 2020. https://sociable.co/technology/futures-gonna-be-weird-musk-memories-downloaded-new-body-robot/.
34. Retalagio A. Elon Musk's neurolink is neuroscience theatre. MIT Technology Review. 2020. https://www.technologyreview.com/2020/08/30/1007786/elon-musks-neuralink-demo-update-neuroscience-theater/.
35. Tarantola A. What neuralink and other BCIs can and can't do, Engadget. https://www.engadget.com/what-neuralink-and-other-bc-is-can-and-cant-do-140014162.html. Accessed 24 Jun 2021.
36. Moriarty JC. Visions of deception: neuroimages and the search for truth. Akron Law Rev. 2009;42:739–60.
37. Allen AL. Dredging up the past: lifelogging, memory, and surveillance. Univ Chic Law Rev. 2008;75:47.
38. Tran L, Kennedy D, Liting Z, Nguyen B, Gurrin C. A virtual reality reminiscence interface for personal lifelogs. In: Multimedia modeling. Lecture notes in computer science, vol. 13142. Cham: Springer; 2022. https://doi.org/10.1007/978-3-030-98355-0_42.
39. Gordon L. Brain-controlled gaming exists, though ethical questions loom over the tech. https://www.washingtonpost.com/video-games/2020/12/16/brain-computer-gaming/. Accessed 16 Dec 2020.
40. Rosca S, Leba M. Design of a brain-controlled video game based on a BCI system. MATEC Web Conf. 2019;290:01019. https://doi.org/10.1051/matecconf/201929001019.
41. Blitz MJ, Bublitz JC. The law and ethics of freedom of thought, volume 1: neuroscience, autonomy, and individual rights. New York: Palgrave Macmillan; 2021.
42. Stanley v. Georgia, 394 U.S. 557. 1969.
43. Bublitz JC. Freedom of thought as an international human right: elements of a theory of a living right. In: Blitz MJ, Bublitz JC, editors. The law and ethics of freedom of thought, volume 1: neuroscience, autonomy, and individual rights. London: Palgrave Macmillan; 2021.
44. Marks, M., Cognitive Content Moderation: Freedom of Thought and the First Amendment Right to Receive Subconscious Information, Fla. L. Rev. (forthcoming 2024); 76, available at SSRN: https://papers.ssrn.com/sol3/papers.cfm?abstract_id=4215470
45. Klein E, Rubel A. Privacy and ethics in brain–computer interface research. In: Nam CS, Nijholt A, Lotte F, editors. Brain–computer interfaces handbook. Boca Raton: CRC Press; 2018.
46. Computer Fraud and Abuse Act, 18 United States Code §1030.

47. Van Buren v. United States, 141 S. Ct. 1648, 1652. 2021.
48. Criminal Code, Revised Statutes of Canada 1985, c. C–34, Section 430 (1.1). 1985.
49. Youm K, Park A. The "right to be forgotten" in European Union Law: data protection balanced with free speech? J Mass Commun Q. 2016;93(2):273–96.
50. 'Google Spain v AEPD', ruling of the CJEU on Case C-131/12, from 2014 May 13.
51. Fourneret E. The hybridization of the human with brain implants: the Neuralink project. Cambridge: Cambridge University Press; 2020. https://www.cambridge.org/core/journals/cambridge-quarterly-of-healthcare-ethics/article/abs/hybridization-of-the-human-with-brain-implants-the-neuralink-projec /C8D1E6A47C78A7B1FA159BE5EE7142DE
52. Elliott C. The tyranny of happiness: ethics and cosmetic pharmacology. In: Parens E, editor. Enhancing human traits: ethical and social implications; 1998. p. 177–82.
53. Carter JA. Varieties of (extended) thought manipulation. In: Blitz MJ, Bublitz JC, editors. The law and ethics of freedom of thought, volume 1: neuroscience, autonomy, and individual rights. New York: Palgrave Macmillan; 2021.
54. Burke JF, et al. Brain computer interface to enhance episodic memory in human participants. Front Neurosci. 2015;8:1–10.
55. New York Times Co. v. Sullivan, 376 U.S. 254. 1964.
56. United States vs. White, 401 U.S. 745. 1971.
57. Blitz MJ. Freedom of thought for the extended mind: cognitive enhancement and the constitution. Wis Law Rev. 2010;2010:1049–117.
58. Blitz MJ, Bublitz JC. The law and ethics of freedom of thought, volume 1. New York: Palgrave Macmillan; 2021.
59. Shiffrin SV. A thinker-based approach to freedom of speech. Const Comment. 2011;27:283–307.
60. Blitz MJ. A constitutional right to use thought enhancing technology. In: Dubljevic V, Jotterand F, editors. Cognitive enhancement: ethical and policy implications in international perspectives. Oxford: Oxford University Press; 2016.
61. Mendlow GS. Why is it wrong to punish thought? Yale Law Rev. 2018;127:2345–86.
62. Barfield W. Cyber-humans: our future with machines. Cham: Copernicus Press; 2015.
63. Macklem T. Independence of mind. Oxford: Oxford University Press; 2007.
64. Buckley v. Valeo, 424 U.S. 1. 1976.
65. Perry Education Association v. Perry Local Educators Association, 460 U.S. 37. 1983.
66. American Civil Liberties Union v. Alvarez, 679 F.3d 583 (7th Cir. 2013). 2013.
67. Meyer v. Nebraska, 262 U.S. 390. 1923.
68. Dobbs v. Jackson Women's Health Org., 142 S. Ct. 2228. 2022.
69. Bambauer JY. Sensorship. In: Collins RKL, Skover D, editors. Robotica: speech rights and artificial intelligence. Cambridge: Cambridge University Press; 2018.
70. Blitz MJ. Artificial minds in first amendment borderlands. In: Barfield W, editor. The law of algorithms. Cambridge: Cambridge University Press; 2021.
71. Kolber AJ. Two views of first amendment thought privacy. Univ Pa J Const Law. 2016;18:1381–3.
72. Kolber AJ. Criminalizing cognitive enhancement at the blackjack table. In: Nadel L, Sinnott-Armstrong W, editors. Memory and law. Oxford: Oxford University Press; 2012.
73. Clark A, Chalmers D. The extended mind. In: Clark A, editor. Supersizing the mind: embodiment, action, and cognitive experience, appendix. New York: Oxford University Press; 1998. https://www.nyu.edu/gsas/dept/philo/courses/concepts/clark.html.
74. Darrow JJ, Acorn J, Kesselheim AS. FDA regulation and approval of medical devices: 1976–2020. JAMA. 2021;326(5):420–32. https://doi.org/10.1001/jama.2021.11171.
75. Stern AD, Gordon WJ, Landman AB, et al. Cybersecurity features of digital medical devices: an analysis of FDA product summaries. BMJ Open. 2019;9:e025374. https://doi.org/10.1136/bmjopen-2018-025374.
76. Kramer DB, Baker M, Ransford B, Molina-Markham K, Reynolds MR. Security and privacy qualities of medical devices: an analysis of FDA postmarket surveillance. PLoS One. 2012;7(7):e40200. https://doi.org/10.1371/journal.pone.0040200.

77. Hartzog W, Stutzman F. Obscurity by design. Wash Law Rev. 2013;88:386–418.
78. Hartzog W. Privacy's blueprint: the battle to control the design of new technologies. Cambridge: Harvard University Press; 2018.
79. Lavazza A. Technology against technology: a case for embedding limits in neurodevices to protect our freedom of thought. In: Blitz MJ, Bublitz JC, editors. The law and ethics of freedom of thought, volume 1: neuroscience, autonomy, and individual rights. New York: Palgrave Macmillan; 2021.
80. Zangwill I. The memory clearinghouse. In: The King of Schnorrers: grotesques and fantasies. New York: Macmillan; 1893.
81. Van Vogt AE. The changeling. Astounding Sci Fict. 1944;33(2):7–66.
82. Van Vogt AE. The world of Null-A. New York: Macmillan; 1948.
83. Loftus E, Pickrell JE. The formation of false memories. Psychiatr Ann. 1995;25(12):720–5. https://doi.org/10.3928/0048-5713-19951201-07.
84. Wolfe G. Book of the new sun. Oxford: Oxford University Press; 1998.
85. Dick PK. We can remember it for you wholesale [1966]. In: Dick PK, editor. The collected short stories of Philip K. Dick. Secaucus: Carol Publishing; 1990.
86. Blitz MJ. The right to an artificial reality? Freedom of thought and the fiction of Philip K. Dick. Mich Technol Law Rev. 2021;27:377.
87. Marlone RA. Successful artificial memory has been created. In: The growing science of memory manipulation raises social and ethical questions. New York: Scientific American; 2019.
88. Hadhazy A. Live science: man infects himself with computer virus, live science. 2010. https://www.livescience.com/8290-man-infects-computer-virus.html.

Cyberneurosecurity

Nadine Liv and Dov Greenbaum

1 Introduction

Decades ago, as biological labs came into the internet age, they were subject to increased cybersecurity threats to their computing infrastructure. These attacks often occurred either directly through network infrastructure such as unsecured wifi, through email phishing attacks targeted to unsuspecting lab members, or through infected shared disks. For the most part, these early efforts to infiltrate the computing infrastructure of life science laboratories, both commercial and academic, were either designed to maliciously disable lab computers or to extract information and intellectual property for profit [1].

The area of research that grew out of the need to deal with the issues of cybersecurity as they related primarily to health science research ultimately became known as cyberbiosecurity (or alternatively as biocybersecurity) [2]. Much of the early research in this emerging field focused predominantly on securing the interface between the biosciences and cyberspace, principally in terms of protecting biological research from cybersecurity threats, but also in employing biological methods to the world of cybersecurity [3–5] and in employing cybersecurity methods in the world of biology [6].

N. Liv
Meitar Liquornik Geva Leshem Tal, Ramat Gen, Israel

Harry Radzyner Law School, Reichman University, Herzliya, Israel

D. Greenbaum (✉)
Harry Radzyner Law School, Reichman University, Herzliya, Israel

Zvi Meitar Institute for Legal Implications of Emerging Technologies, Reichman University, Herzliya, Israel

Molecular Biophysics and Biochemistry, Yale University, New Haven, CT, USA
e-mail: dov.greenbaum@runi.ac.il

V. Dubljević, A. Coin (eds.), *Policy, Identity, and Neurotechnology*, Advances in Neuroethics, https://doi.org/10.1007/978-3-031-26801-4_13

Recently, this area of research has become even more relevant to genomic researchers when it was shown that malevolent individuals could target more specific vulnerabilities at the intersection between cyberspace and biology such as genomic engineering. Consider the possibility that malware could masquerade as common academic bioinformatic software such as codon optimization tools. These tools could be employed by unsuspecting researchers to suggest the creation of physical DNA sequences designed to wreak havoc in unwary research systems. Or consider the possibility that a naive researcher's interactions with a commercial DNA producer could be hijacked and the formerly benign DNA code that said researcher intended to order for her experiment is replaced with a malicious sequence, the properties of that malicious sequence potentially further obfuscated via cryptographic tools. Once that DNA strand is returned to our unsuspecting researcher and integrated into her genomic research systems, that DNA, perhaps coding for some toxic protein, could wreak havoc. Proof of concept of such an attack has already been shown [7]. The nature of these types of software and DNA threats are exacerbated by the reality that the necessary tools for their implementation are generally publicly available. Similarly, software masquerading as benign or an upgrade to a BCI may in fact contain malicious code that could be harmful to the user and/or the people around them.

This paper is meant to be an introductory look into the emerging field of cyberneurosecurity (or neurocybersecurity), a subfield within the incipient field of cybersbioecurity. The paper is presented as follows: We first present the field of cyberbiosecurity noting in particular how that term represents a unique field. We then present how the field of cyberneurosecurity is situated within the larger cyberbiosecurity field. Following these definitions, we present brain–computer interfaces (BCIs), the primary source of hackable electronics when discussing cyberneurosecurity. Within the field of BCIs, we discuss issues specific to their security, as well as the neurorights that have arisen as a result of the increasing advancements within BCI technology. We counter some of that discussion of advancements with an acknowledgement of the pervasiveness of neurohype, i.e., neuro related technologies that are currently more fiction than reality. Although not necessarily a technology that will be implemented in the immediate or near future, the possibility of uploading consciousness to an AI machine, an idea that might or might not fall within the aforementioned concept of neurohype, could also conceivably raise many interesting and novel concerns in the field of cyberneurosecurity. Finally, with all of these aspects presented, we provide the reader with a thorough discussion of the actual field of cyberneurosecurity, including discussions of specific cases as well as potential countermeasures to the cyber threats on neurotechnologies.

2 Cyberbiosecurity

Given these fears, one early definition of cyberbiosecurity defined the nascent field as one devoted to the understanding of "the vulnerabilities to unwanted surveillance, intrusions and malicious and harmful activities which can occur *within or at the interface* of comingled life and medical sciences, cyber, cyber-physical, supply chain and infrastructure systems, and developing and instituting measures, to prevent, protect against, mitigate, investigate and attribute such threats as it pertains to security, competitiveness and resilience" [8]. It has been noted by cyberbiosecurity's early promoters that the landscape for cyberbiosecurity would ultimately evolve rapidly and that this definition would eventually need updating.

Part of that updating is the goal of this particular chapter. Concerns in the area of cyberbiosecurity have long grown past just the aforementioned field of genomics. There are equally, if not greater concerns that arise in the area of neurotechnologies. Just like researchers in the field of bioethics found it necessary to include a subfield devoted particularly to neuroethics in the early part of the century [9], we think the time is right to develop and describe a subfield in cyberbiosecurity devoted wholly to neurotechnologies. Overall, this is what we refer to here as the subfield of cyberneurosecurity. Like cyberbiosecurity, we distinguish cyberneurosecurity from the cybersecurity for neuroscience and neurosecurity [10] in that this field is broader, incorporating like cyberbiosecurity, issues relating to neurorights, neuroprivacy, and neuroethics, as well as the potential future uses of neuroscience technologies in the service of cybersecurity and/or hacking. Cyberneurosecurity is not only interested in issues arising at the intersection of brain–computer interfaces, but also with regard to the attacks in future brain to brain (BtB) communications [11] as well as brain to internet communication [12].

In the area of cyberneurosecurity, an unauthorized hack can project force onto an individual, or read the thoughts of an individual [13–15], either locally or remotely [16]. These possibilities [17] take the cybersecurity concerns to a radically different level than those of cyberbiosecurity.

In addition to being somewhat distinct from discussions on cyberbiosecurity, we believe the cyberneurosecurity is a unique and distinct subfield of cyberbiosecurity because of the personal nature of these potential attacks against the human brain, as described herein. Such attacks can lead to both direct and indirect harm with profound ethical and legal implications; many of these harms, as described herein, are unique to the world of neuroscience. For example, researchers have noted that the misuse of neural devices for malicious purposes may not only threaten users' physical security, but also it can influence the user's behavior and alter their sense of identity and personhood. Additionally, in contrast to many other criminal activities associated with biological devices, the attack on neurodevices can create an extreme sense of anxiety and fear and otherwise severely affect the overall mental state of the targeted individual [18]. Such attacks violate centrally human moral values of autonomy, free will, and self-determination [19].

3 Brain–Computer Interfaces

We define this subfield as mostly interested in the neuroscience tools that can read and write to the brain such as brain–machine/computer interfaces (BMI, or BCIs). BCIs have been around for almost half a century, but only recently have there been an uptick in the academic literature relating to their related cybersecurity concerns [20]. This is disconcerting: Brain to Internet (B2I) technology is already available [21], and as some suggest, BCI technology will only become more pervasive within the general public in the coming decade [22]. Already cheap open source devices are readily purchasable.

BCIs can be broadly defined as devices that record, process, analyze, and/or modify brain activity. BCIs directly connect the brain, invasively via surgery, partially invasively via electrocorticography (ECog or iEEG), or wholly non-invasively via EEG or fMRI, to a computer. In some instances, BCIs only report brain activity, in others, BCIs also effectuate events outside of the user's body, typically circumventing peripheral nerves and sending electro-physiological signals directly to a machine, such as a prothesis [23]. BCIs may also mediate incoming signals to the brain. More practically, BCIs can be used for a host of applications [24]. These include gaming [25] and other recreational activities [26], health and medical, and increasingly biometric authentication [27].

BCIs clearly fall within the category of "comingled life sciences… and cyber" set forth in the definition of cyberbiosecurity above. However, as a result of their unique integration of neuroscience and technology, BCIs are at the forefront of the subfield of cyberneurosecurity risks, and as BCIs evolve, cybersecurity concerns relating especially to neuroscience will continue to evolve as well.

4 Neurosecurity

Notably, some have already described components of cyberneurosecurity within what is known as neurosecurity, i.e., the protection of the confidentiality, integrity, and availability of neural devices and their data from malicious third parties [28]. Neurosecurity can be additionally defined as the employment of knowledge as to how the brain functions when employing cybersecurity tools [29, 30]. Neurosecurity has also been defined as the use of neuroscience for national defense: "Creating resilient soldiers (to stress, fatigue, overload)… [or] Developing rapid training and learning techniques" [31].

Some outcomes of the failure to recognize the value of pursuing neurosecurity goals have already been documented to include the creation of software that has been shown to be able to infiltrate BCI systems to extract privacy-related information [32], the hijacking of prosthetic limbs to create damage or limit the motility of a disabled individual (sometimes termed the failure of availability), the malicious programming of neurostimulation therapy to harm a patient [33, 34]—through tools like transcranial magnetic stimulation, transcranial direct current stimulation, deep brain stimulation, sensory prosthetic and orthotic implants transcranial Doppler or direct cortical electrical stimulation, the interception and reprogramming of signaling between a BCI and an external object [35], and the eavesdropping on a brain implant's signals to reveal private information (sometimes termed a breach of integrity) [28], including discrete pieces of private information such as a personal identification code (PIN) [36].

In contrast to neurosecurity, cyberneurosecurity, as we see it, is both broader and more encompassing while also narrower. To wit: more than just cybersecurity attacks on the BCIs to extract information, hijack prosthetic limbs, manipulate output or even to introduce ransomware that serves the connection between a user and their prosthetic, cyberneurosecurity also includes the manipulation of BCIs by their own users. This manipulation is less for malicious purposes, but rather to override safety settings or other built-in limitations. These hacks can be just as dangerous to the user, as well as to those in their vicinity as malicious attacks, but because the hacker isn't unauthorized, they are typically not included in the neurosecurity discussion.

For example, consider the prospect of medical BCIs being modified to provide otherwise unintended and unfair benefits to the consumers of these devices, such as enhancing memory or cognition. Alternatively, a neuromodulation device could be modified to activate the reward circuitry of the brain potentially resulting in the development of addictive behaviors in the pursuit of desirable sensations or experiences. Already BCIs are being developed to provide for some forms of neuromodulation such as creating enhanced memory and cognition in those experiencing mental decline [37]. However, the same technology designed to bring those who are suffering from deficiencies up to par may, hypothetically also someday provide those at par with extra-human abilities. This potential hacking of a medical BCI for non-medical gains of human enhancement [38, 39] will become more likely as BCIs become more commonplace and consumer grade technologies become more promising [40]. Another area of interest within cyberneurosecurity is the topic of neurorights and the corresponding obligations that arise from them.

5 Neurorights

In general, the hacking of neurodevices to manipulate a patient impinges on the patient's autonomy. This manipulation may occur via many paths, including eliciting emotions, manipulating decision-making and preferences, and manipulating memories. In addition, an attack on BCIs can also impinge on other emerging cognitive rights of the user/consumer/patient. These neurorights are a relatively new academic legal area [41]. Notably, neurorights are not explicitly reflected in the vast majority of national constitutions or international legal instruments, with the exception of Chile and potentially Spain [42].

Some have argued that the neurohype regarding products that are far from available has fed the efforts to develop neurorights long before they are necessary. Without knowing exactly was emerging technologies which are capable of the issues relating to neurorights are not yet ripe and perhaps even ultimately misguided depending on how the relevant technologies actually develop and mature. This is especially problematic, according to critics, for countries like Chile and Spain which are zealously taking the broad ideas developing within the neurorights community and turning them into hard and fast legal rights, rules, and regulations. Ultimately, these premature efforts could stifle innovation rather than promoting its development.

One of these rights is the long standing right to cognitive liberty which outlines the right of each individual to be able to think autonomously and independently without outside interference [43]. Cognitive liberty is an umbrella-like right that incorporates many of the standard rights of freedom of speech, freedom of religion, and freedom of choice. Thus, the right to cognitive liberty is the right to make your own choices, unencumbered by the unknown or undesired influence of others: it "guarantees an individual's sovereignty over her mind and entails the permission to both use and refuse neuro-enhancement" [44]. Accordingly, when third parties hack a neurodevice to manipulate the mental states of an individual, they have violated that individual's cognitive liberty.

Notably, the concept of cognitive liberty also suggests that an individual ought to have the right to self-employ mind enhancing neurodevices as well. However, this neuroenhancement is constrained by the obligation not to harm others. This is a real fear when a user of a BCI hacks their own device to operate it outside of manufacturers' safety constraints, one of the potential interests of the field of cyberneurosecurity.

Other neurorights that might be affected by malicious attacks on neurodevices are the right to mental privacy, the right to mental integrity, and the right to psychological continuity [45]. The right to mental privacy grants individuals the right to be free from third parties peering into their thoughts and emotions, e.g., via a hacked BCI.

The right to mental integrity is the individual's right to have control over their own thoughts and the right to prevent third parties from intruding into their brain and introducing fake information. Again, such an intrusion can occur via a hacked BCI that provides input to the brain. Mental integrity can be further impinged via a cyberneurosecurity attack on a BCI that harms neurological tissue. In this case there is a fear that such an attack could manipulate or erase memories that provide individuals with their weltanschauung and their personal autobiographical record.

The right to psychological continuity similarly refers to the right to not have foreign ideas and memories implanted into an individual's mind. Anything that harms or changes an individual's particular mental sense of self is a potential violation of this right. Any alterations in mental states may affect areas critical to a person's identity and personality [45]. Some have countered that this right particularly highlights many of the concerns with neurorights proposals: Broad statements of rights like these can be misleading or confusing and even vulnerable to counterclaims. Consider the reality that humans are always having foreign ideas and memories placed in our minds simply through daily interactions with reality. Any new idea can affect our mental sense of self. Arguably, the whole process of education would seem to be a problem with regard to the concept of psychological continuity and yet few would suggest that we should disincentivize education and the learning of new ideas that can change our outlook and mindset.

However, even within the scope of these rights, states can also arguably limit an individual's right to waive other countervailing rights, for example, by enforcing the right of cognitive liberty to prevent a user from hacking their own BCI. Legally, while most modern states allow for self-determinism and the ability of each citizen to decide for themselves who they are, there are also aspects of paternalism within the modern state that will typically step in to prevent self-harm or activities that can harm others.

6 Neurohype?

In many areas of neurotechnologies, there is a concern that much of the ethical, legal, and social issues raised are associated with technologies that are improbable and unlikely; i.e., hype. This concern is often referred to as neurohype, the idea that the lay public is often presented with technological claims regarding neuroscience that are beyond the actual capabilities of the technology, or that authors of various articles on the subject often buy into the hype and pontificate about technologies and their concerns that are years away from reality, if ever reality at all [46]. To some degree, some of this neurohype narrative arises out of a failure in science communication. Many researchers in the neuroscience field are themselves influenced by science fiction to pursue and create new neuroscience realities emulating the fictional accounts and they could end up communicating their research as similar to the fictional technologies. And, to their credit, much of what neuroscientists can accomplish today was arguably science fiction a decade ago [47].

To this end, there may be a concern that many of the issues raised by cyberneurosecurity are themselves unripe, resulting from technology that is merely neurohype, conflating science fiction for reality [46, 48]. In particular, some may think that claims about what is and what isn't possible in regard to hacking technology like BCIs overstate the concerns and create problems where none yet exists. As such while many of the issues mentioned herein are associated with proofs of concept, neurohype still remains a potential caveat on the following assessment of cyberneurosecurity.

7 Cyberneurosecurity: How Cyberbiosecurity Specifically Applies to BCIs

Their 50 years of development notwithstanding, the modern version of brain–computer interface technologies (EEGs themselves were developed nearly a century ago [49]) continues to evolve. As they do, BCIs will continue to bring benefits to the field of medicine, where they are used to diagnose medical conditions, aid in rehabilitation, or control prostheses [50]. Data from these devices can become accessible or manipulatable through a hack. In other cases, the hacker can potentially control, the movement, emotion, or even the brain functions of the target. In worst case scenarios, hacking these devices can cause long-term damage to the brain or the individual, and even death.

Data transmitted from the brain can be collected and interpreted to provide an increasing amount of actionable information about a patient. In general, medical devices are often not the best protected against cyberattacks as they often offer little encryption and employ default passwords to allow for easy interfacing with existing hospital infrastructure [51, 52]. Customers and patients typically are uninterested in encrypting their data: "Given a choice between dancing pigs and security, users will pick dancing pigs every time" [53].

This is especially the case in neurotechnologies where latency between BCI and the end result (e.g., the movement of a prosthetic) is already high. Encryption and decryption would only serve to increase that undesirable latency. Further, encryption draws power that would further limit the battery life of a remote device. Because of these reasons and more, neurotechnological devices in particular and medical devices in general are often seen as vulnerable weak spots in hospital networks [52], making the already increasingly profitable hack of a hospital [54] all the more enticing through providing additional types of personal information: neurological.

Arguably even non-neurotechnological medical technologies also provide opportunities to hack the brain. An insulin pump, for example, or any device that is somehow associated with the peripheral nervous system, is itself potentially an opportunity for a side-channel attack, i.e., by using relevant data collected on the induvial rather than exploiting a design flaw, on the human brain.

Regardless as to whether the BCI is inputting to the brain or collecting output, there are ample opportunities for malicious activities. These activities can result in numerous negative impacts relating to the integrity of medical data collected from the brain or transferred to the brain, the confidentiality of that data, resulting in private and personal information being transferred to third parties, the availability of the data that is generated to manipulate a device, and of course the safety of the user who may suffer from long- or short-term psychological and/or physical damage.

Integrity can also refer to the possibility that third party hackers can effect behavior changes on a person with a BCI by stimulating pleasure and pain sensors every time an activity is desired or undesired. Such technologies are not yet thought to have been developed, but are not necessarily beyond the technological limitations of the current state of the art. In an extreme case, one could imagine the user of a BCI having their pleasure and pain sensors triggered surreptitiously via geolocation sensors, perhaps even creating a situation wherein a person with a brain–machine interface might be limited in where they can and cannot go due to third parties triggering pain regions in the brain every time the individual moves beyond a certain point.

Similarly, availability concerns can also relate to accessibility of the BCI, the devices effectuated by the BCI, such as a prosthetic in the case of a hack. In some cases, a BCI hack may even inhibit access to one's own brain; it has been shown that it is possible to stimulate the brain via a BCI as to affect consciousness [55].

As the term suggests, cyberneurosecurity concerns often arise when hackers employ cybersecurity exploits in the area of neurodevices. These include low complexity attacks such as neural flooding which overstimulate neurons via the BCI, and neuronal jamming which is the impeding the information flow from neurons to BCIs. In this instance, a neuronal jamming attack is like a denial of service (DoS) attack, but with biological parts like neurons in contrast to internet infrastructure. Moderate complex attacks can include such hacks as neuronal scanning, which is like port scanning in a cyberattack, but instead of seeking out internet ports, the attack sequentially maliciously stimulates each BCI associated neuron, one at a time and neuronal selective forwarding, which purposefully inhibits only some data from going from the brain to the BCI or vice versa with the intent of incapacitating the information flow. More sophisticated attacks like neuronal spoofing confuse the brain and/or the BCI by replicating an earlier legitimate neuron behavior, but at a different time or location while a neuronal nonce attack modulates the nature of the attack randomly so that the BCI has difficulty in identifying the malicious actions [56]. In addition to these, there are numerous other types of cyberattacks that can exploit the internet-enabled aspects of a BCI [12] as they could any other medical device internet of things (MDIoT) device [57, 58].

8 BCI Data Hacking

Both in health and in employment, the BCI data collected is more than simply a snapshot of some biometric information. An EEG reading can give an indication about the emotional state of the user at a particular point in time. This information can be accumulated over time to create a detailed profile of the user's emotional states [59].

As per standard cybersecurity protocols, data can be maliciously acted upon either when it is at rest, in transit/motion, or in use [60]. Broadly, in either direction of BCI action, there are at least five instances that represent these different aspects of data. When the goal of the BCI is to output data from the brain, these five instances include:

1. Neural data acquisition wherein neural signals are generated, representing the data in its rawest form
2. Data capture from one of the electrodes associated with the brain
3. The conversion of analog neural signals to digital data. This conversion often also includes the reduction of noise from the raw data, resulting in cleaner and more useful signal
4. Processing and decoding digitized signal, in some cases by way of artificial intelligence, in an effort to extract actionable information from the initial neural impulses and
5. The use of any actionable data is put into practice by way of any external device, such as a prosthetic, or a display showing the processed signal [12]

A BCI system can also go in the reverse direction, effecting an external input on the brain wherein

1. An external input is collected through sensors or other inputs
2. That input is then collected and analyzed, and in some cases, converted into a neural firing pattern
3. That firing pattern is then optimized by assessing which neurons ought to be stimulated and by how much voltage
4. Those parameters are passed on to the device that is in physical contact with the brain
5. And finally, that BCI physically stimulates the brain according to the determined parameters

Potential cyberneurosecurity attacks can include attacks in each of the aforementioned steps, both in acquiring signal and in signaling the brain. Here researchers have documented numerous different types of attacks.

During the data acquisition phase (first or last depending on whether the BCI is for input or output), for example, a hack could falsify external stimuli, or the electrodes could be tricked into receiving inaccurate data such as through subliminal stimulations [61]. In the latter neurostimulation of the brain, it is theorized that the nature of the stimulation could disrupt the parameters of the firing pattern increasing/decreasing the quantity of spikes, their voltage, their dispersion, or other modifications [62].

In some cases, this type of hacking can even cause long-term tissue damage [63]. In other cases changing the parameters of the firing pattern, or even introducing novel firing patterns unrelated to external stimuli can create false perceptions that can result in psychological [64] and even physical [65] concerns.

Other cyberneurosecurity concerns in data acquisition of the BCI signal include jamming attacks which can affect the confidentiality, integrity, and availability of the signal. For example, this malware could prevent the data acquisition component of a BCI from picking up raw signals collected by the electrodes [66]. In conversion of analog to digital signal, there is also the possibility of a cyberneurosecurity attack, especially via malicious malware that could confound the conversion process, or extract the data regarding private thoughts of the user and provide it to a third party [67]. Such technology, while still in its infancy, is already able to decode images from BCI outputs [68–72], or extract other information [32]. Notably different from other digital communications that can be hijacked, often times those using BCIs are particularly vulnerable, e.g., mentally and physically handicapped, and the data that might be transferred between BCIs and other devices may also include particularly private and informative information. Finally, malware can also intervene in the digital to analog conversion of brain stimulation signal, as well as extract data regarding the nature of the neurological treatments.

9 Specific Cases of Cyberneurosecurity Concerns in Medical, Recreational, and Employment Uses of BCI

A typical hacking case that comes up considerably in the literature is that of malicious activity performed against a brain–computer interface in the context of medical care. Prominent in these cases is the fear of a hostile takeover by a cyber-attacker against a brain–computer interface that operates a patient's neuro-prostheses, against their will [73].

Another commonly described concern stems from the misdiagnosis of neurological diseases when neurodevices are hacked and the integrity of information is disrupted and/or misrepresented information is provided in its place [51]. The hacking of this data can also severely impact the privacy and autonomy of the patient. Notably, this hacking need not require that the victim of the hacking even be physically connected to a brain–machine interface such as an EEG. Research has shown that remote brain access is a near-future reality [74].

Private data from BCIs can also be extracted from non-medical contexts. Consider, for example, SmartCap an Australian company that manufactures wearable technologies for monitoring the fatigue of workers in various industries such as truck drivers, miners, and commercial workers [75]. Another similar technology, Life, an EEG-based headband that provides real-time feedback and allows users—e.g., truck drivers—to manage their alertness by sending alerts delivered via a dedicated app linked via Bluetooth and thus reducing the risk involved in their work [76]. Similarly, the Chinese government is funding a project to scan the brain data of workers in various industries. Production line workers, state-owned companies' employees, and high-speed train drivers are required to use headgear with EEG technology that purports to detect changes in emotional states [77]. The project scans brain data to identify signs of depression, anxiety, or anger through artificial intelligence (AI) and businesses adjust themselves accordingly [78]. These are just a few of the many similar technologies available [79].

There is no doubt that the use of BCIs for employer or state surveillance purposes is one of the most worrying dystopian scenarios regarding this technology. However, even in their best light, these applications, while ostensibly monitoring employee engagement in order to improve safety during high-risk tasks and alert employees or supervisors to dangerous physical or mental situations [80], can also be hacked to expose the sensitive and private data collected by the devices to less scrupulous third parties [81].

Regardless of the actual device employed in these employment contexts, in contrast to the medical field with its relatively strict requirements for protecting private patient data, employers have very little if any regulation that requires anything approaching the level of protection within the medical environment, and yet the technology allows them to collect medical grade, or near-medical grade data. That data can be intercepted while it is being collected, transferred, or analyzed. And while the data may be noisier than the data collected in a hospital setting, there is the real possibility of extracting private neurological data.

Even when they are less obligated to limit the nature of data protection, governments and employers are typically regulated with regard to the data that they can collect and analyze. This is not necessarily the case for other industries that are employing nascent BCI technologies. The field of neuromarketing—sometimes known as consumer brain sciences—researches the brain to predict and even manipulate consumer behavior and decision-making [82]. Neurodata is valuable to advertising and marketing bodies due to its potential to identify how and why people respond to different stimuli to better influence consumers [83]. Beyond the concerns that this ability to examine responses and perceptions directly from the brain creates new ethical debates, such as how to set the accepted boundaries of manipulation, the lack of regulation in this recreational use of BCIs is disconcerting, especially as it relates to the safety of collected data.

Similarly, another recreational application for BCIs is the recreational industry itself, specifically in the gaming industry. Third-party brain–computer interface games rely on standard application programming interfaces (API) to gain access to the brain–computer interface. Such application programming interfaces provide unrestricted access to raw EEG signals for brain–computer interface games, and moreover, these games have full control over the stimuli that can be displayed to users. It turns out that attackers can view the content and read the same EEG associated with them [84]. This confidentiality problem is not exclusive to gaming. Most APIs that are used for the development of BCI application grant unrestricted access to data acquired by the brain–computer interface [85].

Specific examples of this technology abound. Aimed at providing a more immersive gaming experience, Valve, a gaming company, has partnered with OpenBCI, a neurotech company responsible for numerous open-source, non-invasive BCI devices [86], and Tobii, an eye-tracking firm [87], to launch a virtual reality (VR) brain–computer interface "Galea" in early 2022 [88]. The company uses brain–computer interface signals to engage the player for a longer period of time by changing the level of difficulty of the game in response to signs of fatigue stress or boredom.

This data can be employed to draw conclusions about the user's preferences. Models of artificial intelligence and machine learning can be trained on the user's brain signals—combined with other biological changes in response to content—allowing organizations to associate specific changes occurring in the user's neural with certain physiological conditions, such as arousal.

Notably, the retail industry has also learned to access neurological information without any devices interacting with the brain. Some refer to aspects of this as biometric psychography, i.e., the use of behavioral and anatomical information such as pupil dilation to measure a person's response to stimuli over time. It can reveal both the physical, mental, and emotional state of a person, and the stimuli that caused him or her to enter that state. In particular, biometric psychography can reveal intimate details about users' preferences and interests. Unlike biometrics, which focuses primarily on identity, biometric psychography focuses on the practice of using biometric data to identify areas of interest, attitudes, and lifestyles related to the user's personality structure [89, 90]. Arguably, although this analysis includes neurological assessments and to some degree, it is based on neurological science, it is likely, by definition, out of the scope of cyberneurosecurity; that is not to say that malicious access to this information need not be protected under a different rubric. However, this neurological information arguably ought still be protected by the aforementioned emerging ideals of neurorights. Regardless as to how data is collected, the underlying principles and morals relating to neurorights stand. One need not interact with the brain biologically to impinge on these rights.

Similarly, while many of the social media platforms we currently use are already influencing user behavior through the implementation of smart algorithms that encourage even without directly interfacing with the brain.

10 Self-Hacking

The use of BCIs for individuals with medical conditions or where their brain function has been impaired is well known [91]. BCIs can also be used to enhance in addition to their restorative powers [92]. And it is not just cognitive enhancement that can be accomplished via a BCI—attached to an exoskeleton it can provide superhuman strength; it can also change a user's mood. Note this hacking could also be used to decrease one's neuroabilities, raising its own set of novel concerns.

There is often a fine line between enhancement/augmentation and therapy [38, 39, 93]. We have long used pharmacological solutions for cognitive enhancement, including caffeine which is readily and widely available. Ought we make a distinction when using a device such as a BCI. Does society believe that it is ethically problematic to appropriate BCI technologies to enhance rather than repair. Some, for example, might question the authenticity of actions that are enabled by enhancing technology [94]. Others might disagree [95].

As such, the hacking of a medical BCI so that it provides additional enhancement not only creates moral concerns, but might also be physically and mentally dangerous for the user herself, and a danger to those around her. As such it is possible that the field of cyberneurosecurity would promote the disincentivization of such hacking to a similar degree that it is not in favor of third-party malicious hacking.

The particular concerns of enhancement via BCIs relate to things like safety and social justice. In terms of social justice, it is likely that availability of BCI technologies and the opportunities to hack them for enhancement will be limited to a small select few with both the skill set, as well as the purchasing power and access to these technologies.

BCI devices that are marketed for recreational use are unlikely to fall under any government oversight vis-à-vis safety [96]. However, the government does provide for oversight of medical devices, and ought to have the ability to prevent those devices from being tampered with unsafely for enhancement purposes. This is especially concerning when medical devices have been tested for limited use, but those who employ those devices for enhancement and recreation are more likely to use the devices more often than they have been clinically trialed for [97], potentially resulting in unforeseen health concerns [98].

11 Countermeasures

The emerging cyberbiosecurity field has also worked to describe and develop putative countermeasures that might begin to deal with some of the concerns raised by this chapter. These include the incorporation of firewalls, antivirus software, whitelists, and blacklists to keep malicious attackers off a BCI's network, cryptographic mechanisms, periodic firmware updates, and even AI technologies that can detect and thwart new and novel attacks. Additionally, some have called for broad use of BCI anonymizer tools that strip all identifying information from BCI data [99]. Regardless of the nature of these countermeasures, practitioners need to develop tools to stress-test and assess the cyber-readiness of various BCIs, especially the increasing number of healthcare devices that employ AI that could obscure or magnify harmful hacks due in part to the lack of transparency and explainability of AI systems [100–102].

The implementations of these countermeasures are non-trivial to implement. Given the aforementioned reference to dancing pigs, security professionals always presume that consumers, including consumers of BCIs, would prefer irrelevant even cosmetic upgrades to their BCIs rather than an upgrade that focused on the security of their devices.

As such, there is a possibility that users who are competent could simply refuse to implement any of these countermeasures and they could not legally be required to upgrade the security of their devices. However, there is also a private law solution: those users could be contractually required to secure their devices with the penalty for failing to upgrade security being the loss of usability of the device. There is precedent with numerous consumer devices wherein the device loses much if not all of their usability if and when the user fails to follow the terms and conditions associated with the use of the device, including the necessary upgrades.

In addition to technological solutions, standards ought to be set that enforce privacy by design [103] and ethical by design [104] products at the manufacturing level [105].

12 Conclusions

We have described herein various aspects of the emerging field of cyberneurosecurity, a subfield of the nearly equally novel field of cyberbiosecurity which is similar but somewhat distinct from the older field of neurosecurity. The further analysis and elucidation of this field are necessary as the state of the art in neuroscience in general and BCIs in particular is advancing quickly. The issues that arise in the field of cyberneurosecurity are also particularly pertinent as they can affect both the general public in addition to the actual user of the BCI, who not only is at risk for physical and mental harm, but could see her emerging neurorights significantly impinged upon. Fortunately, there are many available technological solutions that can be implemented relatively quickly. Unfortunately, there is little overlap between the many different medical and non-medical sectors that employ BCI technology, making it unlikely that we will see broad enforcement, either by government or by the industry itself. Further elucidation of this field will, however, help in promoting necessary oversight as well as additional research into protecting the public.FundingNo industry funding is disclosed.

References

1. Murch R. Security vulnerabilities in the bioeconomy existed prior to synthetic biology. In: Presentation to the NAS National Materials and Manufacturing Board May 1, 2019. https://sites.nationalacademies.org/cs/groups/depssite/documents/webpage/deps_192712.pdf.

2. Potter L, Ayala O, Palmer X-L. Biocybersecurity: a converging threat as an auxiliary to war. In: ICCWS 2021 16th international conference on cyber warfare and security, 2021 p. 291. Academic Conferences Limited.
3. Loohuis K. Dutch researchers build security software to mimic human immune system, May 24, 2021. ComputerWeekly.com.
4. Rauf U. A taxonomy of bio-inspired cyber security approaches: existing techniques and future directions. Arab J Sci Eng. 2018;43(12):6693–708.
5. Pourmoafi S, Vidalis S.. Bio-cyber operations inspired by the human immune system. In: European conference on cyber warfare and security, 2021 (pp. 534–14). Academic Conferences International Limited.
6. Peccoud J, et al. Cyberbiosecurity: from naive trust to risk awareness. Trends Biotechnol. 2018;36:4–7.
7. Puzis R, Farbiash D, Brodt O, Elovici Y, Greenbaum D. Increased cyber-biosecurity for DNA synthesis. Nat Biotechnol. 2020;38(12):1379–81.
8. Randall SM, et al. Cyberbiosecurity: an emerging new discipline to help safeguard the bioeconomy. Front Bioeng Biotechnol. 2018;6:39. https://doi.org/10.3389/fbioe.2018.00039.
9. Farah MJ. Neuroethics: the practical and the philosophical. Trends Cogn Sci. 2005;9(1):34–40.
10. Gladden ME. An axiology of information security for futuristic neuroprostheses: upholding human values in the context of technological posthumanization. Front Neurosci. 2017;11:605.
11. Pais-Vieira M, Chiuffa G, Lebedev M, Yadav A, Nicolelis MA. Building an organic computing device with multiple interconnected brains. Sci Rep. 2015;5(1):1–15.
12. Bernal SL, Celdrán AH, Pérez GM, Barros MT, Balasubramaniam S. Security in brain–computer interfaces: state-of-the-art, opportunities, and future challenges. ACM Comput Surv (CSUR). 2021;54(1):1–35.
13. Boccia M, Piccardi L, Palermo L, Nemmi F, Sulpizio V, Galati G, Guariglia C. A penny for your thoughts! Patterns of fMRI activity reveal the content and the spatial topography of visual mental images. Hum Brain Mapp. 2015;36(3):945–58.
14. Wenzel CH. Can thoughts be read from the brain? Neuroscience Contra Wittgenstein. Synthese. 2022;200(3):1–19.
15. Willett FR, Avansino DT, Hochberg LR, Henderson JM, Shenoy KV. High-performance brain-to-text communication via handwriting. Nature. 2021;593(7858):249–54.
16. Verma R, Swanson RL, Parker D, Ismail AAO, Shinohara RT, Alappatt JA, Doshi J, et al. Neuroimaging findings in US government personnel with possible exposure to directional phenomena in Havana, Cuba. JAMA. 2019;322(4):336–47.
17. Hack C. Meet 10 companies working on reading your thoughts (and even those of your pets). Forbes June 21, 2020. https://www.forbes.com/sites/cathyhackl/2020/06/21/meet-10-companies-working-on-reading-your-thoughts-and-even-those-of-your-pets/?sh=4e4c92ab427c.
18. Ienca M, Haselager P. Hacking the brain: brain–computer interfacing technology and the ethics of neurosecurity. Ethics Inf Technol. 2016;18(2):117–29.
19. Ienca and Haselager, supra note 19.
20. Vidal JJ. Toward direct brain–computer communication. Annu Rev Biophys Bioeng. 1973;2(1):157–80.
21. Khan AA, Laghari AA, Shaikh AA, Dootio MA, Estrela VV, Lopes RT. A blockchain security module for brain–computer interface (BCI) with multimedia life cycle framework (MLCF). Neurosci Inform. 2021;2:100030.
22. Bernal G, Montgomery SM, Maes P. Brain–computer interfaces, open-source, and democratizing the future of augmented consciousness. Front Comput Sci. 2021;3:23.
23. Sambana B, Mishra P. A survey on brain–computer interaction. 2022. https://arxiv.org/abs/2201.00997.
24. Bonaci T, Calo R, Chizeck HJ. App stores for the brain: privacy and security in brain–computer interfaces. In: 2014 IEEE International symposium on ethics in science, technology and engineering. Pittsburgh: IEEE; 2014. p. 1–7.

25. Paszkiel S. Using BCI and VR technology in neurogaming. In: Analysis and classification of EEG signals for brain–computer interfaces. Cham: Springer; 2020. p. 93–9.
26. Pal D, Palit S, Dey A. Brain computer interface: a review. In: Computational advancement in communication, circuits and systems. Singapore: Springer; 2022. p. 25–35.
27. Zhang S, Sun L, Mao X, Hu C, Liu P. Review on EEG-based authentication technology. Comput Intell Neurosci. 2021;2021:5229576.
28. Denning T, Matsuoka Y, Kohno T. Neurosecurity: security and privacy for neural devices. Neurosurg Focus. 2009;27(1):E7.
29. Anderson BB, Kirwan CB, Jenkins JL, Eargle D, Howard S, Vance A. How polymorphic warnings reduce habituation in the brain: Insights from an fMRI study. In: Proceedings of the 33rd annual ACM conference on human factors in computing systems, pp. 2883–2892. 2015.
30. Vance A, Jenkins JL, Anderson BB, Bjornn DK, Kirwan CB. Tuning out security warnings: a longitudinal examination of habituation through fMRI, eye tracking, and field experiments. MIS Q. 2018;42(2):355–80.
31. Yonas G. NS2 NeuroScience for national security. In: Presented at the end of the beginning—improving warfighter information intake under stress (DARPA augmented cognition mission accomplished Meeting) January 25, 2007. https://www.osti.gov/servlets/purl/1724532.
32. Martinovic I, Davies D, Frank M, Perito D, Ros T, Song D. On the feasibility of side-channel attacks with brain–computer interfaces. In: 21st {USENIX} security symposium ({USENIX security 12), pp. 143–158. 2012.
33. Markosian C, Taruvai VS, Mammis A. Neuromodulatory hacking: a review of the technology and security risks of spinal cord stimulation. Acta Neurochir. 2020;162(12):3213–9.
34. Pycroft L, Boccard SG, Owen SL, Stein JF, Fitzgerald JJ, Green AL, Aziz TZ. Brainjacking: implant security issues in invasive neuromodulation. World Neurosurg. 2016;92:454–62.
35. Cusack B, Sundararajan K, Khaleghparast R. Neurosecurity for brainware devices. Perth Edith Cowan University; 2017.
36. Lange J, Massart C, Mouraux A, Standaert FX. Side-channel attacks against the human brain the PIN code case study. In: International workshop on constructive side-channel analysis and secure design. Cham: Springer; 2017. p. 171–89.
37. Belkacem AN, Jamil N, Palmer JA, Ouhbi S, Chen C. Brain computer interfaces for improving the quality of life of older adults and elderly patients. Front Neurosci. 2020;14:692.
38. Greenbaum D. Ethical, legal and social concerns relating to exoskeletons. ACM SIGCAS Comput Soc. 2016;45(3):234–9.
39. Greenbaum D, Cabrera LY. ELSI in human enhancement: what distinguishes it from therapy? Front Genet. 2020;11:618.
40. Keskin C, et al. Changing an application state using neurological data. US Patent 9, 864,431. 2018 Jan 9.
41. Yuste R, Goering S, Bi G, Carmena JM, Carter A, Fins JJ, Friesen P, Gallant J, Huggins JE, Illes J, Kellmeyer P. Four ethical priorities for neurotechnologies and AI. Nature. 2017;551(7679):159–63.
42. Strickland E, Gallucci M. First win for the Neurorights campaign: Chile plans to regulate all neurotech and ban the sale of brain data. IEEE Spectr. 2022;59(1):26–58.
43. Sententia W. Neuroethical considerations: cognitive liberty and converging technologies for improving human cognition. Ann N Y Acad Sci. 2004;1013(1):221–8.
44. Bublitz J-C. My mind is mine!? Cognitive liberty as a legal concept. In: Cognitive enhancement. Dordrecht: Springer; 2013. p. 233–64.
45. Ienca M, Andorno R. Towards new human rights in the age of neuroscience and neurotechnology. Life Sci Soc Policy. 2017;13(1):1–27.
46. Dadia T, Greenbaum D. Neuralink: the ethical 'Rithmatic of reading and writing' to the brain. AJOB Neurosci. 2019;10(4):187–9.
47. Shemma A, Meirom R, Greenbaum D. The impact of the humanities in science and technology research: a multidisciplinary approach to the ethical, social, and legal impacts of science and innovation. Am J Bioeth. 2016;14(4):20–8.
48. Is this a fact? There's no citation for this, and it seems like a generalization.

49. Zhu G, Huang Y, Wang X. Basic theory of EEG. In: Multi-modal EEG monitoring of severely neurologically ill patients. Singapore: Springer; 2022. p. 3–25.
50. Greenberg J, et al. Privacy and the connected mind—understanding the data flows and privacy risks of brain–computer interfaces. https://fpf.org/wp-content/uploads/2021/11/FPF-BCI-Report-Final.pdf.
51. Greenbaum D. Cyberbiosecurity: an emerging field that has ethical implications for clinical neuroscience. Camb Q Healthc Ethics. 2021;30(4):662–8.
52. Greenbaum D. Avoiding regulation in the medical internet of things. In: Cohen IG, editor. Big data, health law, and bioethics. Cambridge: Cambridge University Press; 2018.
53. Schneier B. Security in the real world: how to evaluate security technology. Comput Secur J. 1999;15:1–14.
54. Muthuppalaniappan M, Stevenson K. Healthcare cyber-attacks and the COVID-19 pandemic: an urgent threat to global health. Int J Qual Health Care. 2021;33(1):mzaa117.
55. Liu J, Lee HJ, Weitz AJ, Fang Z, Lin P, Choy M, Fisher R, et al. Frequency-selective control of cortical and subcortical networks by central thalamus. elife. 2015;4:e09215.
56. Bernal SL, Celdrán AH, Pérez GM. Eight reasons why cybersecurity on novel generations of brain–computer interfaces must be prioritized. 2021. https://arxiv.org/abs/2106.04968.
57. Sherman M, Idan Z, Greenbaum D. Who watches the step-watchers: the ups and downs of turning anecdotal citizen science into actionable clinical data. Am J Bioeth. 2019;19(8):44–6.
58. Greenbaum D. Avoiding regulation in the medical internet of things. In: Cohen IG, Lynch HF, Vayena E, Gasser U, editors. Big data, health law, and bioethics. Cambridge: Cambridge University Press; 2018.
59. Hildt E. Affective brain–computer music interfaces—drivers and implications. Front Hum Neurosci. 2021;15:711407. https://doi.org/10.3389/fnhum.2021.711407/full.
60. Solterbeck A. Protecting data at rest and in motion. Netw Secur. 2006;2006(9):14–7.
61. Frank M, Hwu T, Jain S, Knight RT, Martinovic I, Mittal P, Perito D, Sluganovic I, Song D. Using EEG-based BCI devices to subliminally probe for private information. In: Proceedings of the 2017 on workshop on privacy in the electronic society, pp. 133–136. 2017.
62. Bernal SL, Celdrán AH, Maimó LF, Barros MT, Balasubramaniam S, Pérez GM. Cyberattacks on miniature brain implants to disrupt spontaneous neural signaling. IEEE Access. 2020;8:152204–22.
63. Parastarfeizabadi M, Kouzani AZ. Advances in closed-loop deep brain stimulation devices. J Neuroeng Rehabil. 2017;14(1):1–20.
64. Marin E, Singelée D, Yang B, Volski V, Vandenbosch GA, Nuttin B, Preneel B. Securing wireless neurostimulators. In: Proceedings of the eighth ACM conference on data and application security and privacy, pp. 287–298. 2018.
65. Polania R, Nitsche MA, Ruff CC. Studying and modifying brain function with non-invasive brain stimulation. Nat Neurosci. 2018;21(2):174–87.
66. Landau O, Puzis R, Nissim N. Mind your mind: EEG-based brain–computer interfaces and their security in cyber space. ACM Comput Surv (CSUR). 2020;53(1):1–38.
67. Bonaci T, Calo R, Chizeck HJ. App stores for the brain: privacy and security in brain–computer interfaces. In: In 2014 IEEE international symposium on ethics in science, technology and engineering. Pittsburgh: IEEE; 2014. p. 1–7.
68. Shen G, Horikawa T, Majima K, Kamitani Y. Deep image reconstruction from human brain activity. PLoS Comput Biol. 2019;15(1):e1006633.
69. VanRullen R, Reddy L. Reconstructing faces from fMRI patterns using deep generative neural networks. Commun Biol. 2019;2(1):1–10.
70. Burr C, Cristianini N. Can machines read our minds? Mind Mach. 2019;29(3):461–94.
71. Ren Z, Li J, Xue X, Li X, Yang F, Jiao Z, Gao X. Reconstructing seen image from brain activity by visually-guided cognitive representation and adversarial learning. NeuroImage. 2021;228:117602.
72. Huang W, Yan H, Wang C, Yang X, Li J, Zuo Z, Zhang J, Chen H. Deep natural image reconstruction from human brain activity based on conditional progressively growing generative adversarial networks. Neurosci Bull. 2021;37(3):369–79.

73. Li Q, Ding D, Conti M. Brain–computer interface applications: security and privacy challenges. In: 2015 IEEE conference on communications and network security (CNS) Pittsburgh: IEEE; 2015. p. 663–6.
74. Canham M, Sawyer BD. Neurosecurity. Am Intell J. 2019;36(2):40–7.
75. Patel K, Shah H, Dcosta M, Shastri D. Evaluating NeuroSky's single-channel EEG sensor for drowsiness detection. In: International conference on human–computer interaction. Cham: Springer; 2017. p. 243–50.
76. http://www.smartcaptech.com.
77. Kılıç B, Aydın S. Classification of contrasting discrete emotional states indicated by EEG based graph theoretical network measures. Neuroinformatics. 2022;20:863–77.
78. https://www.technologyreview.com/2018/04/30/143155/with-brain-scanning-hats-china-signals-it-has-no-interest-in-workers-privacy/.
79. LaRocco J, Le MD, Paeng DG. A systemic review of available low-cost EEG headsets used for drowsiness detection. Front Neuroinform. 2020;14:42.
80. https://www.nytimes.com/2020/02/06/business/drowsy-driving-truckers.html.
81. Krausová A. Legal aspects of brain–computer interfaces. Masaryk Univ J Law Technol. 2014;8:199–208.
82. https://hbr.org/2019/01/neuromarketing-what-you-need-to-know.
83. Matthews S, Bernal SL, Celdrán AH, Pérez GM. What is it and is it a threat to privacy? In: Clausen J, Levy N, editors. Handbook of neuroethics; 2015. p. 1627–45. https://doi.org/10.1007/978-94-007-4707-4_154.
84. Martinovic I, Davies D, Frank M, Perito D, Ros T, Song D. On the feasibility of side-channel attacks with brain–computer interfaces. USENIX Secur. 2012;12:143–58.
85. Takabi H, Bhalotiya A, Alohaly M. Brain computer interface (BCI) applications: privacy threats and countermeasures. In: IEEE 2nd international conference on collaboration and internet computing. Pittsburgh: IEEE; 2016. p. 102–11.
86. https://openbci.com.
87. https://www.tobii.com/.
88. https://www.roadtovr.com/valve-openbci-immersive-vr-games/.
89. Heller B. Human Rights and Immersive Technology, Carr Center for human rights policy. Cambridge: Harvard Kennedy School; 2020.
90. Heller B. Watching androids dream of electric sheep: immersive technology, biometric psychography, and the law. Vanderbilt J Entertainment Technol Law. 2020;23:1.
91. Mane R, Chouhan T, Guan C. BCI for stroke rehabilitation: motor and beyond. J Neural Eng. 2020;17(4):041001.
92. Buch ER, Santarnecchi E, Antal A, Born J, Celnik PA, Classen J, Gerloff C, et al. Effects of tDCS on motor learning and memory formation: a consensus and critical position paper. Clin Neurophysiol. 2017;128(4):589–603.
93. Parens E. Enhancing human traits: ethical and social implications. Washington, DC: Georgetown University Press; 2000.
94. Erler A. Does memory modification threaten our authenticity? Neuroethics. 2011;4(3):235–49.
95. Coin A, Dubljević V. The authenticity of machine-augmented human intelligence: therapy, enhancement, and the extended mind. Neuroethics. 2021;14(2):283–90.
96. Wexler A. A pragmatic analysis of the regulation of consumer transcranial direct current stimulation (TDCS) devices in the United States. J Law Biosci. 2016;2(3):669–96.
97. Wexler A. Who uses direct-to-consumer brain stimulation products, and why? A study of home users of tDCS devices. J Cogn Enhancement. 2018;2(1):114–34.
98. Goering S, Klein E, Specker Sullivan L, Wexler A, Agüera y Arcas B, Bi G, Carmena JM, et al. Recommendations for responsible development and application of neurotechnologies. Neuroethics. 2021;14(3):365–86.
99. Chizeck H, Bonaci T. Brain–computer interface anonymizer. US Patent App. 14/174818. 2014 Aug 14. http://www.google.com/patents/US20140228701.

100. Zhang X, Ma Z, Zheng H, Li T, Chen K, Wang X, Liu C, Xu L, Wu X, Lin D, Lin H. The combination of brain–computer interfaces and artificial intelligence: applications and challenges. Ann Transl Med. 2020;8(11):712.
101. Olsen S, Zhang J, Liang KF, Lam M, Riaz U, Kao JC. An artificial intelligence that increases simulated brain–computer interface performance. J Neural Eng. 2021;18(4):046053.
102. Aggarwal S, Chugh N. Review of machine learning techniques for EEG based brain computer interface. Arch Comput Methods Eng. 2022;29:3001–20.
103. Cavoukian A. Privacy by design: the 7 foundational principles. Inf Priv Commiss Ontario Can. 2009;5:12.
104. Mulvenna M, Boger J, Bond R. Ethical by design: a manifesto. In: Proceedings of the European conference on cognitive ergonomics 2017. Umea: ITWIL; 2017. p. 51–4.
105. Neuralink. 2023. https://neuralink.com/.

Perspectives of Current FDA Guidance on BCI Technology

Michael Pflanzer

1 Regulatory Role and Scope of the FDA and BCI Regulation

The FDA is a United States federal regulatory body responsible for protecting public health by ensuring the safety, efficacy, and security of drugs, biological products, and medical devices. The agency has a variety of methods for enforcing its regulations including warning letters, seizure of assets, injunctions, criminal prosecution, and fines [1]. Recently the FDA has turned to "regulatory shaming" to exert socio-political power on pharmaceutical companies, most recently in an effort to shame big pharmaceutical companies who were using unethical practices to suppress competition from generic pharmaceutical manufacturers [2]. The FDA's guidance regarding the development of brain computer interface (BCI) technology, by contrast, represents an effort to collaborate with developers to communicate early the kinds of regulatory expectations that will govern clinical and non-clinical trials. The goal of this guidance, facilitated by feedback from a panel of 15 experts, is to develop a framework for improving the mobility and independence of patients with paralysis or amputations through controllable prostheses. The FDA refers to this as "leapfrog guidance," a novel regulatory strategy in which it "share[s] initial thoughts regarding emerging technologies that are likely to be of public health importance" [3]. This form of guidance strives to seek a balance between safety and expediency in a competitive international market. Competent guidance is important from the very beginning of device development as negative public sentiment can lead to a moratorium on BCI [4]. This form of regulatory guidance is meant to facilitate

M. Pflanzer (✉)
Science, Technology, and Society Program, North Carolina State University, Raleigh, NC, USA

Department of Communications, Rhetoric, and Digital Media, North Carolina State University, Raleigh, NC, USA
e-mail: mspflanz@ncsu.edu

V. Dubljević, A. Coin (eds.), *Policy, Identity, and Neurotechnology*, Advances in Neuroethics, https://doi.org/10.1007/978-3-031-26801-4_14

rather than to hinder the development of BCI by communicating to the developers of the technology the expectations the FDA will likely have regarding clinical and non-clinical trials.

Medical devices, including BCI, are regulated by the FDA's Center for Radiological Health (CDRH) [5]. Medical devices are classified as either Class I, II, or III, and regulatory control increases from Class I to Class III. The classification of neurological devices by the FDA is contained in the Code of Federal Regulations, Chapter 21 (21CFR882) [6]. Given the classification of similar neurological technology and the long-term invasive nature of implanted BCIs, it is likely that future implanted BCI will be Class III devices. This speculation is further supported by the fact that the Neuropace RNS device, designed for therapeutic applications in adults with epilepsy, is the only existing implanted BCI with FDA classification and it is classified as a Class III device [7]. Class III devices must obtain Premarket Approval (PMA) from the FDA before they may be marketed. See Table 1 for examples of the different classifications of medical devices. The PMA application process is very strict; the FDA requires "sufficient, valid scientific evidence" that the device is safe and effective for its intended use [8]. Such evidence will be collected by firms in both non-clinical and clinical trials.

The current FDA guidance on BCI is an effort by the FDA to communicate early with BCI stakeholders the anticipated requirements for *Q-Submissions* and *Investigational Device Exemptions* (IDEs). These pre-submission documents are voluntary, but the FDA argues that such communications between the FDA and developers will "improve the quality of subsequent submissions, shorten total review times, and facilitate the development process for new devices" [9]. As such, the current FDA guidance for BCI technology contains early expectations about the kinds of information and evidence the FDA will require as firms consider non-clinical and clinical trials for the technology. It is non-binding and intended to facilitate the successful development and future PMA of BCI technology.

Clinical performance regulations are designed not only to ensure that the final products are safe and effective, but that the clinical trials themselves do not pose a threat to patients. Clinical trials begin with IDEs, a process in which developers can demonstrate that the device itself is safe. IDEs are designed to demonstrate the safety and feasibility of the technology, ensuring that the BCI device continues to receive and transmit high-quality brain signals without unacceptable risk to the patient. IDE records are not publicly available until after the FDA has approved an application for PMA or until it has received notification of a completed Product

Table 1 FDA medical device classifications with examples

Class	Risk level	Examples
I	Low	Bandages, surgical instruments, non-electric wheelchairs, electric toothbrushes
II	Intermediate	Surgical gloves, pregnancy tests, syringes, blood pressure cuffs
III	High	Pacemakers, breast implants, defibrillators, implanted prosthetics

Development Protocol (PDP) [10]. Furthermore, the FDA will not disclose the existence of an approved IDE unless such information has already been made public.

The guidance, *Implanted brain–computer interface devices for patients with paralysis or amputation-non-clinical testing and clinical considerations* [6], is limited in scope for several reasons [11]. First, the guidance only applies to implanted BCI technology. Common *non-invasive* BCI technologies, such as electroencephalogram (EEG) technology (Class II devices), are therefore not governed by this guidance. Furthermore, the guidance only applies to therapeutic purposes of restoring "normality" to patients with paralysis or amputated limbs. This not only excludes "enhancement" and "augmentative" functionalities, but also excludes technologies that target sensory capabilities such as vision and hearing. While there are many societal implications of enhancement applications of BCI technology (e.g., concerns of implications to identity, autonomy, and impact on society), such applications are beyond the regulatory scope of the FDA [4].

Because the FDA frames this guidance in ways that suggest that the technology is still early in development and that PMAs or human trials will not occur any time soon, there are several firms already conducting clinical trials with animals [12]. Furthermore, BCI developer Synchron has received permission from the FDA to proceed with an IDE in a human trial [13]. The technology is already here and FDA attempts at regulating BCI technology will shape guidance and regulation of therapeutic applications of implanted BCI technology, but also in all forms of BCI. While the ethical implications of BCI are discussed thoroughly in other chapters of this book, the final sections of this chapter will introduce some of the ethical concerns regarding the FDA's role in shaping the future of BCI technology.

2 FDA Guidance of BCI

After a brief introduction regarding the motivations and limitations of scope of the FDA guidance on BCIs (both of which have been addressed in Sect. 1 of this chapter), the guidance is divided into a series of subsections that represent different risks toward future users of BCI devices. Recall that a role of the FDA, and the CDRH specifically, is to ensure the safety and effectiveness of medical devices. While every section of the guidance is relevant to the issue of safety, criteria that influence device effectiveness are also clearly important in some sections (e.g., human factors). Rather than merely reciting the recommendations contained within each subsection, this section of the paper will briefly summarize the most salient recommendations while explaining the importance of each subsection (what *is* pyrogenicity, for example). It may therefore be constructive to follow along with the guidance to facilitate understanding of contents not thoroughly recited in this chapter. Chapter subsections therefore match the lettering and title of each subsection contained in the guidance.

2.1 Device Description

This section requires a detailed description of the component parts of the BCI technology including such parameters as power sources, number of output channels, physical description of hardware and software features, and the ways in which the components communicate internally to the patient, the programming, and the environment. It further suggests flowcharts and diagrams that demonstrate the ways in which each component interacts with other components, the patient or user, and/or the environment. The guidance also recommends a detailed description of all component safety features, alarm conditions, and user communications.

This section of guidance appears to be based upon the presumption that BCI devices are demonstrated to be safe after establishing the reliability of component parts that constitute the technology. The entire guidance contains links to other FDA forms both for reference by developers and for usage in the case that one or more components of a BCI prototype are already approved by the FDA. This is consistent with FDA (and other U.S. regulatory bodies) efforts to streamline the application process and to make it less burdensome for both regulators and developers, therefore getting new therapies into the hands of patients, providers, and consumers without sacrificing the standards of safety and effectiveness that the FDA is expected to maintain.

2.2 Risk Management

The topic of risk management is important for any medical device, but this is particularly true for BCI devices given their invasive, permanent nature, and proximity to the human brain. Many of the subsequent subsections delineate the kinds of risks that the FDA anticipates as well as the recommended detailed descriptions of how these risks are mitigated. The FDA expects device manufacturers to provide a detailed account of the potential risks from the perspective of the user and the ways in which these risks are diminished to an "acceptable level." The FDA further requests that device applicants define what they perceive as acceptable and to justify why the residual risks are both acceptable and necessary. One common source of unexpected risk is human error, a concern addressed in Sect. 2.5. The guidance further suggests that risk analyses include the hazards that precipitate single-fault conditions and assurances that such failure of any single component of the BCI device will not present an unacceptable risk. These recommendations ensure that not only are BCI device usage risks mitigated to an acceptable level during proper operation, but that if a component fails that the user is not suddenly subject to unacceptable risks.

From a neuroethical perspective, it is likely that users, device manufacturers, and regulators all have a different interpretation of what risks are acceptable. There will likely be considerable variance in the kinds and severity of risks that users are willing to accept. What one user is willing to risk in anticipation of significant therapeutic improvements may be too much for another user. The subjective component of risk management means that, while it is a relatively small section compared to the rest of the guidance, this section is likely to be implicated in future BCI usage and ethics debate.

2.3 Software

BCI device software is important not only because it is "the brains" behind the hardware's normal operation, but also because it is expected to mitigate risks when the device does not operate as intended. This section of the FDA guidance refers readers to the FDA's *Guidance for the Content of Premarket Submissions for Software Contained in Medical Devices*, a guidance which contains frameworks for determining the level of concern for medical devices [14]. In the BCI guidance, the FDA anticipates that the software contained in BCI devices will present a "major" level of concern. Per the software guidance, this level of concern is met if the use, misuse, or failure of the medical device software could result in serious injury or death to the patient or operator. The FDA recommends that developers provide a detailed description of the software that includes its programming language and its compatible hardware platforms and operating systems. Detailed operational instructions are encouraged, and software performance testing data is also recommended. Documentation included with the BCI devices is required to contain sufficient information to describe the role of the software and address software-related concerns.

This section of the guidance also contains further recommendations for early feasibility studies, particularly with respect to the safe operation of the hardware. It is recommended that cybersecurity concerns are addressed. This section links to several other FDA guidance documents related to off-the-shelf software, cybersecurity management, and early feasibility studies. Such documentation will enable the FDA to consider such risks and their mitigation while determining whether the BCI device software is both safe and effective. This will be particularly important in future neuroethical debates as some of the concerns for BCI devices include the potential that the devices can be hacked [15, 16].

2.4 Human Factors

Human factors refer to those actions, by humans, that either deliberately or accidentally subvert the intended use or operation of a product or technology. One of the most cited human factors regarding medical devices is user frustration. A frustrating calibration process reduces device effectiveness as the perceived benefits of the technology may not outweigh the tedious or complicated operation of the medical devices [17]. Human factors may also introduce mistakes, for example, if calibration and usage are too complicated for the user or if device behavior may be inadvertently modified during normal operation. Additionally, the learning curve for new devices is one of the greatest predictors of device effectiveness [4]. This highlights the importance of the FDA's recommendation to consider contextual information such as the intended users, the intended behavior of users, and the environmental conditions of usage when considering human factors validation.

While the FDA does not typically require human factors validation and evaluation for early feasibility studies, it recommends considering including human factors in such studies when the insights developed from their inclusion will benefit the device development. Their inclusion is further recommended because this data may be necessary to support future marketing submissions to the FDA. The FDA recommends an investigational protocol that examines the usability of the device, both by patients and practitioners, and that developers implement a plan to modify either the BCI device or the operational instructions such that usability may be improved. This protocol should also identify and explore any use-related risks associated with the usage of the BCI device. The FDA is especially concerned with "critical tasks" those tasks which, when performed incorrectly or not at all, will or are likely to cause harm to the patient or to compromise the patient's healthcare [18].

2.5 Biocompatibility

This subsection contains recommendations to ensure the biocompatibility of BCI devices and their components. Biocompatibility minimizes the risks of a harmful biological response to component materials. FDA guidance contains four categories of biocompatibility, ranging from Category 1 in which components have permanent contact (>30 days) with neural tissue, cerebrospinal fluid, or blood to Category 4 in which components have only limited (<24 h) or prolonged (>24 h–30 days) contact with intact skin. Endpoints are outcomes or events which can objectively demonstrate whether an intervention is beneficial. While the biocompatibility evaluations require more endpoints for the more sensitive categories (e.g., Category 1 requires 12 endpoints, including hemocompatibility tests) all categories share some endpoints in common (e.g., cytotoxicity, sensitization, and irritation or intracutaneous reactivity). Applicants are recommended to consult the FDA guidance *Use of International Standard ISO-10993-1, "Biological evaluation of medical devices—Part 1: Evaluation and testing within a risk management process"* [19] for guidance regarding biocompatibility assessments. Applicants are further recommended to determine whether materials contained within BCI components have already been approved by the FDA for use in other medical devices. These approvals can be found in Letters of Authorization (LOAs) or device Master Files (MAFs). Section 2.13 of the guidance contains additional documentation regarding device MAFs.

2.6 Sterility

The importance of sterility requirements for implanted BCI devices is rather intuitive. Devices must be sterilized to minimize infections and other complications. The current guidance contains recommendations for documentation that establish sterilization procedures. First, the sponsor is encouraged to provide a thorough description of the sterilization process and environment and to conform, as applicable, with FDA guidance regarding radiation and chemical sterilization. Second, sponsors should provide data regarding sterilization validation including all standards utilized and the extent to which those standards were satisfied. The sterility of medical devices is defined by the probability, represented by its sterility assurance level (SAL), of a viable microorganism after it has been sterilized. BCI sponsors should disclose the SAL noting that sterility requirements are higher for devices intended for internal implantation (10^{-6} as compared with 10^{-3} for those that only contact intact skin). The SAL corresponds with the level of sterility achieved. While the SAL can never reach 0 (a condition in which 100% sterility is assured), a lower SAL corresponds to a greater certainty of sterility.

The proper sterilization of medical devices, particularly implanted devices, is intuitive. Failure to properly sterilize a device can lead to a variety of potentially lethal bacterial infections. This is particularly true with the evolution of bacterial pathogens to resist traditional medical interventions. It is important to remember that immunocompromised patients may also utilize BCI technology and that they are more susceptible to bacterial infection. Multidrug-resistant bacteria necessitate greater attention to detail to sterilization procedures to mitigate the life-threatening risks that they present [20].

2.7 Pyrogenicity

Pyrogens are substances that cause a fever response when introduced into a living body. From a pharmaceutical perspective, the most common pyrogenic threats are gram negative bacterial (GNB) endotoxins. GNBs are highly resistant to antibiotics and present a significant threat in hospital settings due to their high risk of death or severe disease [21]. Chemicals that leach from medical devices can also cause a febrile reaction or an increase in host body temperature. The FDA therefore understandably limits the risk of both GNB endotoxins and material-mediated pyrogens with pyrogen limit specifications. Currently the number of pyrogenic assays is very limited. The FDA-approved assay for medical devices (RPT) requires in vivo testing of rabbits [22], thus necessitating animal trials, a concern addressed in Sect. 2.14 of this chapter. This subsection of the guidance therefore refers to several FDA links containing information regarding testing, limiting, and reporting pyrogenic data.

2.8 Shelf Life and Packaging

This subsection simply recommends that applicants submit data demonstrating that the packaging of BCI devices and their components does not inadvertently undermine the sterility of the devices below the minimum required sterility levels. Device applications should contain a description of the packaging and the protocols implemented to establish packaging and sterility integrity. The guidance further recommends bench data demonstrating that the devices or their components do not suffer any detriment to safety or effectiveness before they reach the patient, even in the case that they are sitting on a shelf (e.g., in a warehouse). Details and data regarding shelf-life testing protocols should also be included in device applications.

While not explicitly addressed by this guidance, the FDA requires the following labeling information: (1) the name and place of business of the manufacturer, packer, or distributor, (2) the quantity of contents, if appropriate, and (3) the statement, "CAUTION: Investigational device. Limited by Federal (or United States) law to investigational use." The labeling must also include all "relevant contraindications, hazards, adverse effects, interfering substances or devices, warnings, and precautions" [10, 12]. These labeling requirements are specifically for IDE applications only; more stringent labeling requirements will follow for an approved application of the technology.

2.9 Electrical Safety and Electromagnetic Compatibility

This is another intuitive subsection of the guidance that stipulates that device developers should address the electrical and electromagnetic hazards associated with use of the BCI technology. These hazards include not only the possibility that the device will fail to operate safely and effectively but also that the device will not interfere with other electrical medical devices (e.g., pacemakers). This short section recommends that implanted BCI devices be tested in their intended environments in accordance with a variety of linked FDA standards.

2.10 Wireless Technology

Due to their nature as implanted medical devices, implanted BCI must communicate with both internal and external devices and technologies and therefore typically utilize wireless technology to communicate neural signals meant to control assistive technology or to excite or inhibit neural pathways. Medical data and information may also be communicated externally to operators or providers to ensure the safe and effective operation of the technology. This wireless technology must therefore be compatible with other medical devices and not interfere with or risk interference by other devices and signals. The FDA recommends that applicants consider the kinds of radiofrequency devices and other wireless technology that may operate in the vicinity of implanted BCI and to assess and mitigate the existence of any risk factors. Applicants are again referred to FDA guidance that specifically includes considerations for devices intended for in-home use.

Conspicuously absent from the discussion of Wireless Technology (Sect. 2.8) is any discussion for countermeasures against efforts to hack the BCI technology. Given often-cited fears and recent research suggesting that "neurohacking" is possible [15], it would be prudent to include guidelines that require applicants to demonstrate the BCI technology is safe from attempts to hack its software. One hazard of any networked technologies is the prevalence of potential exploits that are unknown to both the public and software vendors at the time they are introduced to the market. These exploits present lucrative opportunities for biohackers to leverage wireless security flaws for financial or other nefarious benefit [23].

2.11 Magnetic Resonance (MR) Compatibility

The vibration, movement, and heat generation associated with magnetic resonance imaging can potentially cause tissue damage in patients with implanted BCI devices. This short subsection therefore refers applicants to the FDA guidance *Testing and Labeling Medical Devices for Safety in the Magnetic Resonance (MR) Environment.* This guidance defines three levels of MR compatibility: (1) MR Safe devices are electrically non-conductive, nonmetallic, and nonmagnetic and are at no risk from MR exposure, (2) MR Unsafe devices are those which pose an unacceptable risk to patients, medical staff, or others in an MR environment, and (3) MR Conditional devices which is demonstrated to be safe within pre-established MR environments [24].

A perfectly MR compatible BCI will be one in which it can operate concurrently with "simultaneous MRI and electrode recording or stimulation without artifacts in imaging" [25]. It is likely that there will be tradeoffs between function and susceptibility, but efforts shall be made to demonstrate that the BCI device is both safe and effective. Zhang and colleagues [25] explain that there are significant potential concerns for MR testing of patients with implanted devices including risk of force on implanted medical devices, interference with device signaling, and distortion of the magnetic fields by the BCI implant.

2.12 Non-clinical Bench Testing

The FDA defines bench testing as the non-clinical process of performance testing on living or dead animal or human tissue. Subsections previously discussed (e.g., biocompatibility, sterility, electrical safety, etc.) are excluded from this definition of bench testing; bench testing instead refers to biological and mechanical performance testing intended to determine the expected life of the device in "worse-case *in vivo* conditions." The FDA recommends submitting detailed test report summaries, test protocols, and completed test reports in accordance with established FDA guidance. Components listed in this subsection of the FDA guidance include eight categories of components, and each of these categories contains a variety of recommendations for bench testing that demonstrates the safety and integrity of the components. The final category of this subsection is system level testing. This category importantly recommends that BCI device components integrate safely and effectively, even when some or all components are manufactured by different companies.

2.13 Referencing Master Files (MAF) and Other FDA Premarket Submissions

Device master files are classes of files in which things subject to FDA oversight (e.g., drugs, medical devices, food, etc.) have been previously supported to conform with FDA regulations. These master files (MAFs) are referenced when confidentiality of designs or other specifications must be considered (e.g., when a medical device uses proprietary components from an external manufacturer) and also for efficiency when the safety and effectiveness of components (a probe, for example) have been supported. When MAFs are referenced for the purpose of proprietary confidentiality, they must also be accompanied by a Letter of Authorization (LOA).

2.14 Non-clinical Animal Testing

Non-clinical animal testing is recommended to establish the in vivo safety of medical devices. This testing is appropriate for performance testing that cannot be evaluated in bench testing or in clinical studies. The FDA has guidance regulating the ethical treatment of experimental animals. The guidance is based upon the 3Rs—to "reduce, refine, and replace" animal testing when feasible. Applicants are further encouraged to consider best practices for animal research and to utilize the Q-Submission Program for FDA validation of study designs. There are several sets of requirements listed in this subsection that provide the level of detail expected in study designs. Some of these factors include the purpose of the study, detailed methods sections, the species and strain of the animal model, and why this strain was selected. The section further outlines some expectations about testing procedures (e.g., sedation for acute stimulation tests, and the usage of surgical tools designed for human implantation during animal surgery).

2.15 Clinical Performance Testing

The final subsection of the current FDA guidance on implanted BCI technology is arguably the most important. It outlines FDA expectations regarding clinical trials and performance testing. This subsection begins with application requirements for both IDEs and early feasibility studies. In both cases, any clinical data from previous trials must be provided. In the case that such data is unavailable for the proposed intended use, detailed background clinical information such as data or publications on similar technology must also be included in application materials. The FDA notes that such information may come from outside of the United States. The first FDA-approved IDE for implanted BCI, for example, follows a clinical trial in Australia in which the device was successfully implanted into four patients [26]. Supporting documentation shall include a brief narrative but include details such as the purpose, methodology, sample size, and summary of the study and number of investigational sites in and outside of the United States.

The FDA generally assumes that all implanted BCI will be classified as significant risk (SR) devices and will thus be governed by the FDA guidance for SR devices (21 CFR 812), for institutional review boards (IRBs) (21 CFR 56), and for informed consent (21 CFR 50) [6]. Those familiar with the IRB and informed consent process will likely find little of the following considerations required in IDE submissions surprising.

According to FDA guidance, it is vital to consider the patient populations. There are a wide variety of conditions that may qualify an individual for participation in a clinical trial governed by the BCI guidance. These conditions, which include paralysis or loss of function due to spinal cord injuries, stroke, or neuromuscular disorders, vary in severity, as does the perceived distress caused by such conditions. The FDA advises applicants to consider such information when recruiting participants: clinical trials require that the potential benefits of participation outweigh the potential risks and these considerations are largely subjective. What constitutes an acceptable risk for one individual may be unacceptable to another.

Applicants must also consider that BCI devices will largely be operated in a home environment. A laboratory setting not only limits external validity and the evaluation of human factors, but it also does not necessarily reflect the same level of risk and benefit that a patient might experience in their home environment. A trained caretaker may also be willing and able to assist a trial participant and in such cases the applicant must demonstrate that such caretakers are sufficiently available and adequately trained.

Finally, IDE applications must provide a sufficiently detailed investigational plan. This plan is constituted by the usual considerations for explaining a study methodology: the study purpose, design, duration, and inclusion and exclusion criteria must all be defined. It is likely, for example, that the goal of the study design will be to demonstrate the safety and effectiveness of the technology. The FDA recommends a duration of at least 1 year to adequately demonstrate the long-term integrity of device components such as probes. Certain conditions such as a medical history of seizures, psychotic disorders, or intellectual impairment are among others listed by the FDA as likely exclusion criteria. Other details, such as patient demographics, the treatment protocol (e.g., post-surgical recovery time and regiment), and study endpoints should also be included. Applicants must conduct risk analyses and consider adverse events severe enough to justify device removal or the discontinuation of patient participation. While the FDA considers both primary and secondary endpoints, it also leverages subjective experience and welcomes participant input.

3 Implications of the FDA Guidance

The FDA is widely regarded as a regulatory model for medical devices and interventions. While the FDA is charged with ensuring the safety and effectiveness of therapeutic medical devices and the BCI guidance is specifically limited in scope to implanted devices, the FDA's early recommendations regarding BCI regulation will therefore be examined by many. The FDA has a long history of regulating medical neurological devices, necessary because of the constant risk of diminishing utility and increasing risk factors presented by such technologies, particularly those that are invasive and permanent such as implanted BCI devices [4]. The kinds of recommendations and suggestions that the FDA is making are relevant to any developer of BCI—whether devices are designed for therapeutic or enhancement purposes or whether they are internal or external.

As discussed previously, the FDA guidance constitutes a "leapfrog" guidance, a novel attempt by the FDA to communicate early the expectations the agency will have for applicants hoping to conduct early feasibility studies, and non-clinical and clinical trials for BCI devices. The hope is that this will facilitate device development by streamlining the regulatory approval process because there will be fewer surprises for developers and because applications will therefore better model the recommended guidelines. It may however become necessary to distinguish between actions intended to alleviate burden in the regulatory approval process and those designed simply to reduce regulatory oversight.

The Cures Act is a bipartisan bill signed into law in 2016 that is designed to accelerate the development and deployment of medical products and interventions. This law authorized $500 million over 9 years to the FDA. FDA Commissioner Scott Gottlieb announced that the agency had "embraced the concept of least burdensome regulation as clarified and expanded in the Cures Act," promising that such developments had reduced review times and improved the quality of device applications [27]. While this innovation-friendly approach may foster the delivery of novel medical technologies to the patients who need them, it also follows a trend of government deregulation and suggests that potential conflicts of interest may undermine the FDA's primary objective of ensuring the safety and effectiveness of medical products.

It is, however, unclear to what extent industry regulations are even modeled after the current FDA guidance. For example, Elon Musk, CEO of BCI developer Neuralink, revealed in a 2020 press conference that its internal regulations are far stricter than the FDA guidance [28]. Communication between the FDA and developers or applicants is not always public record, however, as an approved IDE will only be acknowledged by the FDA after it receives notice that the applicant has already made such news public [9].

Synchron, a rival neurotechnology firm, claims to have beaten Neuralink to clinical trials, announcing in 2021 that it had been the first company to receive an IDE from the FDA to conduct clinical trials of implanted BCI technology [13]. While research evidence takes, on average, 17 years to reach clinical practice [29], the clinical trial director David Putrino is optimistic that the pace of BCI technology implementation will be much quicker [26]. Has this time been reduced so drastically because of the pace of innovation or because the FDA has pushed for more efficient and less burdensome regulation? Should we be concerned that attempts at deregulation will undermine the FDA's mission of ensuring the safety and effectiveness of implanted BCI and other related technology?

While United States firms are leading the way in BCI development, they do face international competition. As of 2021, Chinese BCI research was nearly 10 years behind in development, but the Chinese government has recently begun funneling money into BCI development with the intention that the "China Brain Project" soon leads the rest of the world in applied neuroscience developments [30, 31]. Moreover, Chinese media reported in 2020 that China's Zhejiang University successfully implanted BCI technology in a paralyzed, 72-year-old man, thus restoring some motor function and allowing him to shake hands, pick up drinks, and play mahjong [32]. Additionally, Synchron conducted the first ever human clinical trial of implanted BCI in 2019, successfully implanting the technology in a patient with severe paralysis [26, 33]. This trial later recruited and successfully implanted the technology in three more patients with the same disorder, demonstrating improved motor function and quality of life for the participants [26]. This international competition places greater pressure upon the FDA to streamline the application process while finding the proper balance between facilitating technological innovation and preserving the regulatory pressure that ensures that approved devices are both safe and effective.

On the other hand, some critics may question what role the FDA should have in regulating medical devices. Recall that the current guidance is limited in scope to therapeutic applications of implanted BCI technology. This raises the question: How do we draw the line between therapeutic and enhancement applications of medical devices and technology? The debate continues and some researchers are optimistic the enhancement potential of implanted BCI might create the "extended mind," capable of keeping up with artificial intelligence [34]. The extended mind hypothesis (EM), originally posed by Andy Clark and David Chalmers, posits that the cognitive processes that make up the human mind extend beyond the brain's physical boundaries [35]. If one can argue that the absence of such an extended mind constitutes a disadvantage, does enhancement oriented BCI then become therapeutic? Does a device identical to therapeutic application devices but marketed for enhancement deserve any less scrutiny or regulation?

This contention between ethicists and developers has led some critics to question the extent to which the FDA is qualified to make normative sociological judgments [36]. There are some within the disabled community, for example, who not only do not consider their conditions as deficits or liabilities to be corrected but find such normative judgments to be harmful. These normative judgments reinforce what sociologists refer to as the "deficit model of disability"—critics suggest that such framing stigmatizes patients who prefer not to see their disabilities as deficits and have instead embraced their conditions as a form of diversity. Who then is responsible for deciding whether such technology is a form of augmentation, enhancement, restoration, or therapy? Does the guidance therefore become non-binding if the technology is developed and marketed to disabled patients as an augmentation or enhancement device? These are the kinds of questions that may arise as BCI technology develops. One might argue that therapeutic applications end, and enhancement begins, when the technology increases human neural potential beyond the normal, expected level but this once again relies upon normative judgments outside of the purview of the FDA. It may be difficult however for the FDA to disentangle therapeutic and enhancement applications of BCI [4]. It would seem far less messy for the FDA to regulate all medical devices (therapeutic and enhancement oriented) rather than for the agency to distinguish between therapeutic and enhancement applications of medical technology. One potential solution proposes that the FDA creates a special subsection of Class IV Devices for enhancement applications of medical technology (the author suggests "Class IV–E Devices") that require greater scrutiny over therapeutic applications [4]. This solution would also require the FDA to expand its regulatory scope—in an era in which partisans on both sides of the political aisle are pushing for deregulation.

4 Conclusions

The anticipatory motivations of the FDA's "leapfrog" guidance on BCI technology are potentially frustrated by the speed with which the technology has developed. Synchron, for example, already claiming to be the first BCI developer to reach clinical trials in humans, has predicted that their technology will reach production in the next 5–7 years [13]. And while United States regulatory bodies may yet again struggle to keep pace with that of technological progress, the guidance will nevertheless be informative to stakeholders of both clinical and therapeutic applications for BCI technology. While developers of enhancement BCI devices may not be subject to the scrutiny and regulatory oversight of the FDA, the guidance, designed to ensure that devices are both safe and effective, will provide a benchmark with which to compare internal guidance. At a recent press conference, for example, Elon Musk claimed that Neuralink's internal guidance was far stricter than that proposed by the FDA—a claim which, if true, is only possible if the company had previously referenced and were well familiar with the FDA's guidance [28].

The FDA takes advantage of the fact that many of the components within BCI are already regulated in some other form of medical technology [37]. This facilitates the development and subsequent premarket approval process both because the FDA already has a reasonable idea about the kinds of expectations it will have regarding each component, but also because it can both reference existing FDA documentation and encourage developers to utilize MAFs for quicker approval. But it does appear that the existing guidance could have included greater foresight for potential hazards of the technology. While the FDA does not make ethical judgments about the appropriateness or philosophical implications of BCI technology, it also fails to address the potential for biohacking or concerns over how medical and personal information will be secured when internal devices communicate with external hardware. These potentials represent serious safety risks which arguably undermine the FDA's ability to evaluate the safety of the device.

FDA Commissioner Gottlieb suggests that "leapfrog" guidance, as discussed in this chapter, will facilitate quicker development and deployment of life-saving and other powerful therapeutic interventions and devices. And while the FDA may be commonly associated with bureaucratic delays, it must prioritize safety and effectiveness of medications and medical devices over the efficiency with which they are approved. These devices hold tremendous potential, but public acceptance will rely upon both the ethical considerations discussed in the literature on BCI ethics and appropriate regulatory oversight that ensures that therapeutic applications will not result in unacceptable risks to their users. The FDA is often looked to as an exemplar of medical regulation, but only time will tell if even novel approaches such as "leapfrog" guidance will be enough for it to keep up with the international pace of technological innovation.

References

1. Food and Drug Administration. Types of FDA enforcement actions. Silver Spring: U.S Food and Drug Administration; 2022. https://www.fda.gov/animal-veterinary/resources-you/types-fda-enforcement-actions
2. Yadin S. Shaming big pharma. Yale J Regul. 2019;36:131. https://www.yalejreg.com/bulletin/shaming-big-pharma/
3. Food and Drug Administration. Investing in advanced manufacturing to support public health preparedness. Silver Spring: U.S Food and Drug Administration; 2020. https://www.fda.gov/news-events/fda-voices/investing-advanced-manufacturing-support-public-health-preparedness
4. Chan E. The FDA and the future of the brain–computer interface: adapting FDA device law to the challenges of human–machine enhancement. UIC John Marshall J Inf Technol Priv Law. 2007;25(1):4. https://repository.law.uic.edu/jitpl/vol25/iss1/4
5. Food and Drug Administration. Overview of device regulation. Silver Spring: U.S Food and Drug Administration; 2020. https://www.fda.gov/medical-devices/device-advice-comprehensive-regulatory-assistance/overview-device-regulation
6. Food and Drug Administration. Neurological devices, CFR—code of federal regulations title 21. Silver Spring: U.S. Food and Drug Administration; 2022. https://www.accessdata.fda.gov/scripts/cdrh/cfdocs/cfcfr/CFRSearch.cfm?CFRPart=882&showFR=1&subpartNode=21:8.0.1.1.28.5
7. Food and Drug Administration. Premarket approval [neuropace RNS system]. Silver Spring: U.S Food and Drug Administration; 2013. https://www.accessdata.fda.gov/scripts/cdrh/cfdocs/cfpma/pma.cfm?id=P100026
8. Food and Drug Administration. Premarket approval (PMA). Silver Spring: U.S Food and Drug Administration; 2019. https://www.fda.gov/medical-devices/premarket-submissions-selecting-and-preparing-correct-submission/premarket-approval-pma. https://www.fda.gov/medical-devices/investigational-device-exemption-ide/faqs-about-investigational-device-exemption
9. Food and Drug Administration. Requests for feedback and meetings for medical device submissions: the Q-submission program. Silver Spring: U.S Food and Drug Administration; 2021. https://www.fda.gov/media/114034/download
10. Food and Drug Administration. FAQS about investigational device exemption. Silver Spring: U.S Food and Drug Administration; 2019. https://www.fda.gov/medical-devices/investigational-device-exemption-ide/faqs-about-investigational-device-exemption
11. Food and Drug Administration. Implanted brain–computer interface devices for patients with paralysis or amputation-nonclinical testing and clinical considerations; guidance for industry and food and drug administration staff. Silver Spring: U.S Food and Drug Administration; 2021. https://www.federalregister.gov/documents/2021/05/20/2021-10622/implanted-brain-computerinterface-devices-for-patients-with-paralysis-or-amputation-non-clinical.

12. Coin A, Mulder M, Dubljević V. Ethical aspects of BCI technology: what is the state of the art? Philosophies. 2020;5(4):31. https://doi.org/10.3390/philosophies5040031.
13. Synchron. About us: Radically outpacing traditional BCI. 2022. https://synchron.com/about-us.
14. Food and Drug Administration. Content of premarket submissions for device software functions. Silver Spring: U.S Food and Drug Administration; 2021. https://www.federalregister.gov/documents/2021/11/04/2021-24061/content-of-premarket-submissions-for-device-software-functions-draft-guidance-for-industry-and-food
15. Ienca M, Haselager P. Hacking the brain: brain–computer interfacing technology and the ethics of neurosecurity. Ethics Inf Technol. 2016;8(2):117–29. https://doi.org/10.1007/s10676-016-9398-9.
16. Sample M, Sattler S, Blain-Moraes S, Rodríguez-Arias D, Racine E. Do publics share experts' concerns about brain–computer interfaces? A trinational survey on the ethics of neural technology. Sci Technol Hum Values. 2020;45(6):1242–70. https://doi.org/10.1177/0162243919879220.
17. Saha S, Khondaker MA, Ahmed K, Mostafa R, Naik GR, Darvishi S, Khandoker AH, Baumert M. Progress in brain–computer interface: challenges and opportunities. Front Syst Neurosci. 2021;15:578875. https://doi.org/10.3389/fnsys.2021.578875.
18. Food and Drug Administration. Human factors studies and related clinical study considerations in combination product design and development. Silver Spring: U.S Food and Drug Administration; 2016. https://www.fda.gov/files/about%20fda/published/Human-Factors-Studies-and-Related-Clinical-Study-Considerations-in-Combination-Product-Design-and-Development.pdf
19. Food and Drug Administration. Use of international standard ISO 10993-1, "biological evaluation of medical devices—part 1: evaluation and testing within a risk management process". Silver Spring: U.S Food and Drug Administration; 2020. https://www.fda.gov/regulatory-information/search-fda-guidance-documents/use-international-standard-iso-10993-1-biological-evaluation-medical-devices-part-1-evaluation-and
20. Josephs-Spaulding J, Singh OV. Medical device sterilization and reprocessing in the era of multidrug-resistant (MDR) bacteria: issues and regulatory concepts. Front Med Technol. 2021;2:587352. https://doi.org/10.3389/fmedt.2020.587352.
21. Oliveira J, Reygaert WC. Gram negative bacteria. In: National Institutes of Health. Treasure Island (FL): StatPearls Publishing; 2022.
22. Franco E, Garcia-Recio V, Jiménez P, Garrosa M, Girbés T, Cordoba-Diaz M, et al. Endotoxins from a pharmacopoeial point of view. Toxins. 2018;10(8):331. http://www.mdpi.com/2072-6651/10/8/331
23. Guo M, Hata H, Babar A. Revenue maximizing markets for zero-day exploits. In: Baldoni M, Chopra AK, Son TC, Hirayama K, Torroni P, editors. PRIMA 2016: principles and practice of multi-agent systems. Cham: Springer International Publishing; 2016. p. 247–60. https://doi.org/10.1007/978-3-319-44832-9_15.
24. Food and Drug Administration. Testing and labeling medical devices for safety in the magnetic resonance (MR) environment. Silver Spring: U.S Food and Drug Administration; 2021. https://www.fda.gov/media/74201/download
25. Zhang Y, Le S, Li H, Ji B, Wang M-H, Tao J, et al. MRI magnetic compatible electrical neural interface: from materials to application. Biosens Bioelectron. 2021;194:113592.
26. Field H. This brain–computer interface company just hit a major milestone in the US. Emerging Tech Brew. 2022. https://www.emergingtechbrew.com/stories/2022/08/1*/this-brain-computer-interface-company-just-hit-a-major-milestone-in-the-us.
27. Food and Drug Administration. Implementing the 21st century cures act: a 2018 update from FDA and NIH. Silver Spring: U.S Food and Drug Administration; 2018. https://www.fda.gov/news-events/congressional-testimony/implementing-21st-century-cures-act-2018-update-fda-and-nih-07242018
28. CNET. Watch Elon Musk's entire live neuralink demonstration. 2020. https://www.youtube.com/watch?v=iOWFXqT5MZ4.

29. Morris ZS, Wooding S, Grant J. The answer is 17 years, what is the question: understanding time lags in translational research. J R Soc Med. 2011;104(12):510–20.
30. Poo MM, Du JL, Ip NY, Xiong ZQ, Xu B, Tan T. China brain project: basic neuroscience. Neuron. 2016;92:591–6. https://doi.org/10.1016/j.neuron.2016.10.050.
31. Rao P. China's brain–computer interface landscape in 2021: has the dragon woken up to neurotech? 2021. https://www.from-the-interface.com/China-BCI-neurotech/.
32. Hong'e M. China completes successful brain–computer interface implant. China News Service. 2020. http://www.ecns.cn/cns-wire/2020-01-17/detail-ifzsuknk2867059.shtml.
33. Oxley TJ, Yoo PE, Rind GS, et al. Motor neuroprosthesis implanted with neurointerventional surgery improves capacity for activities of daily living tasks in severe paralysis: first in-human experience. J Neurointervent Surg. 2021;13:102–8. https://jnis.bmj.com/content/neurintsurg/13/2/102.full.pdf
34. Coin A, Dubljević V. The authenticity of machine-augmented human intelligence: therapy, enhancement, and the extended mind. Neuroethics. 2020;14:283–90. https://doi.org/10.1007/s12152-020-09453-5.
35. Clark A, Chalmers D. The extended mind. Analysis. 1998;58(1):7–19. https://doi.org/10.1093/analys/58.1.7.
36. Binkley CE, Politz MS, Green BP. Who, if not the FDA, should regulate implantable brain–computer interface devices? AMA J Ethics. 2021;23(9):E745–9. https://journalofethics.ama-assn.org/article/who-if-not-fda-should-regulate-implantable-brain-computer-interface-devices/2021-09
37. Food and Drug Administration. Advancing regulatory science at FDA: focus areas of regulation (FARS). Silver Spring: U.S Food and Drug Administration; 2021. https://www.fda.gov/media/145001/download.

Neurotechnology, Stakeholders, and Neuroethics: Real Decisions and Trade-Offs from an Insider's Perspective

Adam Molnar, David Stanley, and Davide Valeriani

1 Introduction

For now, most public discourse around the advancement of neurotechnology comes from academia, certain large organizations, and, primarily, startups. The term neurotechnology usually refers to brain–computer interface (BCI) technology, which is defined "as a computer-based system that acquires brain signals, analyzes them, and translates them into commands that are relayed to an output device to carry out a desired action" [1]. We think it is critical for pioneers in the BCI space, among other critical technology areas, to think holistically about what their work means in terms of its impact on humanity. Since larger companies tend to be more reserved in statements made publicly, have relatively fewer resources dedicated to specific emerging technologies, and acquire smaller companies, we emphasize the need to map and understand decisions made from these smaller organizations that set ripple effects for larger technological involvement, use, and adoption.

1.1 Why Focus on Smaller Companies?

Smaller organizations, such as startups, are able to move more flexibly and are often the first to give momentum to new technologies. Reasons for focusing on the role that startups play in setting the tone for new technologies are that they (1) can disrupt larger organizations; (2) are small enough to change how and why their companies operate to accommodate a specific change they envision for the world; and (3) have unique challenges with regard to making decisions that have specific trade-offs due to resource limitations.

A. Molnar (✉) · D. Stanley · D. Valeriani
Neurable, Inc., Boston, MA, USA
e-mail: adam@neurable.com; david.stanley@neurable.com; davide.valeriani@neurable.com

V. Dubljević, A. Coin (eds.), *Policy, Identity, and Neurotechnology*, Advances in Neuroethics, https://doi.org/10.1007/978-3-031-26801-4_15

Technological advances made in recent years have evolved BCIs from assistive devices to technologies with a variety of applications, spanning basic neuroscience research to human augmentation [2–4]. BCIs have also broadened their scope from technologies mainly used in clinical or research lab settings to consumer devices that can be used in our everyday lives, which has spawned more neurotechnology companies today than in any past period. For clarity, the word *neurotechnology* has been introduced to represent all technologies that use neural activity as input, such as a brain-controlled prosthesis, or output, such as brain stimulation devices, identifying BCIs as a particular type of neurotechnology. For the rest of this paper, we will refer to neurotechnology in the broad sense.

1.2 How Hard Is Neurotechnology Development?

Really hard. A general consensus in the academic community is that neurotechnology capabilities, especially with consumer products, tend to be exaggerated and overhyped. Reliably collecting data from the brain, analyzing them in real-time to extract specific patterns associated with a given mental task, and building end-to-end systems that work across multiple people are among the greatest challenges faced by neurotechnology devices. The high cost and technical competency required to develop new neurotechnology represent additional burden. In other words, the neurotechnology business is really hard.

As a result, to date, the majority of developments in neurotechnology have been academic and driven by laboratories around the world. Nonetheless, as neurotechnology receives more attention, they generally see an increase in funding, which accelerates development addressing some of the critical choke points preventing mainstream adoption.

The opportunities to significantly improve our understanding of ourselves provided by neurotechnology go hand in hand with novel questions and concerns related to the ethical considerations relevant to the development, application, and commercialization of these technologies. The lack of regulation in this field leaves small and large organizations with a key decision making spectrum; ignore such ethical concerns and be fully driven by profit-oriented stakeholders, or use their expertise and experience to direct and educate the field on how to tackle them. It is important to acknowledge that these decisions are often not clear cut. Nonetheless, as a small startup in the field, we embraced this second route. The rest of the paper will share our unique perspective in this domain.

2 What Is Neuroethics?

The first use of the term neuroethics, although not necessarily aligned with the current field of neurotechnology, was in a 1973 paper entitled "Neuro-ethics of 'walking' in the newborn" by Harvard physician Anneliese A. Pontius [5]. However, writer William Safire is widely credited with giving the word its current meaning in 2002, defining it as "the examination of what is right and wrong, good and bad about the treatment of, perfection of, or unwelcome invasion of and worrisome manipulation of the human brain [6]." In another definition of neuroethics, Martha J. Farah, a pioneer in the field and researcher at the Center for Neuroscience and Society at the University of Pennsylvania, makes and contextualizes a relevant comparison to neuroethics with the field of genetics: "Like the field of genetics, neuroscience concerns the biological foundations of who we are, of our essence. The relation of self to brain is, if anything, more direct than that of self to genome. Perhaps more important, neural interventions are generally more easily accomplished than genetic interventions. Yet until recently there has been little awareness of the ethical issues arising from neuroscience [7]."

Neuroethics encompasses a large and varied set of issues. Some of these concern the practical implications of neurotechnology for individuals and society. Technological progress is making it possible to monitor and manipulate the human mind with ever more precision through a variety of neuroimaging methods and interventions [7]. For the first time, it may be possible to breach the privacy of the human mind and judge people not only by their actions, but also by their thoughts and predilections. The alteration of brain function in normal humans, with the goal of enhancing psychological function, is increasingly feasible and, indeed, increasingly practiced. The sooner neuroethical concerns are addressed, the earlier companies can start translating new findings in neuroscience into products. As Farah states, "progress in basic neuroscience is illuminating the relation between mind and brain, a topic of great philosophical importance. Our understanding of why people behave as they do is closely bound up with the content of our laws, social mores, and religious beliefs. Neuroscience is providing us with increasingly comprehensive explanations of human behavior in purely material terms. Although the field of neuroethics is young, the time seems ripe for a review in which the key issues of neuroethics, both practical and philosophical, are surveyed and placed in relation to one another [7]."

2.1 Neurable: Who Are We?

Neurable is a leading BCI company that commercializes sensor, signal processing, and algorithm advances into more practical, everyday form factors. We spun the company out of the University of Michigan's Direct-Brain Interface (UM-DBI) Laboratory in 2015 and have since gone on to commercialize BCIs in the consumer space. Having worked in this domain for nearly a decade, we have seen maturation and development in the field, alongside key technologies and considerations.

As pioneers in the space, we believe it is paramount to get ahead of potential ethical issues. This falls under the field of "neuroethics" and, to this end, we have been involved in relevant conversations and initiatives as a way to proactively further the space in regard to neuroethical responsibility and policy. To date, we have presented and published on the topic, openly tried to inform the community, promised not to sell user data, and consulted with the United Nations, the United States' Cybersecurity and Infrastructure Security Agency (CISA), and multiple world-renowned academic institutions.

3 Stakeholders

Knowing who influences decision-making allows one in and/or outside of an organization to understand how to best understand, influence, and/or shape outcomes, especially when it pertains to neuroethics. We define "stakeholder" in this chapter as an entity, either specifically or conceptually, that affects how decisions pertaining to neurotechnology and relevant neuroethical considerations are made. A stakeholder could be a group or a single individual that directly or indirectly influences how decisions are made. The following presents a breakdown of key stakeholders involved in decision-making for neurotech companies with a special emphasis on neuroethics. It should be noted that this is not an exhaustive list but a summary of key influencing forces.

3.1 Companies

Companies or organizations are directly involved in the planning, development and commercialization of neurotechnology devices. They play a critical role in deciding which technological advances are translated into products and are made available to the end user. Neurotech companies are also composed of three main subcategories of stakeholders: founders, employees, and advisors.

Founders establish directions, priorities, and goals of the company, having a direct impact on what problems to solve through neurotechnology. They also communicate with and receive input and feedback from other stakeholders, such as investors and end users, ensuring the company's direction is aligned with their needs.

Employees are the key players in implementing neurotechnology, translating advances in science and engineering into products and services. They include engineers, scientists, assistants, interns, and managers. They are the ones responsible for ensuring that data are collected rigorously and ethically, that algorithms are fair and robust across different user populations, and that personal data are transmitted and stored safely and securely. As such, employees play a critical role in effectively implementing neurotech solutions.

Advisors help founders and employees with setting feasible directions and goals for the success of the company. They usually are world-leading experts in a specific area related to the company, such as neuroscience, engineering, or manufacturing. Advisors use their broad view of the market and scientific breakthroughs to counsel the company on opportunities that might not be evident internally.

3.2 Investors

BCI or other neurotechnology companies tend to be considered "Moonshot" initiatives that likely require investment capital to help move promising technology toward commercialization [8]. For this reason, it is important to understand who the investors are and what their role is in the space. Investors tend to be high-net worth individuals, venture-capital corporations, and/or other forms of equity-based investment vehicles that generally invest money into a company in exchange for a percentage of ownership.

In addition to investment, they may also act in an advisory or support capacity, helping the company navigate opportunities, networks, and other benefits.

They are relevant when considering neuroethics because they play a strong role in how the company develops, matures, and ultimately exits. An exit is when a company returns value back to its shareholders either through an acquisition, initial public offering (IPO), or some other financial structure that allows equity holders to "cash out."

Investors also have a responsibility to their fund. For this reason, investors generally invest in companies they expect to make a return on their investment and advise toward protecting said investment.

3.3 Academia

Academia represents the community of people concerned with the pursuit of research and education. It includes students, postdocs, researchers, scientists, professors, as well as professional organizations aimed at promoting the collaboration and exchange of knowledge among community members. The main role of academia is pushing the frontier of a field forward through the scientific method. In the context of neurotechnology, academia is particularly concerned with extending our knowledge of how the brain functions and what technologies can be built to interface with it. Academia is also broadly involved in the development of neuroethics and new neural rights, advocating for rigorous protection of user's data and fair and equitable access to neurotechnology [9]. Another key role of academia is guaranteeing rigor via the strict application of the scientific method to ensure unbiased and well-controlled experimental design and methodology, as well as analysis, interpretation, and reporting of results.

Professional organizations in the neurotech field include the Society for Neuroscience and the Organization for Human Brain Mapping, which are particularly focused on connecting researchers in basic neuroscience. The BCI Society and the Institute of Electrical and Electronics Engineers (IEEE) mainly focus on promoting research around the development and use of neurotechnology. NeuroTechX and BrainMind facilitate the advancement of neurotechnology development via professional training opportunities. Scientific journals are also part of these organizations, with the main goal of making new knowledge available to the whole community. Together, these organizations play a critical role in scientific discovery by providing tools and opportunities for exchanging ideas, validating discoveries via peer reviews, and making new knowledge available to key stakeholders in society.

Overall, academia's interests in neurotechnology development are to ensure that scientific rigor is maintained throughout product development, and that the field keeps advancing and innovating with new discoveries that can enable new products.

3.4 General Public

Although members of the general public may not be directly involved with neurotechnology or neuroscience, they are nonetheless important stakeholders from an ethical standpoint. The reason for this is twofold. First, members of the general public may become more involved with the field in the future, perhaps in the form of end users or patients. Second, it is the general public's perception of the field that determines its reputation as a whole. Without a positive reputation, neurotechnology will be harder to move forward.

Advanced neurotechnology has the potential to reshape society, which could affect the general public even if they are not end users. Social media provides a good analogy. Although many people today abstain from using social media, they are nonetheless affected by its presence. Therefore, like social media and other transformative technologies, neurotechnology's broad effect on the general public should be considered.

3.5 Customers

Customers are individuals, entities, or organizations who ultimately pay for a product or service. They generally evaluate these relative to alternatives and make a decision, such as whether or not to make a purchase. It can be the company, organization, or individual who buys the product. This often represents one of the largest considerations, since companies strive for growth through sales, which means successfully bringing a product or service to customers. They give feedback on how the product is working, both directly and indirectly. When it comes to thinking about the customer's role in neuroethics as a stakeholder, we think it is also beneficial to think of it as a direct and indirect influence. In regard to direct influence, some customers have explicit needs or requirements on how their data is collected, managed, and ultimately used. We can think of the customer's indirect influence before and after purchasing the product or service and how their sentiments influence a greater perspective on the field as a whole.

3.6 End Users

Quite often, the customers and the end users are the same. However, a company may purchase a batch of neurotech devices for use by its employees. In this case, the employees would be the end users but not the customers. The end users are the individuals who are actually interacting with the neurotechnological device. The end users are of primary ethical importance because they are the ones who are at risk if the device is unsafe or has other negative effects, or if privacy is violated. The end users play a role in consuming the outputs that the neurotechnological device produces, providing feedback and serving as the primary source of neurological data. To this end, it is imperative that their informed consent be protected and that their input be considered throughout the course of product design.

3.7 Government

The government is an important stakeholder, as it is ultimately beholden to the general public. It is responsible for defining the regulations within which neurotechnology must operate so as to be both safe and beneficial to the public. Its role in regulating neurotechnology is similar to its role in other fields, such as biotechnology and medical devices. For example, the government defines various classes of medical devices, each with increasingly stringent safety criteria as devices become more invasive. A similar process should exist for neurotech devices but with additional criteria for user privacy, given the sensitive personal nature of neurological data. The government's role is critical in neuroethics because many of the policies they implement will derive from neuroethical principles. Similarly, government agencies will be important players in shaping neuroethics literature and discourse.

4 Case Studies

4.1 Building a Profitable Company Versus Building a Company We Want

In 2017, Neurable showcased technology to the world for the first time at a marquee technology conference called SIGGRAPH. We debuted a demo wherein one could control a virtual reality experience using brain signals, specifically, P300 event related potentials (ERPs). The demo garnered us much attention and was illustrative of never-before-seen capabilities made possible through intellectual property (IP) in signal processing and algorithm development. We even found ourselves on the cover of the New York Times with a headline, "A Game You Can Control With Your Mind" [10]. Neurable was catapulted to the front of the BCI space and then tried to commercialize its product, which was built in alignment with our vision of creating a world without limitations, i.e., allowing people of all different states of ability to control technology leveraging neural activity. It was an exciting time.

In 2018, virtual and augmented reality began to falter and the momentum that carried us through our first investment in 2016 was starting to slow down. Immersive systems were struggling to find the elusive "product-market fit," a term used in entrepreneurship to define when a company has found, built, and established a product or service that successfully addresses a customer's need.

Neurable could no longer depend on virtual reality as a market to commercialize, since virtual reality's install base was too low and the technology was still struggling with other key pain points, which needed to be solved before the addition of a neural interface could be considered. For this reason, our company needed to "pivot" or reconsider its path to market in order to continue raising capital to survive and grow. One of the benefits of being catapulted to the front of BCI attention was that we were able to conduct some of the best customer discovery, a term used to identify potential problems to be solved from potential customers, which we continued to do in 2018 as we were considering our pivot. One of the markets that we looked at was neuromarketing, a field that uses various brain-imaging and interpretation modalities to reveal greater insights from customers. This field has been subject to dubious claims and questionable technology, but indeed has a well-defined business problem and paying customers. Specifically, neuromarketing seeks to identify marketing perspectives, such as buyer intent, decision-making, interest and, to some degree, measurements along the arousal-valence spectrum, the former being more psychologically rooted and the latter being a more questionable and less proven estimate of emotion.

In doing customer discovery, speaking with companies big and small from the consumer-packaged goods (CPG) space, marketing, and more, we actually received opportunities to commercialize and start generating revenue. We had found a path for product-market fit! We were even offered contracts with Fortune 500 companies to work with them. However, herein lay a predicament. Neurable's vision is to help people transcend their limitations and create technologies that ultimately empower the end user. In weighing this aspect, we concluded that the path toward neuromarketing, albeit potentially lucrative, went against the company's core values.

We had to wrestle with two options

1. Do we take the paying opportunities, which would allow us to keep the proverbial and literal lights on, continue to pay our employees, grow as a business, and progress toward our investors' interest of de-risking our success? *or*
2. Do we stick true to our vision, find a way to product-market fit that better aligns with our vision and principles, which may jeopardize the livelihoods of our employees, increase the risk to our investors, and potentially waste all the blood, sweat, and tears spent to make the most out of this opportunity?

In reflecting, speaking with our team, advisors, and investors, we ultimately decided to go with option two. Proceeding as a neuromarketing company, although potentially de-risking the future of the company, was not a company we wanted to be a part of or build. Our founders, investors, and team had no interest in creating tools to help engineer better ads, effectively taking advantage of people and potentially breaching neuroethical principles by invoking an unwelcome manipulation of the human brain. While there may be potentially "good" or ethical applications of neuromarketing, our company's vision was not one that wanted to be spent building them.

Working on a problem that you are not passionate about is a recipe for disaster since, especially with small companies, there will always be something that goes wrong. It is very hard to push through the tough times for a product or service you are not inherently passionate about, which is why this is often a piece of feedback given to aspiring entrepreneurs when they begin to think about what to work on.

If our vision is to be the company that brings neurotechnology to everyone, the path of neuromarketing did not help; in fact, it would probably hurt in the long run. We wanted to build a brand that the end users, i.e., people who would actually use or be affected by our BCI, had reason to trust. For this reason, we decided to create technology that directly benefited and empowered the end user as opposed to taking advantage of them. This allowed us to stick to our vision, justify the opportunity and risk to our investors, and feel good about saying no to easy revenue.

4.2 Differing Strategies to Research and Development: Agile vs Conservative

A key goal of neurotechnological development is the availability of large amounts of brain data, which enable feature discovery, model training, and validation. Similar to other players in the neurotechnology space, Neurable includes a dedicated team of research engineers who design and conduct experiments by collecting electroencephalography (EEG) data.

A key consideration for startups is focus and prioritization. We would argue that most companies do not *intentionally* do harm but end up prioritizing other aspects that may lead to harm. A difficult decision that we have to make every day at Neurable, for example, is the trade-off between taking a very slow, methodical, well-understood, and safe approach to experimentation and a quick and iterative process, which is a hallmark of lean startup methodology. One area, for example, that we debate at Neurable concerns Institutional Review Board (IRB) oversight, which is not a legal requirement for product development.

There are numerous considerations for neurotechnology development that range from access to implementation but, for the sake of this chapter, we will focus on an area we think is most important: data collection, use, and management. Key requirements for ethical data collection include ensuring that participants understand what type of data is being collected, who is collecting the data, and for what purpose it will be used. This is done using informed consent, both written and oral. Academic research labs and large organizations rely on independent committees (IRBs) to oversee the process of collecting and analyzing data from human populations. In the United States of America, IRBs were established in 1974 by the Department of Health Education and Welfare through regulations on the protection of human subjects engaged in federally funded research [11]. An IRB consists of at least five members of varying backgrounds with professional experience to provide appropriate scientific and ethical review. When academic investigators want to collect and analyze new human data, they first need to seek IRB approval through the submission of a detailed and thorough plan for data collection and analysis [12]. After reviewing the application, the IRB typically provides feedback to the investigators and asks them to make amendments, for example, clarifying certain aspects of the analysis. Obtaining IRB approval is typically a long process that may take between 2 and 9 months, depending on the experience of the investigator and complexity of the study.

At Neurable, we have to weigh these considerations every day. In an ideal world, we would have an IRB approval for every experiment we run. However, that is often not the case. As a startup company, it is critical for Neurable to react to market shifts and to continuously explore and develop new features for its products. For example, while exploring new business opportunities, a stakeholder may ask if we could detect fatigue in automobile drivers from brain activity. While in academic settings, this investigation could represent a multi-year research project, in a startup environment, the timeframe between the proposal and the presentation of a first prototype can be much shorter, for example, 2–4 months or even weeks or days! As such, seeking IRB approval is often unfeasible, as it can slow down innovation and reduce the opportunities for the startups to successfully commercialize.

Generally, at Neurable, our workflow for feature development includes several stages

1. Draft requirements
2. Review scientific literature
3. Develop a new experiment for data collection
4. Recruit participants, gather their written consent, and collect data
5. Develop a prototype solution
6. Test the prototype

We generally go through these steps much faster than researchers in academia would. The pros: we are able to move much more nimbly and react to changes quickly. The cons: we accept more risk that something may go wrong. With that being said, we know that EEG is considered a harmless technology, even by IRB standards, but we still accept greater risk by not consulting with external stakeholders

The reality of a company is that it is extremely impractical to follow the same level of protocol or rigor as an academic institution, particularly so since the push and pull by key stakeholders is different. If experiments were conducted 1–1 in a startup as they would be in academia, we would venture to say that the company would run out of money before being able to bring a product to market, which is especially so for emerging and hardware-based companies. The startup executive team has to keep in mind the well-being of its employees as well as the interests of their funders. If the company does not innovate, grow, sell products, or increase its value, it may not receive further funding. If the company does not receive funding the employees do not get paid, which illuminates the need for lean experimentation This being said, choosing not to pursue IRB approval does not mean ignoring ethical policies for data collection. It now becomes the responsibility of the startup to guarantee and demonstrate that ethical guidelines for data collection are still followed.

Neurable fosters neuroethical considerations into our culture by hiring people who care about using technology for ethical applications in the first place. Our company also fosters discussion on this topic, using it as a point of conversation and consideration for all employees, regardless of domain or title. When situations arise that warrant deeper ethical consideration, such as to procure an IRB or go faster with an experiment, each stakeholder responsible has the ability to voice concerns, and is encouraged to do so. There are multiple avenues in which opinions, criticisms, and objections can be raised including town halls, retrospective meetings, 1–1 s, and anonymized surveys. Neurable breeds a culture of intentional conflict and open communication to allow for these kinds of opinions to flow and better the organization. Lastly, we use our networks and other organizations to help inform and drive ethical decision-making, leveraging network accountability. For example, if the company is quoted as proceeding in a certain fashion, those with whom the company interfaces and shares commitments then make the company beholden to those claims.

These case studies represent a fraction of the types of decision-making problems startups encounter and the trade-offs that must happen, which are affected by the relevant stakeholders who influence decision-making.

5 Conclusion

Neurotechnology is an emerging but disruptive technology. Historically, society has seen signals of negative impact from technology left unaddressed with relevant ethical fallout that could have been mitigated and/or buffeted with some preemptive ethical consideration and policy development. This chapter provided an honest insider's perspective to help illustrate the challenges that startups face in making decisions, especially as they pertain to neuroethical implications. By understanding and being able to empathize with those involved in a small organization, one is able to more effectively recommend policy and/or other interventions to help prevent bigger problems.

However, this responsibility does not solely lie on the small organization but rather on multiple influencing forces, deemed stakeholders, who govern, either directly or indirectly, the outcomes of neurotechnology development. We believe that by understanding both what decisions companies make and why and how they make these decisions, we can more effectively come up with ways to (1) reward organizations setting positive tones for the industry; (2) punish or remediate organizations who take advantage of the system; and (3) contextualize multi-componential decision-making to those who may not have as much experience.

Organizations, both public and private, for-profit and non-profit, can do better to anticipate social needs as they pertain to ethics and apply them when building new technologies or products. This can be aided or made more difficult by external stakeholders especially when it comes to process, incentives, and auditing. It is important to note that this chapter largely deals with the use of non-invasive devices for passive brain recording. There are a number of key topics that were not addressed but should be studied and discussed further in future developments, including: (1) how incentives (especially within the context of capitalistic economies) drive product development and innovation pipelines; (2) the philosophy of neuroethics and how we, as individuals and societies, determine what is right and wrong; (3) how equity and access to technology lead to fair or unfair advantages, especially in regard to even more powerful future neurotechnology and an ever-growing disparity between developed vs. underdeveloped nations; and (4) future capabilities in terms of how humans as a species are able to understand ourselves and each other through the analysis of data for varied benefits or malicious intent.

Driving ethical considerations as a company in an emerging technology field is a challenging endeavor with real implications. Understanding how and why organizations make their decisions is critical to empathizing and assisting in this regard. Neurable, as a for-profit organization, sews ethical values into its cultural fabric to help guide decision-making and invention. This actualization of ethical values happens through formal and indirect conversations, intentional efforts to study and learn from other domains, and frequent communication with customers, partners, and stakeholders. Neurable also consults with various agencies dealing with data to help learn and ensure proper accountability. We aim to set an example by leveraging precedent, momentum, and standards to help move the neurotechnology field toward ethical decision-making. Similarly, we strive to leverage case studies, policy, and the experience of other fields to illuminate growing developments in the neurotechnology domain. Neurotechnology is significant today and will become ever so more important moving forward. Neurable urges stakeholders across the spectrum to heed the considerations made in this chapter, challenge assumptions, and educate themselves to ensure the most ethical future of this new and very important technology.

References

1. Shih JJ, Krusienski DJ, Wolpaw JR. Brain–computer interfaces in medicine. In: Mayo clinic proceedings. Amsterdam: Elsevier; 2012. p. 268–79.
2. Farwell LA, Donchin E. Talking off the top of your head: toward a mental prosthesis utilizing event-related brain potentials. Electroencephalogr Clin Neurophysiol. 1988;70(6):510–23.
3. Wolpaw JR, Birbaumer N, McFarland DJ, Pfurtscheller G, Vaughan TM. Brain–computer interfaces for communication and control. Clin Neurophysiol. 2002;113(6):767–91.
4. Cinel C, Valeriani D, Poli R. Neurotechnologies for human cognitive augmentation: current state of the art and future prospects. Front Hum Neurosci. 2019;13:13.
5. Pontius AA. Neuro-ethics of "walking" in the newborn. Percept Mot Skills. 1973;37(1):235–45.
6. Safire W. Visions for a new field of "neuroethics". In: Neuroethics: mapping the field, conference proceedings, May 13–14, 2002. San Francisco: The Dana Press; 2002. p. 4–9.
7. Farah MJ. Neuroethics: the practical and the philosophical. Trends Cogn Sci. 2005;9(1):34–40.
8. Cattan G. The use of brain–computer interfaces in games is not ready for the general public. Front Comput Sci. 2021;3:20.

9. Ienca M, Andorno R. Towards new human rights in the age of neuroscience and neurotechnology. Life Sci Soc Policy. 2017;13(1):1–27.
10. A Game You Can Control With Your Mind. New York Times. 2017. https://www.nytimes.com/2017/08/27/technology/thought-control-virtual-reality.html.
11. Moon MR. The history and role of institutional review boards: a useful tension. AMA J Ethics. 2009;11(4):311–6.
12. Enfield KB, Truwit JD. The purpose, composition, and function of an institutional review board: balancing priorities. Respir Care. 2008;53(10):1330–6.

GPSR Compliance

The European Union's (EU) General Product Safety Regulation (GPSR) is a set of rules that requires consumer products to be safe and our obligations to ensure this.

If you have any concerns about our products, you can contact us on ProductSafety@springernature.com

In case Publisher is established outside the EU, the EU authorized representative is:

Springer Nature Customer Service Center GmbH
Europaplatz 3
69115 Heidelberg, Germany

Batch number: 10732044

Printed by Printforce, the Netherlands